# Frequency Domain Techniques for $\mathcal{H}_\infty$ Control of Distributed Parameter Systems

## Advances in Design and Control

SIAM's Advances in Design and Control series consists of texts and monographs dealing with all areas of design and control and their applications. Topics of interest include shape optimization, multidisciplinary design, trajectory optimization, feedback, and optimal control. The series focuses on the mathematical and computational aspects of engineering design and control that are usable in a wide variety of scientific and engineering disciplines.

## Series Volumes

Özbay, Hitay, Gümüşsoy, Suat, Kashima, Kenji, and Yamamoto, Yutaka, *Frequency Domain Techniques for $\mathcal{H}_\infty$ Control of Distributed Parameter Systems*

Khalil, Hassan K., *High-Gain Observers in Nonlinear Feedback Control*

Bauso, Dario, *Game Theory with Engineering Applications*

Corless, M., King, C., Shorten, R., and Wirth, F., *AIMD Dynamics and Distributed Resource Allocation*

Walker, Shawn W., *The Shapes of Things: A Practical Guide to Differential Geometry and the Shape Derivative*

Michiels, Wim and Niculescu, Silviu-Iulian, *Stability, Control, and Computation for Time-Delay Systems: An Eigenvalue-Based Approach, Second Edition*

Narang-Siddarth, Anshu and Valasek, John, *Nonlinear Time Scale Systems in Standard and Nonstandard Forms: Analysis and Control*

Bekiaris-Liberis, Nikolaos and Krstic, Miroslav, *Nonlinear Control Under Nonconstant Delays*

Osmolovskii, Nikolai P. and Maurer, Helmut, *Applications to Regular and Bang-Bang Control: Second-Order Necessary and Sufficient Optimality Conditions in Calculus of Variations and Optimal Control*

Biegler, Lorenz T., Campbell, Stephen L., and Mehrmann, Volker, eds., *Control and Optimization with Differential-Algebraic Constraints*

Delfour, M. C. and Zolésio, J.-P., *Shapes and Geometries: Metrics, Analysis, Differential Calculus, and Optimization, Second Edition*

Hovakimyan, Naira and Cao, Chengyu, *$\mathcal{L}_1$ Adaptive Control Theory: Guaranteed Robustness with Fast Adaptation*

Speyer, Jason L. and Jacobson, David H., *Primer on Optimal Control Theory*

Betts, John T., *Practical Methods for Optimal Control and Estimation Using Nonlinear Programming, Second Edition*

Shima, Tal and Rasmussen, Steven, eds., *UAV Cooperative Decision and Control: Challenges and Practical Approaches*

Speyer, Jason L. and Chung, Walter H., *Stochastic Processes, Estimation, and Control*

Krstic, Miroslav and Smyshlyaev, Andrey, *Boundary Control of PDEs: A Course on Backstepping Designs*

Ito, Kazufumi and Kunisch, Karl, *Lagrange Multiplier Approach to Variational Problems and Applications*

Xue, Dingyü, Chen, YangQuan, and Atherton, Derek P., *Linear Feedback Control: Analysis and Design with MATLAB*

Hanson, Floyd B., *Applied Stochastic Processes and Control for Jump-Diffusions: Modeling, Analysis, and Computation*

Michiels, Wim and Niculescu, Silviu-Iulian, *Stability and Stabilization of Time-Delay Systems: An Eigenvalue-Based Approach*

Ioannou, Petros and Fidan, Barış, *Adaptive Control Tutorial*

Bhaya, Amit and Kaszkurewicz, Eugenius, *Control Perspectives on Numerical Algorithms and Matrix Problems*

Robinett III, Rush D., Wilson, David G., Eisler, G. Richard, and Hurtado, John E., *Applied Dynamic Programming for Optimization of Dynamical Systems*

Huang, J., *Nonlinear Output Regulation: Theory and Applications*

Haslinger, J. and Mäkinen, R. A. E., *Introduction to Shape Optimization: Theory, Approximation, and Computation*

Antoulas, Athanasios C., *Approximation of Large-Scale Dynamical Systems*

Gunzburger, Max D., *Perspectives in Flow Control and Optimization*

Delfour, M. C. and Zolésio, J.-P., *Shapes and Geometries: Analysis, Differential Calculus, and Optimization*

Betts, John T., *Practical Methods for Optimal Control Using Nonlinear Programming*

El Ghaoui, Laurent and Niculescu, Silviu-Iulian, eds., *Advances in Linear Matrix Inequality Methods in Control*

Helton, J. William and James, Matthew R., *Extending $\mathcal{H}^\infty$ Control to Nonlinear Systems: Control of Nonlinear Systems to Achieve Performance Objectives*

# Frequency Domain Techniques for $\mathcal{H}_\infty$ Control of Distributed Parameter Systems

## Hitay Özbay
Bilkent University
Ankara, Turkey

## Suat Gümüşsoy
The MathWorks
Natick, Massachusetts

## Kenji Kashima
Kyoto University
Kyoto, Japan

## Yutaka Yamamoto
Kyoto University
Kyoto, Japan

Society for Industrial and Applied Mathematics
Philadelphia

| | |
|---|---|
| *Publications Director* | Kivmars H. Bowling |
| *Executive Editor* | Elizabeth Greenspan |
| *Developmental Editor* | Gina Rinelli Harris |
| *Managing Editor* | Kelly Thomas |
| *Production Editor* | David Riegelhaupt |
| *Copy Editor* | Julia Cochrane |
| *Production Manager* | Donna Witzleben |
| *Production Coordinator* | Cally A. Shrader |
| *Compositor* | Cheryl Hufnagle |
| *Graphic Designer* | Doug Smock |

Royalties from the sale of this book are placed in a fund to help students attend SIAM meetings and other SIAM-related activities. This fund is administered by SIAM, and qualified individuals are encouraged to write directly to SIAM for guidelines.

**Library of Congress Cataloging-in-Publication Data**
Names: Özbay, Hitay, author. |Gümüşsoy, Suat, author. | Kashima, Kenji (Mathematics professor), author. | Yamamoto, Yutaka, 1950- author.
Title: Frequency domain techniques for H [infinity symbol] control of distributed parameter systems / Hitay Özbay (Bilkent University, Ankara, Turkey), Suat Gümüşsoy (The MathWorks, Natick, Massachusetts), Kenji Kashima (Kyoto University, Kyoto, Japan), Yutaka Yamamoto (Kyoto University, Kyoto, Japan).
Description: Philadelphia : Society for Industrial and Applied Mathematics, [2018] | Series: Advances in design and control : 32 | Includes bibliographical references and index. | Description based on print version record and CIP data provided by publisher; resource not viewed.
Identifiers: LCCN 2018033506 (print) | LCCN 2018038453 (ebook) | ISBN 9781611975406 (ebook) | ISBN 9781611975390 (print)
Subjects: LCSH: System analysis. | Distributed parameter systems. | H [infinity symbol] control. | Control theory.
Classification: LCC QA402 (ebook) | LCC QA402 .F7425 2018 (print) | DDC 003--dc23
LC record available at https://lccn.loc.gov/2018033506

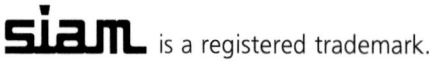 is a registered trademark.

# Contents

# List of Figures

# List of Tables

# Preface

This book presents new computational tools for the $\mathcal{H}_\infty$ control of distributed parameter systems. Transfer functions are considered as input-output descriptions for the plants to be controlled. Detailed examples are given from various applications ranging from computer/communication networks and mechanical systems to biological systems. The main emphasis is on the computation of the controller parameters and reliable implementation. Over the last 10 years new results have been obtained for the computation of the $\mathcal{H}_\infty$ controllers for a large class of distributed parameter systems. This book presents some of these results, taken from articles written by the authors. It is assumed that the plants to be controlled are described by their transfer functions and the weights defining the $\mathcal{H}_\infty$ optimality condition are low order rational functions. First, a special type of factorization is done on the plant transfer function. For this purpose new computational techniques are presented (the general case considers retarded and neutral time delay systems, but for other classes of distributed parameter systems the methods are demonstrated on specific examples). Second, the computation of the optimal controller is reduced to solving a set of finitely many linear equations. Finally, reliable implementation (in terms of series, parallel, and feedback connections of stable transfer functions) is demonstrated. Most of the book is devoted to single-input-single-output plants; in the last chapter, extensions to a certain class of multiple-input-multiple-output plants are demonstrated.

We believe that a first year graduate level control course can be taught from the material presented here. In fact, the first author taught courses at Bilkent University and at The Ohio State University, covering most of the topics included in the book.

Many people have helped us in the production of this book. Historically, we first got interested in the subject matter thanks to Allen Tannenbaum, Ciprian Foias, George Zames, Pramod Khargonekar, Malcom Smith, and Tryphon Georgiou. Some of the earlier results discussed in various parts of the book are based on their contributions. In fact, theoretical foundations of the present work were laid by these pioneers from the mid-1980s to the mid-1990s. One of the first books on this theory is by Foias, Özbay, and Tannenbaum [66].

Parts of the book on fractional order systems are based on discussions with Catherine Bonnet. On several occasions over the last 10 years, the first and last authors have visited INRIA-Saclay and L2S to collaborate with her and with Silviu Niculescu on related topics. We also thank them for providing us an early version of a software, YALTA, developed in their group, which has proven to be very useful in finding the right half plane roots of quasi-polynomials (a crucial step in factorizations used in the book). Likewise, we thank Tomas Vyhlidal for sending us his version of the quasi-polynomial root finding code, QPmR. We also had fruitful discussions with Jie Chen on the delay margin optimization problem (Section 5.1.2). The second author thanks Wim Michiels from K.U. Leuven for high quality research collaboration in his postdoctoral study on numerical methods of $\mathcal{H}_\infty$ norm computation and fixed-order $\mathcal{H}_\infty$ controller design.

As one can easily identify, detailed controller designs in the book are in Chapters 6 and 7. The first section of Chapter 6 is expanded from an old joint work with Onur Toker. The newer part of Chapter 6 (mostly Section 6.2) is based on a joint paper with Mustafa Oğuz Yeğin (more detailed examples are in his MS thesis [265]). The new formula for the optimal Nevanlinna–Pick interpolant (Section 2.4.1) is from a recent work with Veysel Yücesoy. In their MS theses, Erdem Karagül and Ezgi Ateş have worked out examples for $\mathcal{H}_\infty$ control of fractional order systems, and $\mathcal{H}_\infty$-based estimation, respectively. These examples are reworked in the book (Sections 6.3.3 and 6.3.5). All other parts of the book are based on the authors' individual or joint publications.

We should mention that Kirsten Morris was an invisible force behind the conception of this project; she was one of the few vocal people to give us encouragement and suggestions before we considered writing such a book.

Our editor, Elizabeth Greenspan, has been very patient with us throughout the writing process, which was painfully slow due to the first author's involvement with various other activities he could not decline. Completion of the book was possible thanks to a sabbatical leave granted from Bilkent University, and thanks to The Ohio State University for hosting him during this leave.

Many people have read early versions of different parts of the book: we thank all Bilkent students who took EEE644 in Spring 2016 and worked on most of Chapter 2, and OSU students who took ECE5194.11 in Spring 2018 and worked on several examples and exercises throughout the book. We are also thankful to Nazlı Gündeş for reading parts of the book and for providing suggestions and corrections.

Finally, the book would not be possible without constant support from the members our immediate families. This work is dedicated to them:

> Özlem and Olsan E. Özbay
> Bahar and Elfin D. Gümüşsoy
> Satoko, Shinya, Yuri, and Aoi Kashima
> Mamiko, Sho, and Kaoru Yamamoto

Ankara, Columbus
Boston
Kyoto

Spring 2018

# Chapter 1

# Introduction

## 1.1 ▪ Aim and scope

The objective of this book is to provide researchers and practicing engineers with numerical tools for computing $\mathcal{H}_\infty$ controllers for various classes of distributed parameter systems (DPSs). For $\mathcal{H}_\infty$ control of finite dimensional systems, many reliable computational methods exist; and over the last two decades they have been widely used thanks to the Robust Control Toolbox (and its earlier versions and subcomponents) of MATLAB®. On the other hand, for infinite dimensional systems, such as systems with time delays and those represented by partial differential equations, new tools are required to extend existing popular robust control design software packages. Although several tractable solutions for $\mathcal{H}_\infty$ control of various classes of infinite dimensional systems were published more than 20 years ago (between 1987 and 1995; see, e.g., [66]), they have not been incorporated into widely circulated software packages. One of the reasons for the lack of a popular toolbox for the robust control of DPSs is that the computational procedures published earlier are found to be quite technical and difficult to understand by junior graduate students and practicing engineers. Over the last decade, the above mentioned early results on $\mathcal{H}_\infty$ control of infinite dimensional systems have been simplified and new numerical tools have been developed that facilitate the implementation of these computational methods. The aim of the book is to illustrate these new methods and hence provide $\mathcal{H}_\infty$ controller design tools that are easily accessible by practicing engineers as well as researchers in academia.

## 1.2 ▪ Historical perspective

This book is concerned with the design of $\mathcal{H}_\infty$ controllers for single-input-single-output (SISO) linear time invariant (LTI) DPSs; the standard feedback control scheme is as shown in Figure 1.1. It is assumed that the input-output dynamical behavior of a DPS is represented by its transfer function, $P(s)$, called the plant. The goal is to design a controller $C(s)$, achieving design objectives that are specified based on certain properties of the internal signals $(e, u, y)$ in relation to the external signals $(r, v)$; see Chapter 5 for precise statements of the control problems studied in the book.

One of the most interesting properties of a DPS is that its transfer function is an irrational function, which makes the controller design quite challenging, whereas for

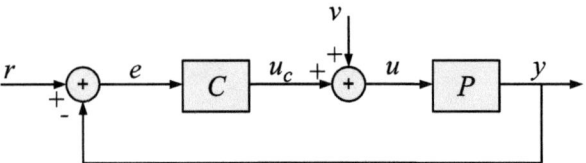

**Figure 1.1.** *Standard feedback system.*

lumped parameter systems the plant has a rational transfer function and it admits a finite dimensional state-space representation, for which there exists a wide range of control design techniques. In Chapter 3, various transfer function examples are given for different DPS models.

The main motivation behind $\mathcal{H}_\infty$ controller design comes from the fact that it captures important robust control problems within this framework [273]. One of the associated robust control problems can be defined as follows: let the "true" plant transfer function be written as

$$P_\Delta(s) = P(s) + \Delta(s),$$

where $P(s)$ is known (called the nominal plant ) and $\Delta(s)$ represents the uncertainty, which is unknown but satisfies certain mild conditions. One of these conditions is expressed in terms of a bound on its magnitude, $|W(j\omega)| > |\Delta(j\omega)|$ for all $\omega \geq 0$, where $W(s)$ is known, and it is called the "additive uncertainty weight." The robust control problem related to the above formulation is to find a fixed controller $C$, from the problem data ($P$ and $W$), such that design objectives are satisfied for all possible $P_\Delta$ as the plant in the feedback loop. In Chapter 6, computational tools are developed for the SISO $\mathcal{H}_\infty$ controller design, where $P(s)$ is irrational and $W(s)$ is rational.

Various design techniques have been developed for $\mathcal{H}_\infty$ control of finite dimensional systems (where $P(s)$ and $W(s)$ are rational functions/matrices) [51, 52, 54, 68, 79, 83, 130, 145, 210, 226, 245, 281, 280], and the corresponding computational tools have been tested in a wide range of applications. Robust control of finite dimensional systems, under other types of uncertainty descriptions, have also been studied extensively; see [19, 24, 45, 237] and their references.

For time delay systems (an important class of infinite dimensional systems), $\mathcal{H}_\infty$ controllers were first obtained in the mid-1980s [63, 67, 282] for the case where the plant is stable. Over the last 20 years there has been significant progress in the extension of these results to larger classes of infinite dimensional systems and more complicated robust control problems; see, e.g., [14, 41, 42, 48, 64, 77, 84, 85, 108, 113, 120, 136, 144, 173, 182, 183, 199, 200, 216, 217, 222, 229, 234, 242]. One of the methods used in the computation of $\mathcal{H}_\infty$ controllers for infinite dimensional systems is the "skew Toeplitz" approach [66]. The role of skew Toeplitz operators in $\mathcal{H}_\infty$ control was first noticed in [23]. An important feature of this approach is that the $\mathcal{H}_\infty$ controllers are computed directly, without approximating the plant, from a set of finitely many linear equations where the number of unknowns depends on the order of the weight(s) and the number of unstable poles of the plant. Chapter 6 of the book is devoted to the exposition of this method with illustrative examples.

For infinite dimensional systems, robust controllers can also be obtained by approximating the plant and then using standard techniques developed for the control of finite dimensional systems, by keeping track of the original approximation error; see, e.g., [15, 40, 116, 137, 138, 168, 169, 202, 220] and their references. See [39] for an early review of $\mathcal{H}_\infty$ control of DPSs in general, [244] for the state-space approach to $\mathcal{H}_\infty$

control of such systems, and [162, 251] for reviews of state-space and operator theoretic approaches to $\mathcal{H}_\infty$ control of time delay systems. For time delay systems and
some other classes of DPSs, $\mathcal{H}_\infty$ controllers can also be derived from a game theoretic
approach [20, 228]. For the most recent results on $\mathcal{H}_\infty$ control of systems with input-
output delays, see [152, 279] and their references. Repetitive control design [105],
under certain performance and robustness conditions, can be posed as a robust control
problem for systems with time delays [99, 209, 231, 252]. Sampled-data controller design, with certain types of optimality conditions, result in an $\mathcal{H}_\infty$ control problem for
time delay systems; see [18, 35, 104, 260] for further references. Robust stability of
time delay systems (within, and outside, the framework of $\mathcal{H}_\infty$ control) is widely studied [33, 53, 55, 71, 72, 74, 89, 100, 106, 111, 127, 128, 132, 133, 134, 139, 178, 179,
165, 207, 208, 235, 236, 243]. See also the recent books [70, 129] for unified treatments
of time domain (state-space–based) methods for time delay systems. Another recent
book on time delay systems is [283]; it uses a small-gain approach (frequency domain
methods) for various robust control problems. Several issues related to robust control of
fractional delay systems are considered in [25]. Stability robustness against small time
delays has been considered for various types of plants; see, e.g., [26, 140, 151, 166]
and their references. Flexible structure models which include internal time delays are
considered in [102, 103]. For spatially invariant DPSs [17], $\mathcal{H}_\infty$ optimal controllers
are obtained from a parameterized family of finite dimensional problems; see also [44].
Robust control of infinite dimensional systems is also covered in the book [43], where
a state-space approach (semigroup theory) is adopted.

## 1.3 ▪ Organization of the book

In Chapter 2, basic mathematical definitions are given. The rest of the book is based
on this preliminary material. For an expanded background material on mathematical
tools used, we refer the reader to [261]. In particular, input-output behavior of a DPS is
represented by its transfer function. An advanced reader who is familiar with the finite
dimensional $\mathcal{H}_\infty$ controller may skip most parts of this chapter. However, the notation
used throughout the book is set up here. So, when in doubt about the notation, the reader
should consult Chapter 2. Chapter 3 includes examples of different types of DPSs and
how transfer functions are constructed. Inner (all-pass) – outer (minimum phase) factorization of a given DPS plays an important role in the $\mathcal{H}_\infty$ controller implementation;
numerical techniques to perform such factorizations are outlined in Chapter 4. Different robust control problems and their relation to the mixed sensitivity minimization are
illustrated in Chapter 5. The $\mathcal{H}_\infty$ controller design algorithm is described in Chapter 6, where different examples are given from various applications. Extension of this
$\mathcal{H}_\infty$ controller design technique to the multi-input-multi-output (MIMO) plants case is
studied in Chapter 7.

The first author taught graduate level one semester courses, based on the contents
of this book, at Bilkent University (2016) and at The Ohio State University (2018). He
also taught an accelerated short course at Beihang University (2014), covering most
of the material from the book. A typical syllabus for such a course is given below; this
assumes that the student has taken the first control course in the undergraduate program.

1. Review of basic feedback control system analysis and design, Nyquist stability
   test (from a reference book such as [50, 167, 191]).

2. Gain, phase, and delay margins for systems with time delay (from [191]).

3. Norms for signals and systems; function spaces; special operators (Sections 2.1, 2.2, 2.3, and 2.7 and supporting material from [261]).

4. Modeling, uncertainty, and robustness (Chapter 3, Section 5.1.1, and supporting material from [51, 191]).

5. Parameterization of stabilizing controllers via factorizations: systems with finitely many poles in $\mathbb{C}_+$ (Sections 4.1–4.4 and 4.6).

6. Stabilization of systems with infinitely many poles in $\mathbb{C}_+$ and repetitive control (Section 4.5).

7. Robust control problems; mixed sensitivity minimization (most of Chapter 5).

8. Model matching problem: finite dimensional case (Sections 2.4.1 and 2.4.2).

9. Model matching problem: infinite dimensional case (Section 2.4.3).

10. Mixed sensitivity minimization: spectral factorization and reduction to one block (Sections 2.5 and 2.6).

11. $\mathcal{H}_\infty$-optimal controller for infinite dimensional systems (Sections 6.1 and 6.2).

12. Examples (Sections 6.3.1–6.3.3).

13. Strong stabilization problems for time delay systems (Section 6.3.4 and supporting materials from [51] and [245]).

14. Extensions to MIMO systems (Chapter 7).

# Chapter 2

# Mathematical Preliminaries

In this chapter, some basic concepts from function spaces and linear operators are reviewed and the notation used is set up. This revision is intended to give a sufficient amount of material to follow the remaining parts of the book. For more detailed treatment, we refer the reader to [46, 66, 261] and their references.

## 2.1 ▪ Signals, systems, and function spaces

In order to put the signals and systems theory into a mathematical framework, some function spaces are defined in this section. Many technical details are omitted here; we strongly recommend consulting [261] for a complete picture.

Throughout the book we consider continuous time systems, which means that the input and output are functions of $t \in \mathbb{R}$. We further assume that the systems considered are causal, linear, and time invariant. If $P$ is such a system whose input is $u$ and output is $y$, as shown in Figure 2.1, then it is represented by its transfer function $P(s)$, which is the Laplace transform of its impulse response.

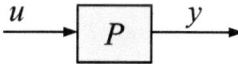

**Figure 2.1.** *Input-output representation of the system $P$.*

When a system is represented by its transfer function, it is assumed to be relaxed at $t = 0$; i.e., internal states are zero initially. In this case, the input $u(t)$, $t \geq 0$, determines the output $y(t)$, $t \geq 0$. All signals we consider are real-valued piecewise continuous functions defined on the positive time interval, $\mathbb{R}_+ = [0, \infty)$. These signals may belong to different function spaces. Some of the function spaces that are of interest to system theory are defined below.

Let $p \geq 1$; then we say that a signal $f$ (which is a function from $\mathbb{R}_+$ to $\mathbb{R}$) is in $\mathcal{L}_p(\mathbb{R}_+)$ if

$$\|f\|_p := \left( \int_0^\infty |f(t)|^p dt \right)^{1/p} < \infty. \tag{2.1}$$

Similarly, $f$ belongs to $\mathcal{L}_\infty(\mathbb{R}_+)$ if it satisfies

$$\|f\|_\infty := \operatorname*{ess\ sup}_{t \geq 0} |f(t)| \; < \; \infty. \tag{2.2}$$

In (2.1), the integration is in the sense of the Lebesgue integral, and in (2.2), note that an essential supremum is used, i.e., the values of $f$ on a set of measure zero do not contribute to the norm. If the signal is discontinuous at only finitely many time instants, then the above mentioned technicalities disappear in the norm computation. From the system theory point of view, most interesting function (signal) spaces are $\mathcal{L}_1(\mathbb{R}_+)$, $\mathcal{L}_2(\mathbb{R}_+)$, and $\mathcal{L}_\infty(\mathbb{R}_+)$. The space $\mathcal{L}_2(\mathbb{R}_+)$ can be seen as the space of *finite energy signals*, and functions in $\mathcal{L}_\infty(\mathbb{R}_+)$ are *essentially bounded signals*. The significance of $\mathcal{L}_1(\mathbb{R}_+)$ will be clear in the next section.

It is also important to note that $\mathcal{L}_2(\mathbb{R}_+)$ is a vector space on which an inner product is defined:

$$\langle f, g \rangle \; := \; \int_0^\infty g(t) f(t) dt \tag{2.3}$$

for all $f, g \in \mathcal{L}_2(\mathbb{R}_+)$.

In system theory, we also deal with functions of a complex variable. Such functions appear when frequency domain techniques are used; they appear as Fourier, or Laplace, transforms of signals. The *right half plane* is denoted by

$$\mathbb{C}_+ := \{ s \in \mathbb{C} \; : \; \operatorname{Re}(s) > 0 \};$$

similarly, the *left half plane* is denoted by

$$\mathbb{C}_- := \{ s \in \mathbb{C} \; : \; \operatorname{Re}(s) < 0 \}.$$

The notation $\mathbb{I}$ will stand for the *imaginary axis*,

$$\mathbb{I} := \{ s \in \mathbb{C} \; : \; \operatorname{Re}(s) = 0 \} = j\mathbb{R}$$

(as is customary in the electrical engineering literature, we use the notation $j = \sqrt{-1}$). For almost all practical applications of interest, we deal with complex-valued functions $F$ satisfying $\overline{F(\bar{s})} = F(s)$ for all $s \in \mathbb{C}$; i.e., these are functions with real coefficients. We restrict ourselves to this class.

A function $F : \mathbb{I} \to \mathbb{C}$ belongs to $\mathcal{L}_2(\mathbb{I})$ if it satisfies

$$\|F\|_2 := \left( \frac{1}{2\pi} \int_{-\infty}^\infty |F(j\omega)|^2 d\omega \right)^{1/2} < \infty. \tag{2.4}$$

The usual inner product in $\mathcal{L}_2(\mathbb{I})$ is defined as

$$\langle F, G \rangle \; := \; \frac{1}{2\pi} \int_{-\infty}^\infty G(j\omega)^* F(j\omega) d\omega, \tag{2.5}$$

where $G(j\omega)^*$ stands for the complex conjugate, i.e., $G(j\omega)^* = \overline{G(j\omega)} = G(-j\omega)$.

A function $F : \mathbb{I} \to \mathbb{C}$ belongs to $\mathcal{L}_\infty(\mathbb{I})$ if it satisfies

$$\|F\|_\infty := \operatorname*{ess\ sup}_{\omega \in \mathbb{R}} |F(j\omega)| < \infty. \tag{2.6}$$

Let $F : \mathbb{C}_+ \to \mathbb{C}$ be an analytic function bounded on $\mathbb{C}_+$; then we say that $F \in \mathcal{H}_\infty(\mathbb{C}_+)$ and define

$$\|F\|_\infty = \operatorname*{ess\ sup}_{\operatorname{Re}(s) > 0} |F(s)|. \tag{2.7}$$

By the maximum modulus principle, for all $F \in \mathcal{H}_\infty(\mathbb{C}_+)$ we have

$$\|F\|_\infty = \text{ess} \sup_{\text{Re}(s)>0} |F(s)| = \text{ess} \sup_{\omega \in \mathbb{R}} |F(j\omega)|. \tag{2.8}$$

In other words, the norms $\| \cdot \|_\infty$ in (2.6) and (2.7) are the same when $F \in \mathcal{H}_\infty(\mathbb{C}_+)$.

On the other hand, a function in $F \in \mathcal{L}_\infty(\mathbb{I})$ may not have a bounded analytic extension in $\mathbb{C}_+$. Therefore, the meaning of the norm $\| \cdot \|_\infty$, whether it is (2.2), or (2.6), or (2.7), depends on the function space, and we will assume that this will be clear from the context. Now consider an analytic function $F : \mathbb{C}_+ \to \mathbb{C}$ for which $\|F\|_2$ defined in (2.4) is finite; in this case, we say that $F$ belongs to $\mathcal{H}_2(\mathbb{C}_+)$. In other words, $\mathcal{H}_2(\mathbb{C}_+)$ consists of functions in $\mathcal{L}_2(\mathbb{I})$ that have analytic extensions in $\mathbb{C}_+$. We also define $\mathcal{H}_2(\mathbb{C}_-)$ as the subspace of $\mathcal{L}_2(\mathbb{I})$ consisting of functions admitting analytical extensions in $\mathbb{C}_-$.

Let $f$ be a signal in $\mathcal{L}_2(\mathbb{R}_+)$; then its Laplace transform $F(s)$ belongs to $\mathcal{H}_2(\mathbb{C}_+)$. By using the time-reversal properties of the Laplace transform, we have that for a function $g \in \mathcal{L}_2(\mathbb{R}_-)$ its Laplace transform $G(s)$ belongs to $\mathcal{H}_2(\mathbb{C}_-)$. Moreover, any function $f \in \mathcal{L}_2(\mathbb{R})$ can be decomposed as

$$f = f_+ + f_-,$$

where $f_+ \in \mathcal{L}_2(\mathbb{R}_+)$ and $f_- \in \mathcal{L}_2(\mathbb{R}_-)$, with the property that

$$\|f\|_2^2 = \|f_+\|_2^2 + \|f_-\|_2^2.$$

The regions of convergence for the Laplace transforms of $f$, $f_+$, $f_-$ (namely $F$, $F_+$, and $F_-$, respectively) include $\mathbb{I}$, $\mathbb{C}_+$, and $\mathbb{C}_-$, respectively. By Parseval's theorem [261], we have that $\|f\|_2 = \|F\|_2$, $\|f_+\|_2 = \|F_+\|_2$, and $\|f_-\|_2 = \|F_-\|_2$ (the right-hand sides are norm computations in $\mathcal{L}_2(\mathbb{I})$, $\mathcal{H}_2(\mathbb{C}_+)$, and $\mathcal{H}_2(\mathbb{C}_-)$, respectively). The inner products on $\mathcal{H}_2(\mathbb{C}_+)$ and $\mathcal{H}_2(\mathbb{C}_-)$ are inherited from the inner product definition on $\mathcal{L}_2(\mathbb{I})$.

In the rest of this book, $\mathcal{H}_2$ stands for $\mathcal{H}_2(\mathbb{C}_+)$, $\mathcal{H}_2^\perp$ stands for $\mathcal{H}_2(\mathbb{C}_-)$, and $\mathcal{H}_\infty$ stands for $\mathcal{H}_\infty(\mathbb{C}_+)$. We will also use the shorthand notations $\mathcal{L}_\infty$ for $\mathcal{L}_\infty(\mathbb{I})$, and $\mathcal{L}_2$ for $\mathcal{L}_2(\mathbb{I})$ as well as $\mathcal{L}_2(\mathbb{R}_+)$, whenever the meaning is clear from the context.

**Exercise 2.1.**

1. Let $u(t)$ denote the *unit step function* defined by

$$u(t) = \begin{cases} 0 & \text{if } t < 0, \\ 1 & \text{if } t \geq 0. \end{cases}$$

Determine whether the functions given below belong to any of the function spaces $\mathcal{L}_1(\mathbb{R}_+)$, $\mathcal{L}_2(\mathbb{R}_+)$, $\mathcal{L}_\infty(\mathbb{R}_+)$, and compute their associated norms.

$$f(t) = e^{-2t}u(t), \quad f(t) = 2t\,(u(t) - u(t-1)) + \frac{1}{t+1}u(t-1),$$

$$f(t) = \begin{cases} 0 & \text{if } t < 0, \\ t^{-3/4} & \text{if } 0 < t \leq 1, \\ e^{-2t} & \text{if } t > 1, \end{cases} \quad f(t) = \begin{cases} 0 & \text{if } t < 0, \\ t^{-1/4} & \text{if } 0 < t \leq 1, \\ t^{-4}\sin(2t) & \text{if } t > 1. \end{cases}$$

2. Define $f(t) = e^{-t}u(t)$ and $g(t) = e^{2t}(u(t) - u(t-1))$, and compute $\langle f, g \rangle$. Verify Parseval's theorem by computing the inner product $\langle F, G \rangle$, where $F$ and $G$ are Laplace transforms of $f$ and $g$, respectively.

3. Show that the function $F$ given below is in $\mathcal{L}_\infty$ and $\mathcal{L}_2$,

$$F(s) = \frac{(s-5)+2(s+3)e^{-hs}}{(s-1)(s+1)}, \quad \text{where} \quad h = \ln(2).$$

By using partial fraction expansions, find $F_1$ and $F_2$ such that $F = F_1 + F_2$, and such that $F_1 \in \mathcal{H}_2$ and $F_2 \in \mathcal{H}_2^\perp$. Compute the $\mathcal{L}_2$ norms of $F$, $F_1$, and $F_2$, and verify that $\|F\|_2^2 = \|F_1\|_2^2 + \|F_2\|_2^2$. Note that $F_2 = 0$ in this case. Repeat the question for $h = 1$ sec.

## 2.2 ▪ System norms

Consider an LTI system as shown in Figure 2.1. From the input-output point of view, we say that the system is *stable* if all inputs $u$ with $\|u\| < \infty$ lead to outputs $y$ with $\|y\| < \infty$, and

$$\|\mathbf{P}\| = \sup_{u \neq 0} \frac{\|y\|}{\|u\|} < \infty. \tag{2.9}$$

The *system norm*, $\|\mathbf{P}\|$, can be seen as the largest signal amplification through the system. Depending on which norm we select in measuring the signal strength in $u$ and $y$, we have different ways of determining system stability.

Let $h(t)$, $t \geq 0$, be the *impulse response* of the system $P$; i.e., $h$ is the output, $y$, when the input, $u$, is the Dirac delta function, denoted by $\delta(t)$. In a slightly awkward manner we use the notation $P(s)$ for the Laplace transform of $h$; $P(s)$ is called the *transfer function* of the system, and it satisfies

$$Y(s) = P(s)U(s), \tag{2.10}$$

where $U(s)$ and $Y(s)$ are Laplace transforms of $u(t)$ and $y(t)$, respectively. The identity (2.10) determines the input-output behavior of the system.

The following are well-established facts [51]:

$$\|\mathbf{P}\| = \sup_{u \neq 0} \frac{\|y\|_\infty}{\|u\|_\infty} = \int_0^\infty |h(t)|dt = \|h\|_1, \tag{2.11}$$

$$\|\mathbf{P}\| = \sup_{u \neq 0} \frac{\|y\|_2}{\|u\|_2} = \operatorname*{ess\,sup}_{\operatorname{Re}(s)>0} |P(s)| = \|P\|_\infty. \tag{2.12}$$

It is also an interesting exercise to show that, [46],

$$\|\mathbf{P}\| = \sup_{u \neq 0} \frac{\|y\|_1}{\|u\|_1} = \int_0^\infty |h(t)|dt = \|h\|_1.$$

The equality (2.11) says that the system is bounded-input-bounded-output (BIBO) stable (or $\mathcal{L}_\infty(\mathbb{R}_+)$ to $\mathcal{L}_\infty(\mathbb{R}_+)$ stable) if and only if its impulse response is absolutely integrable, i.e., the $\mathcal{L}_1(\mathbb{R}_+)$ norm of the impulse response is finite. The importance of the signal space $\mathcal{L}_1(\mathbb{R}_+)$ is now clear from this fact.

On the other hand, (2.12) says that from the energy amplification point of view the system is stable (or $\mathcal{L}_2$ to $\mathcal{L}_2$ stable) if and only if its transfer function $P(s)$ is in $\mathcal{H}_\infty$. Therefore, the space $\mathcal{H}_\infty$ can be seen as the set of all $\mathcal{L}_2$ to $\mathcal{L}_2$ stable systems. Moreover, by (2.8), when $P \in \mathcal{H}_\infty$ the system norm is computed as the peak value of the frequency response magnitude. Throughout the book we use this stability definition, which is formally stated as follows.

**Definition 2.2.** *We say that an LTI system represented by the transfer function $P(s)$ is* stable *if $P \in \mathcal{H}_\infty$.*

When dealing with finite dimensional systems, where $P(s)$ is rational, the system is $\mathcal{L}_\infty(\mathbb{R}_+)$ to $\mathcal{L}_\infty(\mathbb{R}_+)$ stable if and only if it is $\mathcal{L}_2$ to $\mathcal{L}_2$ stable. It is easy to see this because in this case the impulse response is a linear combination of functions in the form $t^{m_i - 1}e^{\lambda_i t}$, $t \geq 0$, where $\lambda_i$ is a pole of $P(s)$ with multiplicity $m_i$. So we have $\|h\|_1 < \infty$ if and only if $\|P\|_\infty < \infty$, if and only if $\operatorname{Re}(\lambda_i) < 0$ for all $i$. This statement is not true when we deal with infinite dimensional plants, where $P(s)$ is not rational. There are examples where all of the poles of $P(s)$ have negative real parts, yet the system is not stable; see the recent papers [176, 177] and their references.

**Exercise 2.3.**
1. Consider an impulse response $h \in \mathcal{L}_1(\mathbb{R}_+)$. Recall that $H(s) = \int_0^\infty h(t)e^{-st}dt$, so, in general, we have $\|H\|_\infty \leq \|h\|_1$. Prove that if $h(t) \geq 0$ for all $t \geq 0$, then we have $\|H\|_\infty = H(0) = \|h\|_1$.
2. Show that the function given below is in $\mathcal{H}_\infty$ for all $h \geq 0$:

$$F(s) = \frac{(s+1) + \frac{2}{3}(s-4)e^{-h(s-1)}}{(s-1)}.$$

Compute $F(1)$ for $h = \frac{1}{4}$. Is $F \in \mathcal{L}_2$?
3. The purpose of this exercise is to confirm an example given in [176], where a transfer function is given in the form $G(s) = n_g(s)/d_g(s)$, where $n_g$ is a polynomial and $d_g$ is a quasi-polynomial

$$d_g(s) = (s^3 + 3s^2 + 2s) + (2s^3 + 2s^2 - s + 0.5)e^{-s} + (s^3 - s^2 + 0.5s)e^{-2s}.$$

In [176] it was shown that all the finite roots of the equation $d_g(s) = 0$ are in $\mathbb{C}_-$, and there are infinitely many roots with the real parts of the roots converging to zero from below. That is, $G$ has at least one chain of poles converging to $\mathbb{I}$.
As a first step, observe this fact by using the MATLAB based function QPmR.m [249], which can be downloaded from *http://www.cak.fs.cvut.cz/algorithms*. Plot the roots of $d_g$ in the region $-0.002 \leq \operatorname{Re}(s) \leq 0.0001$, $-200 \leq \operatorname{Im}(s) \leq 200$.
Then, let $n_g(s) = 5$ and plot the frequency response magnitude $|G(j\omega)|$ (Bode magnitude plot of $G$) as a function of $\omega$. For this purpose, you may have to take 2,000 logarithmically spaced frequency points between 0.01 and 10 and then 200,000 points between 10 and 100 (only a very large number of data points will illustrate the behavior as $\omega \to \infty$). Observing this behavior, you may want to prove that as $\omega \to \infty$ we have $|G(j\omega)| \leq 10$; that is, $|d_g(j\omega)| \geq 0.5$ for all large $\omega$ values.
Now take $n_g(s) = 5\,(s+1)$, obtain $|G(j\omega)|$, and plot as above. Observe that there is a sequence $\{\omega_k\}_{k=1}^\infty$ such that $|G(j\omega_k)| \to \infty$ as $\omega_k \to \infty$. In other words, it is not possible to find a constant $k$ such that $|d_g(j\omega)| \geq k\sqrt{\omega^2 + 1}$ for $\omega \to \infty$. Thus, in this case we have $\|G\|_\infty = \infty$, although all its poles are in $\mathbb{C}_-$.

## 2.3 ▪ Some special operators

In this section we define some operators that play an important role in the solution of robust control problems. Again, we keep the discussion at a minimum level here and refer the reader to [261] for further details.

From the discussion above, we see that a transfer function $P \in \mathcal{H}_\infty$ represents a bounded operator from $\mathcal{H}_2$ to $\mathcal{H}_2$; more precisely, for all $U \in \mathcal{H}_2$ we get that $Y(s) = P(s)U(s)$ in $\mathcal{H}_2$, and

$$\|Y\|_2 \leq \|P\|_\infty \|U\|_2. \tag{2.13}$$

Note that $P \in \mathcal{H}_\infty$ can also be seen as a bounded operator from $\mathcal{H}_\infty$ to $\mathcal{H}_\infty$ because for any $U \in \mathcal{H}_\infty$ we have $Y = PU \in \mathcal{H}_\infty$, and

$$\|Y\|_\infty \leq \|P\|_\infty \|U\|_\infty. \tag{2.14}$$

**Definition 2.4.** *We say that a transfer function $P(s)$ is* proper *if there exists $d \in \mathbb{R}_+$ such that as $|s| \to \infty$ we have $|P(s)| \leq d$; it is said to be* strictly proper *if this property holds with $d = 0$. If $P(s)$ is not proper, then it is* improper. *If $P(s)$ is proper but not strictly proper, then $P$ is called* biproper; *in this case $P^{-1}(s)$ is also proper.*

A transfer function $M \in \mathcal{H}_\infty$ is said to be *inner* if $|M(s)| \leq 1$ for all $s \in \mathbb{C}_+$ and $|M(j\omega)| = 1$ almost everywhere, $\omega \in \mathbb{R}$. Note that $M$ is a norm preserving operator. More precisely, if $G = MF$ for $F \in \mathcal{H}_2$ (respectively, $F \in \mathcal{H}_\infty$), we have that $|G(j\omega)| = |F(j\omega)|$ almost everywhere, $\omega \in \mathbb{R}$, which implies that $\|G\|_2 = \|F\|_2$ (and, respectively, $\|G\|_\infty = \|F\|_\infty$). Inner functions are also called *all-pass* transfer functions because the Bode magnitude plots of the input and the output are the same (what comes in goes out without any attenuation or amplification, but there may be some phase shift between the input and the output).

A function $N_o \in \mathcal{H}_\infty$ is called *outer* if the closure of $\{N_o F : F \in \mathcal{H}_2\}$ in $\mathcal{H}_2$ is the whole space $\mathcal{H}_2$. In particular, this implies that $N_o(s)$ does not have any zeros in $\mathbb{C}_+$. In a less formal way, we can say that $N_o \in \mathcal{H}_\infty$ is outer if it satisfies the following property: for all $N \in \mathcal{H}_\infty$ satisfying (i) $\text{sign}(N(\epsilon)) = \text{sign}(N_o(\epsilon))$ for all $\epsilon > 0$ sufficiently small and (ii) $|N(j\omega)| = |N_o(j\omega)|$ almost everywhere $\omega \in \mathbb{R}$, we have that

$$\angle N_o(j\omega) \geq \angle N(j\omega) \quad \text{a.e.} \ \omega \in \mathbb{R}, \tag{2.15}$$

where $\angle$ represents the phase of the function. The condition (i) matches the signs of $N$ and $N_o$ near $s = 0$ (i.e., signs of DC gains are matched; there may be some pathological outer functions where $s = 0$ corresponds to an *essential singularity*, in this case, the signs are matched at another point on the positive real axis). The condition (ii) says that the Bode magnitude plots of $N$ and $N_o$ are indistinguishable. The inequality (2.15) means that the phase drop through the system $N_o$ is smaller than the phase drop through $N$ at all frequencies. For this reason, outer functions are called *minimum-phase* functions. In fact, *minimum-phase shift* is a better terminology; otherwise, (2.15) suggests that outer functions have *maximum phase* among all functions having the same magnitude, with matched sign.

For any given $N \in \mathcal{H}_\infty$ we can perform an *inner-outer factorization*, so that $N = MN_o$, where $M$ is inner and $N_o$ is outer. Moreover, this factorization is unique up to multiplication by a constant $\alpha$, where $|\alpha| = 1$. Note that among all functions satisfying the standing assumption $\overline{F(\bar{s})} = F(s)$, the functions $F_0(s) = 1$ and $F_1(s) = -1$ are the only ones which are both inner and outer.

**Example 2.5.** Consider the following transfer function in $\mathcal{H}_\infty$:

$$N(s) = \frac{e^{-hs}(a - s)}{s^2 + 2s + 2}, \quad h \geq 0, \ a > 0.$$

We can rewrite it as $N(s) = M(s)N_o(s)$, where $M$ is inner and $N_o$ is outer:

$$M(s) = \frac{e^{-hs}(a-s)}{(a+s)}, \quad N_o(s) = \frac{(a+s)}{(s+2+2s+2)}.$$

The inner function contains all transport delays and the zeros in $\mathbb{C}_+$. The inverse of the outer function does not contain any poles in $\mathbb{C}_+$, but it may be improper, as in this example, and may also have poles on $\mathbb{I}$. More interesting inner-outer factorization examples will be given in Chapter 4.   ■

Finite Blaschke products of the form

$$B_N(s) = \prod_{k=1}^{N} \frac{a_k - s}{\overline{a}_k + s},$$

where $a_k \in \mathbb{C}_+$ appear in *complex conjugate pairs*, $k = 1, \ldots, N$, are typical factors of inner functions. The infinite Blaschke product defined by a sequence $\{a_k\}_{k=1}^{\infty}$, where the elements appear in complex conjugate pairs with $\text{Re}(a_k) > 0$,

$$B_\infty(s) = \prod_{k=1}^{\infty} \frac{a_k - s}{\overline{a}_k + s}, \tag{2.16}$$

converges uniformly in $\mathbb{C}_+$ [4], and thus is an inner function, provided that

$$\sum_{k=1}^{\infty} \frac{\text{Re}(a_k)}{1 + |a_k|^2} < \infty. \tag{2.17}$$

Note that the product of two inner functions is also inner, and, similarly that the product of two outer functions is also outer.

For a biproper outer function $N_o \in \mathcal{H}_\infty$ with no zeros on $\mathbb{I}$, we see that $N_o^{-1} \in \mathcal{H}_\infty$ and hence

$$N_o \mathcal{H}_2 := \{N_o(s)F(s) \ : \ F \in \mathcal{H}_2\} = \mathcal{H}_2,$$

and similarly

$$N_o \mathcal{H}_\infty := \{N_o(s)F(s) \ : \ F \in \mathcal{H}_\infty\} = \mathcal{H}_\infty.$$

On the other hand, for a nonconstant inner function $M \in \mathcal{H}_\infty$ we have

$$M\mathcal{H}_2 := \{M(s)F(s) \ : \ F \in \mathcal{H}_2\} \subset \mathcal{H}_2.$$

Note that $M\mathcal{H}_2$ is an invariant subspace of $\mathcal{H}_2$, and we can define its orthogonal complement in $\mathcal{H}_2$ as $\mathcal{H}(M) := \mathcal{H}_2 \ominus M\mathcal{H}_2$. This means that for every $F \in \mathcal{H}_2$ we have a unique decomposition

$$F = G + MH, \quad \text{where} \quad G \in \mathcal{H}(M) \quad \text{and} \quad H \in \mathcal{H}_2. \tag{2.18}$$

It can be shown that for $M = M_1 M_2$, where $M_1$ and $M_2$ are two inner functions, we have $\mathcal{H}(M) = \mathcal{H}(M_1) \oplus M_1 \mathcal{H}(M_2)$.

As implicit in the above discussion, we associate a function $F \in \mathcal{H}_\infty$ with a *multiplication operator* on $\mathcal{H}_2$, i.e., $\mathbf{M}_F \ : \ \mathcal{H}_2 \to \mathcal{H}_2$ defined by $\mathbf{M}_F(U) = FU$ for all $U \in \mathcal{H}_2$.

With the help of the inner product on $\mathcal{L}_2$, we can define *projection operators* $\mathbf{\Pi}_-$, $\mathbf{\Pi}_+$, and $\mathbf{\Pi}_{\mathcal{H}(M)}$ as follows: let $F \in \mathcal{L}_2$ be decomposed as $F = F_- + F_+$, where $F_- \in \mathcal{H}_2^\perp$ and $F_+ \in \mathcal{H}_2$. Then

$$\mathbf{\Pi}_- \;:\; \mathcal{L}_2 \to \mathcal{H}_2^\perp, \quad \mathbf{\Pi}_-(F) = F_-, \tag{2.19}$$

$$\mathbf{\Pi}_+ \;:\; \mathcal{L}_2 \to \mathcal{H}_2, \quad \mathbf{\Pi}_+(F) = F_+, \tag{2.20}$$

$$\mathbf{\Pi}_{\mathcal{H}(M)} \;:\; \mathcal{H}_2 \to \mathcal{H}(M), \quad \mathbf{\Pi}_{\mathcal{H}(M)}(F) = G, \tag{2.21}$$

where $G \in \mathcal{H}(M)$ is defined by the decomposition (2.18). For a given $F \in \mathcal{H}_2$ how do we compute $G$ and $H$ in (2.18)? Clearly, $G = F - MH$, so it suffices to find $H$. Since $M^*G \in \mathcal{H}_2^\perp$, we have that

$$H = \mathbf{\Pi}_+ M^* F.$$

Let $G(s)$ be a given transfer function such that $G \in \mathcal{L}_\infty$. Then, $\mathbf{M}_G$ is a bounded multiplication operator from $\mathcal{L}_2$ to $\mathcal{L}_2$ defined as $\mathbf{M}_G(F) = GF$, recalling that we have $\|\mathbf{M}_G(F)\|_2 \le \|G\|_\infty \|F\|_2$. Next we define two special operators from a given $G \in \mathcal{L}_\infty$:

$$\mathbf{\Gamma}_G \;:\; \mathcal{H}_2 \to \mathcal{H}_2^\perp, \quad \mathbf{\Gamma}_G(F) := \mathbf{\Pi}_- GF, \tag{2.22}$$

$$\mathbf{\Upsilon}_G \;:\; \mathcal{H}_2 \to \mathcal{H}_2, \quad \mathbf{\Upsilon}_G(F) := \mathbf{\Pi}_+ GF. \tag{2.23}$$

The operator $\mathbf{\Gamma}_G$ is called the *Hankel operator with symbol $G$*, and the operator $\mathbf{\Upsilon}_G$ is called the *Toeplitz operator with symbol $G$*. They play a crucial role in the solution of the $\mathcal{H}_\infty$ control problems.

Let $W(s)$ be a stable transfer function, i.e., $W \in \mathcal{H}_\infty$, and let $M \in \mathcal{H}_\infty$ be an inner function such that $W(s)$ and $M(s)$ do not have common zeros in $\mathbb{C}_+$. Then, we define

$$W(\mathbf{T}) \;:\; \mathcal{H}(M) \to \mathcal{H}(M), \quad W(\mathbf{T})(F) = \mathbf{\Pi}_{\mathcal{H}(M)} \mathbf{M}_W F \text{ for } F \in \mathcal{H}(M). \tag{2.24}$$

In the computation of $\mathcal{H}_\infty$ controllers for infinite dimensional systems, $W(\mathbf{T})$ (which we call the *Sarason's operator*), is known to be very useful.

## 2.4 ▪ The model matching problem

As we shall see later in the book, many interesting robust control problems are transformed to the following *one-block problem*: given $W \in \mathcal{H}_\infty$ and an inner function $M \in \mathcal{H}_\infty$, find

$$\gamma_{\text{opt}} = \inf_{Q \in \mathcal{H}_\infty} \|W - MQ\|_\infty \tag{2.25}$$

and the corresponding optimal $Q_{\text{opt}} \in \mathcal{H}_\infty$ achieving $\|W - MQ_{\text{opt}}\|_\infty = \gamma_{\text{opt}}$.

It is known that, under certain mild assumptions that hold in practical control problems, $\gamma_{\text{opt}} = \|W(\mathbf{T})\|$ (norm of the Sarason operator) and $Q_{\text{opt}}$ is computed from the singular vector of $W(\mathbf{T})$ corresponding to the largest singular value (which is the norm). This is formally stated below.

**Theorem 2.6 (Sarason).** *Assume that $\gamma_{\text{opt}} = \|W(\mathbf{T})\|$ is achieved at the discrete spectrum. Accordingly, let $G_o$ be a nonzero function in $\mathcal{H}(M)$ satisfying*

$$\left( \gamma_{\text{opt}}^2 \mathbf{I} - W(\mathbf{T})^* W(\mathbf{T}) \right) G_o = 0. \tag{2.26}$$

*Then*

$$F_{\text{opt}} := (W - MQ_{\text{opt}}) = \frac{W(\mathbf{T})G_o}{G_o}. \tag{2.27}$$

*Moreover, the function $F_{\text{opt}}(s)$ is equal to $\gamma_{\text{opt}}$ times an inner function.*

See [65, 66, 221, 261] for the proof and more detailed discussion on Sarason's theorem and its relation to interpolation problems and the "commutant lifting theorem" of [232].

Let us concentrate on the finite dimensional case for the moment, where $W$ and $M$ are rational functions. In this case $\mathcal{H}(M)$ is a finite dimensional subspace of $\mathcal{H}_2$, and hence $W(\mathbf{T})$ can be represented by a finite size matrix; computation of its singular values and vectors is straightforward.

**Example 2.7.** For two distinct numbers $\alpha_1, \alpha_2 \in \mathbb{R}_+$, consider the inner functions

$$M_i(s) = \frac{(s - \alpha_i)}{(s + \alpha_i)}, \quad i = 1, 2, \quad \text{and} \quad M(s) = M_1(s)M_2(s).$$

Then $G \in \mathcal{H}(M)$ if and only if $\langle M^*G, F \rangle = 0$ for all $F \in \mathcal{H}_2$. This means that each

$$G_i(s) = \frac{\sqrt{2\alpha_i}}{s + \alpha_i}$$

is a normal basis for $\mathcal{H}(M_i)$. Then $\{G_1, G_2\}$ forms a basis for $\mathcal{H}(M)$. But $G_1$ and $G_2$ are not orthogonal to each other; so an orthonormal basis for $\mathcal{H}(M)$ can be obtained by the Gram–Schmidt procedure as follows. Let $\widehat{G}_1 = G_1$, and let $\widehat{G}_2 = E_2/\|E_2\|_2$, where

$$E_2 = G_2 - \langle G_2, \widehat{G}_1 \rangle \widehat{G}_1 = \alpha_{12} \, M_1(s) \, G_2(s), \quad \text{with} \quad \alpha_{12} := \left( \frac{\alpha_2 - \alpha_1}{\alpha_2 + \alpha_1} \right).$$

Since $\|E_2\|_2 = \alpha_{12}$, we have

$$\widehat{G}_2(s) = M_1(s)G_2(s).$$

Thus, $\{G_1, M_1G_2\}$ is an orthonormal basis for $\mathcal{H}(M)$.

Now consider $W(s) = (s + \varepsilon)^{-1}$. One can compute the matrix representation of $W(\mathbf{T})$ by finding the action of this operator on the basis functions:

$$W(\mathbf{T})\widehat{G}_1 = \mathbf{\Pi}_{\mathcal{H}(M)}W(s)\widehat{G}_1(s) = \left( \frac{1}{\alpha_1 + \varepsilon} \right) \widehat{G}_1(s) + \left( \frac{-2\sqrt{\alpha_1\alpha_2}}{\alpha_1 + \varepsilon} \right) \left( \frac{1}{\alpha_2 + \varepsilon} \right) \widehat{G}_2(s),$$

$$W(\mathbf{T})\widehat{G}_2 = \mathbf{\Pi}_{\mathcal{H}(M)}W(s)\widehat{G}_2(s) = 0 \, \widehat{G}_1(s) + \left( \frac{1}{\alpha_2 + \varepsilon} \right) \widehat{G}_2(s).$$

In conclusion, we have a triangular matrix representation of $W(\mathbf{T})$ in the orthonormal basis $\{G_1, M_1G_2\} = \{\widehat{G}_1, \widehat{G}_2\}$:

$$W(\mathbf{T})\, [\widehat{G}_1 \ \widehat{G}_2] = [\widehat{G}_1 \ \widehat{G}_2] \begin{bmatrix} 1 & 0 \\ 0 & \alpha_2 + \varepsilon \end{bmatrix}^{-1} \begin{bmatrix} 1 & 0 \\ -2\sqrt{\alpha_1\alpha_2} & 1 \end{bmatrix} \begin{bmatrix} \alpha_1 + \varepsilon & 0 \\ 0 & 1 \end{bmatrix}^{-1}.$$

According to Sarason's theorem, $\gamma_{\text{opt}} = \|W(\mathbf{T})\| = \sqrt{\sigma_{\max}}$, where $\sigma_{\max}$ is the largest eigenvalue of $(\mathsf{X}^\mathsf{T} \mathsf{X})$ where

$$\mathsf{X} = \begin{bmatrix} 1 & 0 \\ 0 & \alpha_2 + \varepsilon \end{bmatrix}^{-1} \begin{bmatrix} 1 & 0 \\ -2\sqrt{\alpha_1\alpha_2} & 1 \end{bmatrix} \begin{bmatrix} \alpha_1 + \varepsilon & 0 \\ 0 & 1 \end{bmatrix}^{-1}. \tag{2.28}$$

It is a simple exercise to check that $\sigma_{\max}$ is the largest $\sigma$ satisfying

$$\sigma^2 \, (\alpha_1 + \varepsilon)^2(\alpha_2 + \varepsilon)^2 - \sigma \left( (\alpha_2 + \varepsilon)^2 + (\alpha_1 + \varepsilon)^2 + 4\alpha_1\alpha_2 \right) + 1 = 0. \tag{2.29}$$

The optimal solution (2.27) can be determined from the following MATLAB commands:

```
[U,S,V]=svd(X); gopt=S(1,1); Vp=V'; Go=Vp(:,1);
XGo=X*Go; d1=Go(1); d2=Go(2); n1=XGo(1); n2=XGo(2);
```

The final result is expressed in the form

$$F_{\text{opt}}(s) = \frac{W(\mathbf{T})G_o}{G_o} = \frac{\text{n1}\ G_1(s) + \text{n2}\ M_1(s)G_2(s)}{\text{d1}\ G_1(s) + \text{d2}\ M_1(s)G_2(s)}$$

$$= \frac{\text{n1}\ \sqrt{\alpha_1}(s+\alpha_2) + \text{n2}\ \sqrt{\alpha_2}(s-\alpha_1)}{\text{d1}\ \sqrt{\alpha_1}(s+\alpha_2) + \text{d2}\ \sqrt{\alpha_2}(s-\alpha_1)} = \lambda\ \left(\frac{s-\phi_0}{s+\phi_0}\right),$$

where

$$\lambda = \frac{\text{n1}\ \sqrt{\alpha_1} + \text{n2}\ \sqrt{\alpha_2}}{\text{d1}\ \sqrt{\alpha_1} + \text{d2}\ \sqrt{\alpha_2}},$$

$$\phi_0 = \left(\frac{\text{d1}\ \sqrt{\alpha_1}\alpha_2 - \text{d2}\ \sqrt{\alpha_2}\alpha_1}{\text{d1}\ \sqrt{\alpha_1} + \text{d2}\ \sqrt{\alpha_2}}\right) = -\left(\frac{\text{n1}\ \sqrt{\alpha_1}\alpha_2 - \text{n2}\ \sqrt{\alpha_2}\alpha_1}{\text{n1}\ \sqrt{\alpha_1} + \text{n2}\ \sqrt{\alpha_2}}\right) > 0,$$

and $\gamma_{\text{opt}} = |\lambda|$.  ∎

**Exercise 2.8.**
1. Compute $\gamma_{\text{opt}}$ and $F_{\text{opt}}(s)$ for $\alpha_1 = 5$, $\alpha_2 = 2$, and $\varepsilon = \frac{1}{10}$.
*Hint*: The answer is $\gamma_{\text{opt}} = 0.7742$ with $\phi_0 = 8.3917$ and $\lambda = -\gamma_{\text{opt}}$.

2. Let $\alpha_1 = 3 + j$ and $\alpha_2 = 3 - j$. Define $M(s) = M_1(s)M_2(s)$, with

$$M_1(s) = \frac{s-\alpha_1}{s+\overline{\alpha}_1}, \quad M_2(s) = \frac{s-\alpha_2}{s+\overline{\alpha}_2}.$$

Show that an orthonormal basis for $\mathcal{H}(M)$ is $\{\widehat{G}_1, \widehat{G}_2\}$, where

$$\widehat{G}_1(s) = \frac{\sqrt{2\text{Re}(\alpha_1)}}{s+\overline{\alpha}_1}, \quad \widehat{G}_2(s) = M_1(s)\frac{\sqrt{2\text{Re}(\alpha_2)}}{s+\overline{\alpha}_2}.$$

Compute X and its largest singular value for the case $W(s) = \frac{1}{s+0.1}$.

## 2.4.1 ▪ The Nevanlinna–Pick interpolation

Suppose that the zeros of $M(s)$ are distinct: $\alpha_1, \ldots, \alpha_n \in \mathbb{C}_+$. If we define

$$F(s) = W(s) - M(s)Q(s),$$

then $\gamma_{\text{opt}}$ is the smallest $\gamma > 0$ for which we can find a function $F \in \mathcal{H}_\infty$ with $\|F\|_\infty \le \gamma$ satisfying the interpolation conditions

$$F(\alpha_i) = W(\alpha_i), \quad i = 1, \ldots, n. \tag{2.30}$$

This is the *Nevanlinna–Pick* interpolation problem, and the solution, given in Theorem 2.9 below, is obtained from the *Pick matrix*:

$$P_\gamma := A - \gamma^{-2}B,$$

where

$$[A]_{i,j} = \frac{1}{\overline{\alpha}_i + \alpha_j}, \quad [B]_{i,j} = \frac{\overline{\beta}_i\beta_j}{\overline{\alpha}_i + \alpha_j}, \quad \text{with} \quad \beta_i := W(\alpha_i). \tag{2.31}$$

**Theorem 2.9 (Nevanlinna–Pick).** *The optimal performance level is*

$$\gamma_{\text{opt}} = \sqrt{\lambda_{\max}(\mathsf{A}^{-1}\mathsf{B})},$$

*where $\lambda_{\max}(\cdot)$ stands for the largest eigenvalue. Moreover, for a given $\gamma > \gamma_{\text{opt}}$, all solutions $F \in \mathcal{H}_\infty$ satisfying (2.30) with $\|F\|_\infty \leq \gamma$ are parameterized as*

$$F(s) = \gamma \frac{\Theta_{11}(s)G(s) + \Theta_{12}(s)}{\Theta_{21}(s)G(s) + \Theta_{22}(s)}, \tag{2.32}$$

*with the free parameter $G \in \mathcal{H}_\infty$ subject to $\|G\|_\infty \leq 1$, and the $\Theta_{ij}$'s are computed from*

$$\begin{bmatrix} \Theta_{11}(s) & \Theta_{12}(s) \\ \Theta_{21}(s) & \Theta_{22}(s) \end{bmatrix} = \begin{bmatrix} 1 & 0 \\ 0 & 1 \end{bmatrix} + \mathsf{Q}_1 \mathsf{D}_\alpha(s) \, \mathsf{P}_\gamma^{-1} \mathsf{Q}_2, \tag{2.33}$$

*where*

$$\mathsf{Q}_1 = \begin{bmatrix} \beta_1/\gamma & \cdots & \beta_n/\gamma \\ 1 & \cdots & 1 \end{bmatrix},$$

$$\mathsf{Q}_2 = \begin{bmatrix} -\overline{\beta}_1/\gamma & 1 \\ \vdots & \vdots \\ -\overline{\beta}_n/\gamma & 1 \end{bmatrix},$$

*and $\mathsf{D}_\alpha(s)$ is the diagonal matrix determined by $[\mathsf{D}_\alpha(s)]_{i,i} = (s - \alpha_i)^{-1}$.*

See [16] for a detailed discussion on this result.

From Theorem 2.9 we can obtain a particular $F(s)$ as the central solution (where $G(s) = 0$):

$$F_c(s) = \frac{[\beta_1 \ \cdots \ \beta_n] \, (s\mathsf{I} - \Lambda_\alpha)^{-1} \, \mathsf{B}_\gamma}{1 + [1 \ \cdots \ 1] \, (s\mathsf{I} - \Lambda_\alpha)^{-1} \, \mathsf{B}_\gamma}, \tag{2.34}$$

where

$$\Lambda_\alpha := \text{diag}(\alpha_1, \ldots, \alpha_n), \tag{2.35}$$

$$\mathsf{B}_\gamma := \mathsf{P}_\gamma^{-1} [ \ 1 \ \cdots \ 1 \ ]^{\mathsf{T}}. \tag{2.36}$$

Note that in the above computations, the finite dimensionality of $W(s)$ is not used at all. This means that the Nevanlinna–Pick approach works even if $W(s)$ is not rational. We only need the values of $W$ at $\alpha_i$, $i = 1, \ldots, n$. Another important point to note is that as $\gamma \searrow \gamma_{\text{opt}}$ the matrix $\mathsf{P}_\gamma$ approaches a singular matrix, and hence numerical computations of $\Theta_{i,j}$'s using (2.33) become a challenging issue.

On the other hand, the optimal solution can be computed from the following facts: $F_{\text{opt}} \in \mathcal{H}_\infty$ is such that $F_{\text{opt}}(s)$ is a rational function of order $(n-1)$, and $|F_{\text{opt}}(j\omega)| = \gamma_{\text{opt}}$ for all $\omega \in \mathbb{R}$. Therefore, it can be postulated that

$$F_{\text{opt}}(s) = \lambda \frac{[s^{n-1} \ \cdots \ s \ 1] \, \mathsf{J} \, \Phi}{[s^{n-1} \ \cdots \ s \ 1] \, \Phi} \tag{2.37}$$

for some $\Phi = [\phi_{n-1} \ \cdots \ \phi_0]^{\mathsf{T}} \in \mathbb{R}^n$, where $\lambda \in \mathbb{R}$ with $|\lambda| = \gamma_{\text{opt}}$, and $\mathsf{J}$ is a diagonal matrix with $[\mathsf{J}]_{i,i} = (-1)^{i-1}$, $i = 1, \ldots, n$. Moreover, the roots of the polynomial

$$D_{\text{opt}}(s) := \phi_{n-1}s^{n-1} + \cdots + \phi_1 s + \phi_0$$

must be in $\mathbb{C}_-$. The interpolation conditions (2.30) means that

$$(\lambda \, \mathsf{V}_\alpha \, \mathsf{J} - \mathsf{D}_\beta \, \mathsf{V}_\alpha) \, \Phi = 0, \tag{2.38}$$

where $\mathsf{D}_\beta$ is a diagonal matrix defined by

$$[\mathsf{D}_\beta]_{i,i} = \beta_i = W(\alpha_i),$$

and $\mathsf{V}_\alpha$ is the Vandermonde matrix

$$\mathsf{V}_\alpha = \begin{bmatrix} \alpha_1^{n-1} & \cdots & \alpha_1 & 1 \\ \vdots & \cdots & \vdots & \vdots \\ \alpha_n^{n-1} & \cdots & \alpha_n & 1 \end{bmatrix}.$$

The condition (2.38) can be rewritten as

$$\left(\lambda \mathsf{J} - \mathsf{V}_\alpha^{-1} \mathsf{D}_\beta \, \mathsf{V}_\alpha\right) \Phi = 0. \tag{2.39}$$

Define the coefficients $a_1, \ldots, a_n$ from the polynomial

$$\prod_{i=1}^{n}(s - \alpha_i) =: s^n + a_1 s^{n-1} + \cdots + a_n.$$

Let the associated canonical matrix be

$$\mathsf{A}_o := \begin{bmatrix} -a_1 & 1 & 0 & 0 \\ \vdots & 0 & \ddots & 0 \\ \vdots & 0 & 0 & 1 \\ -a_n & 0 & \cdots & 0 \end{bmatrix}. \tag{2.40}$$

Then, it is a simple exercise to verify that

$$\mathsf{A}_o = \mathsf{V}_\alpha^{-1} \mathsf{D}_\alpha \, \mathsf{V}_\alpha,$$

and since $\beta_i = W(\alpha_i)$, we have that

$$\mathsf{V}_\alpha^{-1} \mathsf{D}_\beta \, \mathsf{V}_\alpha = W(\mathsf{A}_o).$$

In conclusion, (2.39) is equivalent to

$$(\lambda \mathsf{I} - \mathsf{J} \, W(\mathsf{A}_o)) \, \Phi = 0. \tag{2.41}$$

The above discussion leads to the following result.

**Theorem 2.10.** *Consider the eigenvalue–eigenvector problem defined by* (2.41)*. We have that* $\lambda = \pm\gamma_{\mathrm{opt}} = \pm\sqrt{\lambda_{\max}(\mathsf{A}^{-1}\mathsf{B})}$ *is an eigenvalue of the matrix* $\mathsf{J}\,W(\mathsf{A}_o) = \mathsf{J}\,\mathsf{V}_\alpha^{-1}\mathsf{D}_\beta\,\mathsf{V}_\alpha$ *and the corresponding eigenvector* $\Phi$ *defines* $F_{\mathrm{opt}}(s)$ *via* (2.37)*. Once* $\Phi$ *is determined from the eigenvectors of* $\mathsf{J}\,W(\mathsf{A}_o)$*, it can be verified that the roots of* $D_{\mathrm{opt}}(s) = \phi_{n-1}s^{n-1} + \cdots + \phi_1 s + \phi_0$ *are in* $\mathbb{C}_-$*. Finally, the optimal solution of* (2.25) *is given by*

$$Q_{\mathrm{opt}}(s) = \frac{W(s) - F_{\mathrm{opt}}(s)}{M(s)} \in \mathcal{H}_\infty. \tag{2.42}$$

A preliminary version of the above result first appeared in [269]; an expanded version was recently published in [270]. There are many references on the Nevanlinna–Pick interpolation (see, e.g., [16, 51, 66, 167, 261, 276]) and its connections with Sarason's theorem and alternative solutions.

The MATLAB code given below computes the optimal interpolant $F_{opt}(s)$ by using the result summarized in Theorem 2.10.

```
function [gopt,Fopt]=NevPickNew(a,b)
tol=1e-6;  % change this if two eigenvalues are close to each other
n=length(a);
for ii=1:n
    for jj=1:n %construct matrices A and B
        A(ii,jj)=1/(conj(a(ii))+a(jj));
        B(ii,jj)=conj(b(ii))*b(jj)*A(ii,jj);
    end
    Jii(ii)=(-1)^(ii-1);
end
gopt=abs(sqrt(max(eig(inv(A)*B))));
Va=vander(a); Db=diag(b); JJ=diag(Jii);
[EigVec,Lambda]=eig(JJ*inv(Va)*Db*Va);
for ii=1:n
    if abs(abs(Lambda(ii,ii))-gopt)<tol
        kk=ii;
    else
    end
end
PhiVec=EigVec(:,kk);
Fopt=zpk(tf((Lambda(kk,kk)*JJ*PhiVec)',PhiVec'));
end
```

As an example, if we apply the commands

```
>> a=[1+1i,1-1i]; b=[10/(a(1)+0.01), 10/(a(2)+0.01)];
>> [gopt,Fopt]-NevPickNew(a,b)
```

we obtain

$$\text{gopt} = 13.5530 \quad \text{and} \quad F_{opt}(s) = -13.553\,\frac{(s-2.748)}{(s+2.748)}.$$

See Section 2.4.2 and Exercise 2.15 for an alternative derivation of these results.

An important point to note is that the MATLAB code given above uses the eigenvalues of the matrix $J\,V_\alpha^{-1} D_\beta\, V_\alpha$; inversion of a Vandermonde matrix can be numerically problematic if the size of the matrix is large or if there are two interpolation points close to each other (i.e., when $V_\alpha$ is close to being singular). In that case, it might be better to use the eigenvalues and vectors of the matrix $J\,W(A_o)$ for the computation of the optimal interpolant.

**Exercise 2.11.**
Consider $\alpha_1 = 5$, $\alpha_2 = 2$, and $W(s) = \frac{1}{s+0.1}$. Verify that $\gamma_{opt}$ and $F_{opt}(s)$ obtained by the method given in this section match the results obtained in Example 2.7.

## 2.4.2 ▪ The Nehari problem

An alternative solution to the one-block problem is given by the Nehari theorem. To see this, first note that, since $M$ is inner, (2.25) is equivalent to

$$\gamma_{\text{opt}} = \inf_{Q \in \mathcal{H}_\infty} \|M^*W - Q\|_\infty. \tag{2.43}$$

The function $R := M^*W$ is in $\mathcal{L}_\infty$, and we now want to find the nearest $Q \in \mathcal{H}_\infty$ to the given $R \in \mathcal{L}_\infty$. Nehari's theorem [174] says that

$$\gamma_{\text{opt}} = \|\mathbf{\Gamma}_R\|,$$

i.e., the optimal performance level is the norm of the Hankel operator whose symbol is $R$. Moreover, the optimal solution $Q_{\text{opt}} \in \mathcal{H}_\infty$ is determined from the largest singular vector of the Hankel operator. A state-space–based representation of the solution for the finite dimensional case is obtained as follows: first, stable and unstable parts of $R$ are separated:

$$R(s) = R_u(s) + R_s(s),$$

where $R_u(-s)$ and $R_s(s)$ are in $\mathcal{H}_\infty$, with $R_s(\infty) = R(\infty)$. When a finite dimensional $R(s)$ is given, $R_s$ and $R_u$ can be determined from the MATLAB command stabsep.

**Example 2.12.** Let $M(s) = \frac{s^2 - 2s + 2}{s^2 + 2s + 2}$ and $W(s) = \frac{10}{s + 0.01}$. Then,

$$R(s) = \frac{10 \left(s^2 + 2s + 2\right)}{(s + 0.01)(s^2 - 2s + 2)}.$$

The MATLAB commands

```
>> R=tf(10*[1,2,2],conv([1,-2,2],[1,0.01]));
>> [Rs,Ru]=stabsep(R)
```

result in $R_s(s) = \frac{9.802}{s + 0.01}$ and $R_u(s) = \frac{0.198s + 39.6}{s^2 - 2s + 2}$.   ∎

Let us now consider a minimal state-space representation for $R_u(s)$ in the form

$$R_u(s) = C(sI - A)^{-1}B.$$

Since $R_u(-s)$ is in $\mathcal{H}_\infty$, all eigenvalues of $A$ are in $\mathbb{C}_+$. In fact, the eigenvalues of $A$ are the zeros of $M(s)$, i.e., $\alpha_1, \ldots, \alpha_n$, and a particular choice for $A$ would be $A_o$ defined in (2.40) or $\Lambda_\alpha$ defined in (2.35). Below we give a review of the computation of the singular values and vectors of the Hankel operator $\mathbf{\Gamma}_{R_u}$ from [68, 80].

### Singular values of the Hankel operator: Finite dimensional case

When $R = R_u + R_s$, from (2.43) it is clear that

$$\gamma_{\text{opt}} = \inf_{Q \in \mathcal{H}_\infty} \|R_u - (Q - R_s)\|_\infty = \inf_{Q_1 \in \mathcal{H}_\infty} \|R_u - Q_1\|_\infty = \|\mathbf{\Gamma}_{R_u}\|.$$

Let us now examine the action of $\mathbf{\Gamma}_{R_u}$ on an element $G \in \mathcal{H}_2$ using time domain representations.

For this purpose, consider an example: $R_u(s) = \frac{5}{s - 2}$ and $G(s) = \frac{1}{s + 3}$. Let

$$\mathbf{\Gamma}_{R_u}G = \mathbf{\Pi}_- R_u(s)G(s) = \mathbf{\Pi}_- \left(\frac{5}{s - 2}\right)\left(\frac{1}{s + 3}\right) = \mathbf{\Pi}_- \left(\frac{1}{s - 2} - \frac{1}{s + 3}\right) = \frac{1}{s - 2}.$$

Define $Y = \mathbf{\Gamma}_{R_u} G \in \mathcal{H}_2^{\perp}$; then $\mathcal{L}^{-1}\{Y(s)\} = y(t) \in \mathcal{L}_2(-\infty, 0]$. In the above example, we have $Y(s) = \frac{1}{s-2}$ and $y(t) = e^{2t}u(-t)$, where $u(t)$ is the usual unit step function.

For a given $y \in \mathcal{L}_2(-\infty, \infty)$, introducing the notation

$$y_r(t) = y(-t)$$

we can define a reversed Hankel operator $\mathbf{\Gamma}_{R_u}^r(G) = Y_r = \mathcal{L}\{y_r(t)\}$, from $\mathcal{H}_2$ to $\mathcal{H}_2$, or in the time domain from $\mathcal{L}_2[0, \infty)$ to $\mathcal{L}_2[0, \infty)$. Let $f = \mathcal{L}^{-1}\{R_u(s)\} \in \mathcal{L}_1(-\infty, 0]$. Define $h(t) = f(-t)$; then we obtain the representations of $\mathbf{\Gamma}_{R_u}$ and $\mathbf{\Gamma}_{R_u}^r$ in the time domain as follows. First, for a given $g \in \mathcal{L}_2[0, \infty)$, we compute $(R_u G)$ in the time domain from the convolution

$$\mathcal{L}^{-1}\{(R_u G)\} = (f \star g)(t) = \int_{-\infty}^{+\infty} f(t-\tau)g(\tau)d\tau = \int_0^{+\infty} f(t-\tau)g(\tau)d\tau.$$

Then, using $h(t) = f(-t)$ we have

$$y(t) = \mathbf{\Pi}_-(f \star g)(t) = \left( \int_0^{+\infty} h(-t+\tau)g(\tau)d\tau \right) u(-t),$$

$$y_r(t) = \left( \int_0^{+\infty} h(t+\tau)g(\tau)d\tau \right) u(t).$$

When $R_u(s) = C(sI - A)^{-1}B$, we have

$$f(t) = \begin{cases} Ce^{At}B & \text{if } t \leq 0, \\ 0 & \text{if } t > 0, \end{cases}$$

and hence

$$h(t) = \begin{cases} Ce^{-At}B & \text{if } t \geq 0, \\ 0 & \text{if } t < 0. \end{cases}$$

Since $\mathbf{\Gamma}_{R_u}$ is isometrically equivalent to $\mathbf{\Gamma}_{R_u}^r$, we see that

$$\gamma_{\text{opt}} = \|\mathbf{\Gamma}_{R_u}^r\| = \left( \lambda_{\max}(\mathbf{\Gamma}_{R_u}^{r*} \mathbf{\Gamma}_{R_u}^r) \right)^{\frac{1}{2}}.$$

Define $\sigma_i = \lambda_i(\mathbf{\Gamma}_{R_u}^{r*} \mathbf{\Gamma}_{R_u}^r)$ as the $i$th singular value, where $\lambda(\cdot)$ denotes the $i$th eigenvalue. They are computed from the singular value/singular vector equation

$$\mathbf{\Gamma}_{R_u}^{r*} \mathbf{\Gamma}_{R_u}^r g_i = \sigma_i^2 g_i \quad \text{for} \quad g_i \in \mathcal{L}_2[0, \infty); \tag{2.44}$$

here $g_i \neq 0$ is the $i$th singular vector. Recall that

$$\mathbf{\Gamma}_{R_u}^r(g_i) = \begin{cases} \int_0^{+\infty} Ce^{-A(t+\tau)}Bg_i(\tau)d\tau & \text{if } t \geq 0, \\ 0 & \text{if } t < 0. \end{cases}$$

In order to compute the left-hand side of (2.44), we need to understand the action of $\mathbf{\Gamma}_{R_u}^{r*} : \mathcal{L}_2[0, \infty) \to \mathcal{L}_2[0, \infty)$. By definition, we must have

$$\langle \mathbf{\Gamma}_{R_u}^r g, y \rangle = \langle g, \mathbf{\Gamma}_{R_u}^{r*} y \rangle.$$

Now we write the inner products as

$$\langle \mathbf{\Gamma}_{R_u}^r g, y \rangle = \int_0^{\infty} y(t)^T \left( \int_0^{+\infty} Ce^{-A(t+\tau)}Bg(\tau)d\tau \right) dt$$

$$= \int_0^{\infty} \left( \int_0^{\infty} B^T e^{-A^T(t+\tau)}C^T y(t)dt \right)^T g(\tau)d\tau$$

$$= \langle g, \mathbf{\Gamma}_{R_u}^{r*} y \rangle.$$

Thus,

$$\mathbf{\Gamma}_{R_u}^{r*} y(t) = \int_0^\infty B^T e^{-A^T(t+\tau)} C^T y(t) dt.$$

Returning to (2.44), we first note that

$$\mathbf{\Gamma}_{R_u}^r g_i = C e^{-At} x_i, \quad \text{where} \quad x_i := \int_0^\infty e^{-A\tau} B g_i(\tau) d\tau;$$

then

$$\mathbf{\Gamma}_{R_u}^{r*} \mathbf{\Gamma}_{R_u}^r g_i = B^T e^{-A^T t} \left( \int_0^\infty e^{-A^T\tau} C^T C e^{-A\tau} d\tau \right) x_i,$$

with

$$W_o := \left( \int_0^\infty e^{-A^T\tau} C^T C e^{-A\tau} d\tau \right) \tag{2.45}$$

being the observability Grammian. In summary, we have

$$\sigma_i^2 g_i(t) = B^T e^{-A^T t} W_o x_i. \tag{2.46}$$

Multiplying both sides of (2.46) by $e^{-A\tau} B$ and integrating from 0 to $+\infty$, we get a finite dimensional eigenvalue–eigenvector equation in the form

$$\sigma_i^2 x_i = W_c W_o x_i,$$

where

$$W_c := \left( \int_0^\infty e^{-A\tau} B B^T e^{-A^T\tau} d\tau \right) \tag{2.47}$$

is the controllability Grammian of the system. Thus, we see that $\sigma_i^2$ must be an eigenvalue of the matrix $W_c W_o$, and $x_i$ must be the corresponding singular vector. Once we have found the $\sigma_i$'s and $x_i$'s, we can find all the $g_i$'s. The optimal interpolant is given by the following.

**Theorem 2.13.** *Given a minimal realization* $R_u(s) = C(sI - A)^{-1} B$, *with all the eigenvalues of* $A$ *in* $\mathbb{C}_+$, *we have*

$$\gamma_{\text{opt}} = \inf_{Q_1 \in \mathcal{H}_\infty} \| R_u - Q_1 \|_\infty = \sqrt{\lambda_{\max}(W_c W_o)},$$

*where* $W_o$ *and* $W_c$ *are defined in* (2.45) *and* (2.47), *respectively. The optimal interpolant is given by*

$$F_{\text{opt}}(s) = (R(s) - Q_{\text{opt}}(s)) = \frac{(\mathbf{\Gamma}_{R_u} g_{\max})(s)}{g_{\max}(s)},$$

*where* $g_{\max}$ *satisfies* (2.46) *for* $\sigma_i^2 = \gamma_{\text{opt}}^2$, *i.e.,*

$$g_{\max}(t) = \frac{1}{\gamma_{\text{opt}}^2} B^T e^{-A^T t} W_o x_{\max}, \quad t \geq 0,$$

*with* $x_{\max}$ *being the corresponding eigenvector of* $W_c W_o$.

Recall that $W_c$ and $W_o$ are the solutions of the Lyapunov equations

$$A W_c + W_c A^T = B B^T \tag{2.48}$$
$$A^T W_o + W_o A = C^T C. \tag{2.49}$$

By Theorem 2.13, the problem is now reduced to finding $(\mathbf{\Gamma}_{R_u} g_{\max})(s)$:

$$\mathbf{\Gamma}_{R_u} g_{\max}(t) = \text{reverse of } \mathbf{\Gamma}_{R_u}^r g_{\max}(t) = Ce^{At} x_{\max}, \quad t \in (-\infty, 0].$$

Therefore, in the $s$-domain,

$$\mathbf{\Gamma}_{R_u} g_{\max}(s) = C(sI - A)^{-1} x_{\max},$$

which leads us to (see, e.g., [51], [68])

$$Q_{\text{opt}}(s) = R(s) - \gamma_{\text{opt}} \frac{C(sI - A)^{-1} x_{\max}}{B^{\mathsf{T}}(sI + A^{\mathsf{T}})^{-1} y_{\max}}, \tag{2.50}$$

where

$$y_{\max} = \gamma_{\text{opt}}^{-1} W_o x_{\max}.$$

**Example 2.14.** Consider

$$R(s) = \frac{10\,(s^2 + 2s + 2)}{(s + 0.01)(s^2 - 2s + 2)} = \frac{9.802}{s + 0.01} + \frac{0.198s + 39.6}{s^2 - 2s + 2}.$$

The MATLAB commands

```
>> R=tf(10*[1,2,2],conv([1,-2,2],[1,0.01]));
>> [Rs,Ru]=stabsep(R); [A,B,C,D]=ssdata(Ru);
>> Wc=lyap(-A,B*B'); Wo=lyap(-A',C'*C);
>> [xx,gg]=eig(Wc*Wo) %% observe that largest eigenvalue is gg(2,2)
>> x=xx(:,2); gopt=sqrt(gg(2,2)); y=(1/gopt)*Wo*x;
>> QFn=ss(A,x,C,0); QFd=ss(-A',y,B',0); Q=R-gopt*(QFn/QFd);
>> Qopt=zpk(minreal(Q))
```

give us the following result:

$$Q_{\text{opt}}(s) = \frac{13.553\,(s^2 + 2s + 2)}{(s + 2.748)(s + 0.01)}.$$

Note that there are several unstable pole-zero cancellations in the computation of $Q_{\text{opt}}$. The resulting $F_{\text{opt}} = W - M Q_{\text{opt}}$ is obtained from

```
>> W=tf(10,[1,0.01]); M=tf([1,-2,2],[1,2,2]);
>> Fopt=zpk(minreal(W-M*Qopt));
```

As expected, $F_{\text{opt}}(s)$ is an inner function whose magnitude is $\gamma_{\text{opt}} = 13.553$:

$$F_{\text{opt}}(s) = \frac{-13.553\,(s - 2.748)}{(s + 2.748)}.$$

The `minreal` command is used to perform the pole-zero cancellations. ∎

Let us now elaborate on the unstable pole-zero cancellations in the right-hand side of (2.50); in particular, the poles of $R_u(s)$ disappear in $Q_{\text{opt}}(s)$. By comparing the solutions (2.42) and (2.50), we see that

$$F_{\text{opt}}(s) = \lambda \frac{[s^{n-1} \cdots s\ 1]\,\mathbf{J}\,\Phi}{[s^{n-1} \cdots s\ 1]\,\Phi} = \gamma_{\text{opt}} \frac{C(sI - A)^{-1} x_{\max}}{B^{\mathsf{T}}(sI + A^{\mathsf{T}})^{-1} y_{\max}}\,M(s), \tag{2.51}$$

where $\lambda = \pm\gamma_{\text{opt}}$. Another interesting point to note is that in the finite dimensional case we can write $M(s)$ as

$$M(s) = m_\infty \frac{\det(sI - A)}{\det(sI + A^{\mathsf{T}})}, \quad m_\infty := \text{sign}\big(M(\infty)\big).$$

Hence the identity

$$\gamma_{\text{opt}}^{-1} F_{\text{opt}}(s) = \pm \frac{[s^{n-1} \cdots s\ 1]\, \mathsf{J}\, \Phi}{[s^{n-1} \cdots s\ 1]\, \Phi} = m_\infty \frac{C\, \text{adj}(sI - A)\, x_{\max}}{B^{\mathsf{T}}\, \text{adj}(sI + A^{\mathsf{T}})\, y_{\max}} \qquad (2.52)$$

illustrates the connection between the two alternative solutions.

**Exercise 2.15.**
1. Consider $M(s)$ and $W(s)$ given in Example 2.7, with $\alpha_1 = 5$, $\alpha_2 = 2$, and $W(s) = \frac{1}{s+0.1}$. Verify the equivalence of the solutions as described by (2.52).
2. Let $M(s) = \frac{s^2 - 6s + 10}{s^2 + 6s + 10}$ and $W(s) = \frac{1}{s+0.1}$. Verify the identity (2.52).

## 2.4.3 ▪ Infinite dimensional case

The Nehari problem for infinite dimensional systems has been studied extensively by various researchers; see, e.g., [114, 136, 144, 182, 189, 233]. In this section we provide a brief summary of a method in which the optimal solution $Q_{\text{opt}}$ is computed from the singular vectors of the operator $W(\mathbf{T})$ via Theorem 2.6 for the case where $W$ is finite dimensional and $M$ is infinite dimensional.

The singular-value–singular-vector equation is in the form

$$\big(\gamma^2 I - W(\mathbf{T})^* W(\mathbf{T})\big)\, G = 0, \qquad (2.53)$$

where $G \in \mathcal{H}(M)$. It can be shown that the singularity of the operator

$$\mathbf{A}_\gamma := \big(\gamma^2 I - W(\mathbf{T})^* W(\mathbf{T})\big)$$

is equivalent to the singularity of a *skew Toeplitz* operator [66]. By carefully examining the action of $W(\mathbf{T})^* W(\mathbf{T})$ on a typical element of $\mathcal{H}(M)$, we end up getting finitely many equations to solve for finitely many unknowns:

$$\mathsf{R}_\gamma \Phi = 0,$$

where $\mathsf{R}_\gamma$ is a finite dimensional square matrix whose entries are functions of $\gamma$ and where $\Phi$ is a finite size vector. From the singular value decomposition of $\mathsf{R}_\gamma$ as $\gamma$ varies between an upper and a lower bound, we can determine $\gamma_{\text{opt}}$ and the corresponding nonzero $\Phi$. In Chapter 6 these computations are discussed in great detail.

In order to illustrate the basic idea behind this approach, let us consider an arbitrary infinite dimensional $M(s)$ and a first order $W(s) = (s + \varepsilon)^{-1}$. Then, for a candidate singular vector $G \in \mathcal{H}(M)$ we have

$$W(\mathbf{T})G = \mathbf{\Pi}_{\mathcal{H}(M)} WG = WG - M\mathbf{\Pi}_+ W(M^* G) = \frac{1}{s+\varepsilon}\Big(G(s) - M(s)\psi_1\Big),$$

where $\psi_1 = G(-\varepsilon)/M(-\varepsilon)$ is a constant (we assume that $M(s)$ does not have a pole at $s = -\varepsilon$). The next step is to compute

$$W(\mathbf{T})^* W(\mathbf{T})G = \mathbf{\Pi}_{\mathcal{H}(M)} W^* W(G - M\psi_1) = W^* W(G - M\psi_1) - \mathbf{\Pi}_- W^* W(G - M\psi_1).$$

Note that we have

$$\Pi_- W^* W (G - M\psi_1) = \frac{W(\varepsilon)}{\varepsilon - s}\left(\psi_2 - M(\varepsilon)\psi_1\right), \quad \text{where} \quad \psi_2 := G(\varepsilon).$$

Thus, (2.53) is equivalent to

$$\gamma^2 G(s) - \left(\frac{1}{\varepsilon^2 - s^2}\right)(G(s) - M(s)\psi_1) + \frac{W(\varepsilon)}{\varepsilon - s}\left(\psi_2 - M(\varepsilon)\psi_1\right) = 0,$$

and rearranging the above terms, we find that

$$G(s) = \frac{\left(M(s) - M(\varepsilon)\frac{(s+\varepsilon)}{2\varepsilon}\right)\psi_1 + \frac{(s+\varepsilon)}{2\varepsilon}\psi_2}{\gamma^2 s^2 + (1 - \varepsilon^2\gamma^2)}. \tag{2.54}$$

Note that if we put $Q = 0$ in (2.25), we get that $\gamma_{\text{opt}} \leq \varepsilon^{-1} = \|W\|_\infty$. This means that the roots of the denominator of (2.54) are on $\mathbb{I}$ and they are denoted as $\beta_{1,2} := \pm j\sqrt{\gamma^{-2} - \varepsilon^2}$. Since $G \in \mathcal{H}_2$, it must not have poles at $\beta_{1,2}$. This gives us two interpolation conditions:

$$\begin{bmatrix} M(\beta_1) - M(\varepsilon)\frac{(\beta_1+\varepsilon)}{2\varepsilon} & \frac{(\beta_1+\varepsilon)}{2\varepsilon} \\ M(\beta_2) - M(\varepsilon)\frac{(\beta_2+\varepsilon)}{2\varepsilon} & \frac{(\beta_2+\varepsilon)}{2\varepsilon} \end{bmatrix}\begin{bmatrix} \psi_1 \\ \psi_2 \end{bmatrix} = \begin{bmatrix} 0 \\ 0 \end{bmatrix}, \tag{2.55}$$

which is in the form $R_\gamma \Phi = 0$, and the largest $\gamma < \varepsilon^{-1}$ which makes $R_\gamma$ singular is $\gamma_{\text{opt}}$. From the corresponding singular vector we obtain $\psi_1$ and $\psi_2$, which define $G$ by (2.54). We should also point out that the conditions $\psi_2 = G(\varepsilon)$ and $\psi_1 = G(-\varepsilon)/M(-\varepsilon)$ are automatically satisfied by (2.54). Finally, the optimal solution

$$F_{\text{opt}}(s) = \frac{W(\mathbf{T})G}{G} = \frac{W(s)\left(G(s) - M(s)\psi_1\right)}{G(s)} \tag{2.56}$$

can be computed numerically from $G(s)$ defined by (2.54) via $\gamma_{\text{opt}}$, $\psi_1$, and $\psi_2$.

**Exercise 2.16.**
1. Consider $M(s)$ and $W(s)$ given in Example 2.7, with $\alpha_1 = 5$, $\alpha_2 = 2$, and $W(s) = \frac{1}{s+0.1}$. Compute $\gamma_{\text{opt}}$ and $F_{\text{opt}}$ using (2.54), (2.55), and (2.56). Verify that the result matches the ones computed via the Nevanlinna–Pick interpolation and the Nehari methods.
2. Let $M(s) = e^{-hs}$ and $W(s) = \frac{1}{s+0.1}$. Compute $\gamma_{\text{opt}}$ as a function of $h > 0$; i.e., obtain a graph of $\gamma_{\text{opt}}$ versus $h$ for the range $h \in [0,5]$.

3. Consider again $W(s) = \frac{1}{s+\varepsilon}$. Let $\beta_1 := j\omega_\gamma$, where $\omega_\gamma = \sqrt{\gamma_{\text{opt}}^{-2} - \varepsilon^2}$. With this notation, $\gamma_{\text{opt}}$ is obtained from the smallest $\omega_\gamma$ for which the matrix $R_\gamma$ is singular. Show that the set of equations (2.55) is equivalent to having

$$\psi_2 = \left(M(\varepsilon) - \frac{2\varepsilon}{\overline{\beta}_1 + \varepsilon}M(\overline{\beta}_1)\right)\psi_1$$

and

$$\left(\frac{M(\beta_1)}{M(\overline{\beta}_1)} - \frac{\beta_1 + \varepsilon}{\overline{\beta}_1 + \varepsilon}\right)\psi_1 = 0.$$

Clearly, the last equation has a solution $\psi_1 \neq 0$ if and only if

$$\left(M(j\omega_\gamma)\right)^2 = \frac{\varepsilon + j\omega_\gamma}{\varepsilon - j\omega_\gamma}. \tag{2.57}$$

Prove that if the phase of $M(j\omega_\gamma)$ (denoted by $\angle M(j\omega_\gamma)$) is a uniformly decreasing function of $\omega_\gamma$, with $\angle M(j\omega_\gamma) < 0$ for all $\omega_\gamma > 0$, then (2.57) leads to the following result: $\gamma_{\mathrm{opt}}$ is obtained from the smallest $\omega_\gamma$ satisfying

$$-\angle M(j\omega_\gamma) + \tan^{-1}(\omega_\gamma/\varepsilon) = \pi.$$

For example, when $M(s) = e^{-hs}$, the optimal performance level $\gamma_{\mathrm{opt}}$ is obtained from the unique solution $\omega_\gamma > 0$ of the equation

$$h\omega_\gamma + \tan^{-1}(\omega_\gamma/\varepsilon) = \pi.$$

Moreover, in this case we have

$$Q_{\mathrm{opt}}(s) = \frac{\gamma_{\mathrm{opt}}}{1 + H_{\mathrm{opt}}(s)},$$

where $H_{\mathrm{opt}}(s)$ is a stable transfer function whose impulse response is restricted to the finite interval $[0, h]$,

$$H_{\mathrm{opt}}(s) = \frac{2\varepsilon s + \varepsilon^2 - \omega_o^2 + \gamma_{\mathrm{opt}}^{-1}(s + \varepsilon)e^{-hs}}{s^2 + \omega_o^2},$$

where $\omega_o$ denotes the value of $\omega_\gamma$ for $\gamma = \gamma_{\mathrm{opt}}$. As a numerical example, take $\varepsilon = 0.1$ and $h = 0.5$, verify that $\gamma_{\mathrm{opt}} = 0.312$ and $\omega_o = 3.204$. Obtain the impulse response and the Nyquist plot of $H_{\mathrm{opt}}$. Verify that $Q_{\mathrm{opt}} \in \mathcal{H}_\infty$ and that $F_{\mathrm{opt}} = W - MQ_{\mathrm{opt}}$ satisfies $|F_{\mathrm{opt}}(j\omega)| = \gamma_{\mathrm{opt}}$ for all $\omega$.

## 2.5 ▪ Two-block problem

A generalization of the problem (2.25) is the *two-block problem*, which is to find

$$\gamma_{\mathrm{opt}} = \inf_{Q \in \mathcal{H}_\infty} \left\| \begin{bmatrix} R - Q \\ V \end{bmatrix} \right\|_\infty \tag{2.58}$$

for the given data $R \in \mathcal{L}_\infty \cap \mathcal{H}_2^\perp$ and $V \in \mathcal{H}_\infty$. It can be shown that (see, e.g., [66, 274])

$$\gamma_{\mathrm{opt}} = \left\| \begin{bmatrix} \Gamma_R \\ \Upsilon_V \end{bmatrix} \right\|. \tag{2.59}$$

Alternatively, the problem (2.58) can be transformed to a one-block problem using a spectral factorization as follows. Note that for a given $Q \in \mathcal{H}_\infty$ we have

$$\left\| \begin{bmatrix} R - Q \\ V \end{bmatrix} \right\|_\infty \leq \gamma \tag{2.60}$$

if and only if the following inequality holds:

$$(R(j\omega) - Q(j\omega))^*(R(j\omega) - Q(j\omega)) + V(j\omega)^*V(j\omega) \leq \gamma^2 \text{ a.e. } \omega \in \mathbb{R}.$$

Therefore, the candidate $\gamma$ for $\gamma_{opt}$ should be strictly greater than $\|V\|_\infty$. For a fixed $\gamma > \|V\|_\infty$ we can compute $V_\gamma, V_\gamma^{-1} \in \mathcal{H}_\infty$ from the *spectral factorization*

$$V_\gamma(j\omega)^* V_\gamma(j\omega) = \gamma^2 - V(j\omega)^* V(j\omega) \quad \text{a.e. } \omega \in \mathbb{R}. \tag{2.61}$$

Hence, (2.60) holds if and only if

$$\|V_\gamma^{-1} R - Q_1\|_\infty \le 1, \tag{2.62}$$

where $Q_1 := V_\gamma^{-1} Q$. Thus, we conclude that $\gamma_{opt}$ is the smallest $\gamma > 0$ such that $\|\boldsymbol{\Gamma}_{V_\gamma^{-1} R}\| \le 1$. The computation of $\gamma_{opt}$ in this case requires a bisection search, where at each step the norm of the Hankel operator (whose symbol is $V_\gamma^{-1} R$) is computed. The steps followed in this algorithm are as follows:

0. Given the problem data $R$, $V$, tolerance level $\varepsilon$, and an initial guess $\gamma_{opt} \in (\gamma_{min}, \gamma_{max})$. For example, we can take $\gamma_{min} = \|V\|_\infty$ and $\gamma_{max} = \|[R \quad V]^{\mathrm{T}}\|_\infty$.

1. Let $\gamma = (\gamma_{min} + \gamma_{max})/2$, and compute $V_\gamma$.

2. Check whether $\|\boldsymbol{\Gamma}_{V_\gamma^{-1} R}\| \le 1$.
   Yes: define $\gamma_{max} = \gamma$.
   No: define $\gamma_{min} = \gamma$.

3. Check whether $(\gamma_{max} - \gamma_{min}) > \varepsilon \ \gamma_{min}$.
   Yes: return to step 1.
   No: Exit with the statement $\gamma_{min} \le \gamma_{opt} \le \gamma_{max}$.

In the finite dimensional case, where $V_\gamma^{-1} R$ is a rational function, computation of the Hankel operator norm in **step 2** can be done by using the Nehari solution as outlined in Section 2.4.2 (or by using the solution of a Nevanlinna–Pick interpolation as in Section 2.4.1).

The computational procedure given in Chapter 6, for $\gamma_{opt}$ and the corresponding optimal controller, is based on finding the norm of the so-called Hankel+Toeplitz operator, (2.59), or the norm of an operator in the form $W_\gamma(\mathbf{T})$, where $W_\gamma(s) = V_\gamma^{-1} R$. As illustrated above, when the norm of this operator is achieved at the discrete spectrum, we come up with a finite set of linear equations from which $\gamma_{opt}$ is computed; further theoretical details of this approach can be found in [66]. In this book we mainly focus on the computational aspects.

## 2.6 ▪ Spectral factorization

Later, in Section 5.2.4 we will see that, for the mixed sensitivity minimization problems with finite dimensional weights, $V$ appearing in the two-block problem (2.58) is a rational transfer function. Therefore, the spectral factorization (2.61) is a finite dimensional factorization. There are various numerically efficient ways to perform this operation; see, e.g., [281] for state-space formulae involving algebraic Riccati equations. On the other hand, in the SISO case, one can obtain the spectral factor $V_\gamma$ in **step 1** of the above algorithm by computing the roots of a *polynomial* as follows. Let $nV(s)$ and $dV(s)$ be coprime polynomials with all the roots in $\mathbb{C}_-$ such that $V(s) = nV(s)/dV(s)$; it will be clear from the discussion of Section 5.2.4 that, typically, the $V(s)$ we consider is outer and biproper. Therefore, we may assume that $nV$ and $dV$ are in the form

$$nV(s) = a_n s^n + a_{n-1} s^{n-1} + \cdots + a_0, \quad dV(s) = b_n s^n + b_{n-1} s^{n-1} + \cdots + b_0.$$

Then, $V_\gamma(s)$ is in the form $nV_\gamma(s)/dV(s)$, where $nV_\gamma(s)$ is an $n$th order polynomial determined from the polynomial factorization

$$nV_\gamma(s) nV_\gamma(-s) = \gamma^2 dV(s) dV(-s) - nV(s) nV(-s) =: W_\gamma(s)$$

in such a way that all the roots of $nV_\gamma(s)$ are in $\mathbb{C}_-$. Since $W_\gamma(s) = W_\gamma(-s)$, i.e., it is symmetric, we observe that if $z$ is a root of $W_\gamma$, then so is $-z$. Moreover, $\gamma > \|V\|_\infty$ implies that $W_\gamma(s)$ does not have roots on $\mathbb{I}$; thus, $n$ roots $z_1, \ldots, z_n$ in $\mathbb{C}_-$ define $nV_\gamma$. A simple (perhaps not so efficient) MATLAB pseudocode is given below.

```
1. Input nV=[an, ... ,a0]; dV=[bn, ... ,b0]; gg=gamma^2;
2. Define an (n+1)x1 vector J by J(k)=(-1)^(n+1-k), k=1:n+1
3. Define W= gg*conv(dV,dV*diag(J))-conv(nV,nV*diag(J));
4. Let rw=roots(W); and separate them: initially set rwn=[];
       for k=1:2*n; if real(rw(k)) < 0; rwn=[rwn , rw(k)]; else end
5. Extract the gain: Kg=sqrt(abs(gg*dV(1)^2-nV(1)^2));
6. Output nVg=Kg*poly(rwn)
```

**Exercise 2.17.**
Let $V = nV/dV$ be given by $nV(s) = (s+1)(s+2)$ and $dV(s) = s^2 + 2s + 2$.
First, by plotting the Bode magnitude of $V$, show that $\|V\|_\infty = \frac{3}{2}$.
Then, take $\gamma = \sqrt{3}$ and compute $V_\gamma(s)$ by using the above steps.
Verify that

$$V_\gamma(s) = \frac{\sqrt{2}\left(s^2 + \sqrt{\frac{3}{2}}\, s + 2\right)}{s^2 + 2s + 2}.$$

## 2.7 ▪ Computation of the $\mathcal{L}_\infty$ norm

As we have seen in the previous sections, the peak value of the Bode magnitude plot of a SISO transfer function is the $\mathcal{L}_\infty$ norm. When this transfer function is stable, the $\mathcal{L}_\infty$ norm is the $\mathcal{H}_\infty$ norm, which is closely related to the stability robustness and performance of feedback systems (more on these relationships in Chapter 5). In this section, we discuss the issue of numerical computation of the $\mathcal{L}_\infty$ norm for a MIMO system, which may or may not be stable.

Let $G \in \mathcal{L}_\infty$ be a rational transfer function matrix; its $\mathcal{L}_\infty$ norm is defined as

$$\|G\|_\infty := \sup_\omega \sigma_{\max}\{G(j\omega)\},$$

where $\sigma_{\max}(\cdot)$ denotes the largest singular value. By this definition, an estimate of the $\mathcal{L}_\infty$ norm can be obtained by computing the maximum singular value of $G(j\omega)$ over a sufficiently fine grid of frequency points. When $G$ is a rational transfer matrix, which admits a minimal realization in the form

$$G(s) = C(sI - A)^{-1}B + D, \tag{2.63}$$

we can compute the $\mathcal{L}_\infty$ norm by using the associated *Hamiltonian matrix*,

$$H_\gamma = \begin{bmatrix} A + BR_\gamma^{-1}D^\mathsf{T}C & BR_\gamma^{-1}B^\mathsf{T} \\ -C^\mathsf{T}(I + DR_\gamma^{-1}D^\mathsf{T})C & -(A + BR_\gamma^{-1}D^\mathsf{T}C)^\mathsf{T} \end{bmatrix},$$

where $R_\gamma = \gamma^2 I - D^\mathsf{T}D$ for $\gamma > 0$. Formally, the relation between the $\mathcal{L}_\infty$ norm and the Hamiltonian is given by the following [31].

**Lemma 2.18.** *Let $\gamma > \sigma_{\max}\{D\}$. For the transfer function matrix $G(s)$ given by (2.63), we have $\|G\|_\infty < \gamma$ if and only if $H_\gamma$ has no imaginary axis eigenvalues.*

This result can be used for computing the $\mathcal{L}_\infty$ norm of $G$ by a bisection algorithm (see [28]), where explicit upper and lower bounds of $\gamma$ are given. If the Hamiltonian matrix for the mean of the upper and lower bounds has imaginary axis eigenvalues, we set the lower bound to the mean value; otherwise, we set the upper bound to the mean value. We iterate until the upper and lower bounds are within a prespecified tolerance of each other. Faster, quadratically convergent algorithms are presented in [30, 27]. These algorithms require only a lower bound.

In MATLAB, given appropriate size matrices $A, B, C, D$, the transfer function matrix $G$ and its $\mathcal{L}_\infty$ norm can be computed from the following commands:

```
>> sys = ss(A,B,C,D);
>> [sig,w] = norm(sys,inf);
```

where `sig` is the $\mathcal{L}_\infty$ norm and `w` is the frequency where the norm is achieved. An equivalent function is `hinfnorm`, where additionally the stability of the system is checked (whether $A$ is Hurwitz or not); if `sys` is unstable, then `hinfnorm(sys)` gives `Inf` as the output.

When the transfer function is infinite dimensional, the above mentioned techniques are not directly applicable due to irrational terms, e.g., delays. The Hamiltonian *operator* of a transfer function with delays consists of exponential terms. A relation similar to Lemma 2.18 still holds for this case, where $\mathcal{L}_\infty$ norm computation requires finding imaginary axis solutions of a *nonlinear* eigenvalue problem; see [155].

One way to compute the $\mathcal{L}_\infty$ norm of an infinite dimensional system is to compute maximum singular values of the transfer function matrix over a fine grid of frequency points and iterate with a new grid of frequency points around the maximum of the previous iteration step. The key problem here is to choose the frequency points such that they approximate singular value curves well. For general infinite dimensional systems, a visual inspection can be used. For delay systems, it is possible to obtain a linear fractional representation by taking delay terms out. The frequency points are determined based on the bandwidth of the dynamic system without delays and sampling each delay term depending on its active frequency regions. Another approach is to use Padé approximation of delay terms (the related MATLAB command is `pade`); let's say with a $k$th order approximation we obtain a finite dimensional system $G_k$ whose $\mathcal{L}_\infty$ norm is $\gamma_k$: keep increasing the order of the approximation until $|\gamma_k - \gamma_{k-1}|$ is less than a specified level of accuracy; see, e.g., Section 7.2 of [191].

MATLAB has an in-house algorithm to detect critical frequency points of interest of a given transfer function $G$ with delays. The example given below computes the $\mathcal{L}_\infty$ norm for a delay system in a quick and easy way, though there are some pitfalls. The main idea in the algorithm is to use singular values and frequency points returned from the function `sigma` to compute a *predicted* maximum singular value, and the frequency where it is achieved, at each iteration step. Then, around that frequency, an additional 100 points are chosen to do a finer singular value computation. The iteration stops when the maximum singular values from two consecutive iterations are within a specified tolerance value, e.g., $10^{-3}$. The MATLAB code given below computes the $\mathcal{L}_\infty$ norm of the transfer function

$$G(s) = \frac{1}{s + 0.9\, e^{-5.5s}}.$$

It gives svmax=1.9184 as the norm and wmax=1.4146 as the frequency point where the norm is achieved. These results are in agreement with the magnitude plot of $G(j\omega)$, shown in Figure 2.2.

```
s = tf('s');                    % define s
sys = 1/(s+0.9*exp(-5.5*s));    % define the transfer function
tol = 1e-5;                     % tolerance for L-inf norm
[sv,w] = sigma(sys);            % compute singular values and frequencies
                % compress sv and w into 1-D array, add 0 frequency
sv = squeeze(sv); sv = [sigma(sys,0) sv]; w = squeeze(w); w = [0;w];
[svmax,ix] = max(sv);           % compute maximum singular value
wmax = w(ix);                   % frequency attaining max singular value
while true                      % compute the new frequency range
    new_w = linspace(w(max(ix-1,1)),w(min(ix+1,length(w))),100);
    sv_next = sigma(sys,new_w); % next singular values and frqs
    sv_next = squeeze(sv_next); % compress sv and w into 1-D array
    [svmax_next,ix_next] = max(sv_next);
                                % compute next max singular value
    wmax_next = new_w(ix_next); % and its frequency
    if abs(svmax_next-svmax)>tol % if last two sv are not within a
        svmax = svmax_next;      % tolerance of each other: continue
        w = new_w;
        ix = ix_next;
    else                        % last two sv are close : end
        svmax = svmax_next;     % return max sv and its freq
        wmax = wmax_next;
        break;
    end
end
```

The above algorithm gives reliable results on most practical application examples. There are two edge cases that can be considered to make it more robust. First, the frequency points returned from the sigma function may omit boundary frequency at $\infty$, i.e., the case where the norm is achieved at $\infty$. The second case is when two peaks of singular value curves are very close to each other. Then, instead of the max of singular values, singular values close to the maximum within a certain tolerance can be considered. This requires iterating on multiple frequency intervals and increases the code complexity. Also, instead of continuing iterations, one can do a correction step to remove the discretization error of the first iteration by solving a system of nonlinear equations that characterize the singular value curve of $G$, where the initial values are obtained in the first predictor step. The details on the two edge cases and the corrector step approach can be found in [155].

There are other iterative [155] and direct [94] predictor-corrector approaches which guarantee finding the $\mathcal{L}_\infty$ norm of a delay system within a prespecified frequency in-

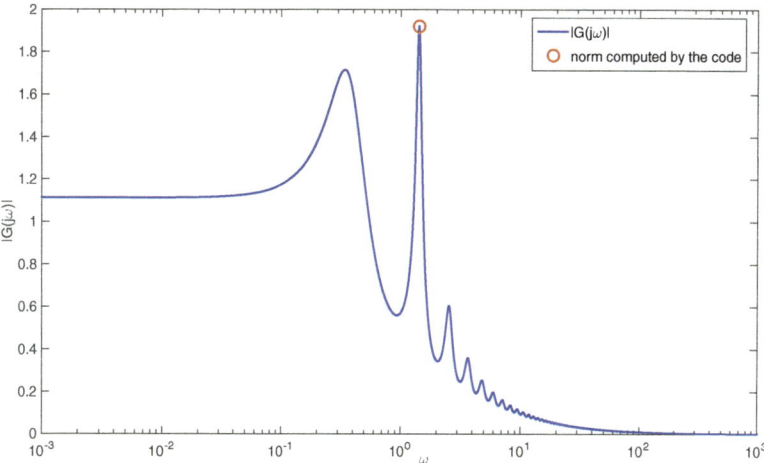

**Figure 2.2.** *Magnitude plot of G.*

terval, unlike the approach mentioned above. In the predictor step, the corresponding operator is discretized using a spectral method, followed by a correction step where the effect of the discretization on computed peak values is removed in the frequency response by local optimizations. Another pitfall to consider may be the sensitivity of the $\mathcal{L}_\infty$ norm with respect to arbitrarily small delay perturbations. In [93], an extension of the $\mathcal{L}_\infty$ norm is defined, the strong $\mathcal{L}_\infty$ norm, which explicitly takes into account small delay perturbations, a common case in any practical control application.

# Chapter 3

# Transfer Functions of DPSs

In this chapter, several examples are given for transfer functions of DPSs. First, retarded and neutral time delay systems are discussed. This is probably the most extensively studied class of DPSs, due to its wide range of applications [70, 89, 129, 156, 178]. The next example is pseudorational systems [258]. Then, other types of infinite dimensional systems, such as those represented by partial differential equations, are given. For an extensive tutorial survey on transfer functions of DPSs represented by partial differential equations, see [38]. Finally, some examples from fractional order systems are considered. For a general theory of transfer functions of infinite dimensional systems, see [43, 284].

## 3.1 ▪ Retarded and neutral delay systems

There is a huge literature on the analysis and control of time delay systems with applications in various fields such as robotics [9, 125], communication networks [110, 148, 212], mechanical systems [8, 37, 267], transportation systems [187], chemical processes [112, 181], and biological systems [196, 107], as well as economics [60, 198, 271, 272]. Many other application examples can be found in books such as [36, 70, 156, 178, 267].

Transport lag is typically due to the physical distance between the controlled variable and the physical process. For example, in oil/gas pipelines, any action taken at the source end (e.g., changing the flow rate) will show its effect at the destination with a transport delay of $\tau = dA/r$, where $d$ is the distance between the two ends, $A$ is the cross-section of the pipeline, and $r$ is the rate of flow, as depicted in Figure 3.1.

A simple mathematical model of the pipeline system is

$$\dot{x}(t) = r_{in}(t) - r_{out}(t), \quad r_{in}(t) = r(t - \tau(t)), \quad \tau(t) = \frac{d\,A}{r(t)},$$

where $r_{out}(t)$ is the outgoing flow rate from the storage; it may vary depending on consumer demand. Note that even for such a simple flow model, we have an implicit function term in the form $r(t - dA/r(t))$. If $r(t)$ is adjusted based on feedback from $x(t)$, then we end up with a closed-loop system involving a *state-dependent delay*. Stability analysis and controller design for such systems are beyond the scope of this book. However, if $r(t)$ is restricted to an interval $[r_-, r_+]$, then the upper and lower bounds

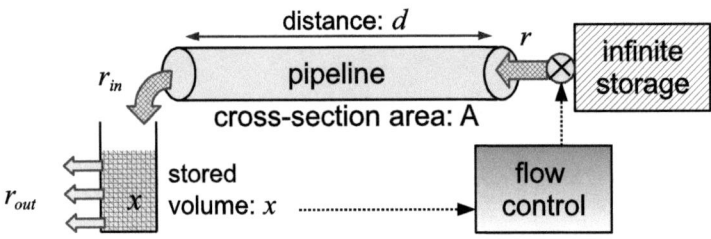

**Figure 3.1.** *Flow control in an oil/gas pipeline.*

of the time delay $\tau$ are determined, $[dA/r_+, dA/r_-]$. This information, along with a
bound on $|\dot{r}(t)|$, can be used to design robust feedback controllers; see, e.g., [212].

In tele-operation applications the command signal generated at the controller's side
reaches the physical system with a time delay (also called propagation delay or communication delay) which is proportional to the distance between the controller and the
plant. For example, the communication time delay is very significant for controlling a
robotic device on Mars from the Earth (on the order of 10 minutes), whereas communication via satellites from Earth to Earth takes several hundreds of milliseconds. A
typical tele-operation scheme, where the feedback control is over a network, is shown
in Figure 3.2 [70, 192], where

$$u(t) = u_c(t - h_f), \qquad y_m(t) = y(t - h_b),$$

with $h_f$ and $h_b$ representing the time delays on the forward and backward paths, respectively. Assuming that $h_f$ and $h_b$ are constant, i.e., ignoring network effects such as delay
jitter, packet drops, and retransmission, the plant seen by the controller in Figure 3.2 is

$$P(s) = e^{-hs} P_0(s), \qquad h = h_f + h_b, \tag{3.1}$$

where $P_0(s)$ is, typically, a rational transfer function.

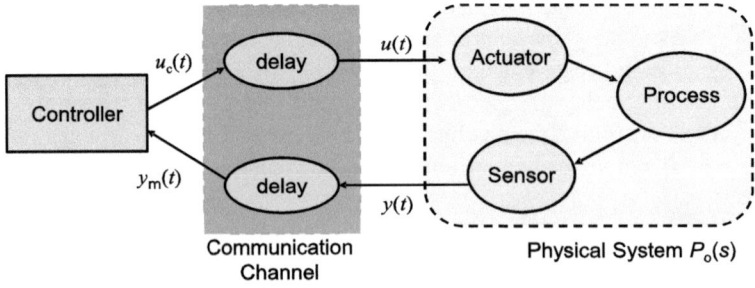

**Figure 3.2.** *General set-up for a typical tele-operation* [192].

The model (3.1) is one of the simplest time delay system models encountered in
practice; in this case, $h > 0$ is the input/output (I/O) time delay; it is also called the
dead-time. Various predictor types of dead-time compensators are proposed in the liter-

ature; see, e.g., [180, 279]. In particular, plants in the form (3.1) with

$$P_0(s) = \frac{K}{s}, \qquad K \in \mathbb{R} \quad \text{(integrator)},$$

$$P_0(s) = \frac{K}{(\tau s + 1)}, \quad K \in \mathbb{R}, \, \tau > 0 \quad \text{(first order stable process)},$$

$$P_0(s) = \frac{K}{(s - a)}, \quad K \in \mathbb{R}, \, a > 0 \quad \text{(first order unstable process)}$$

have been extensively studied; see, e.g., [5, 247] and their references.

Complex systems where series, parallel, and feedback interconnections of several subsystems of the form (3.1) lead to systems with *internal time delays*. As an example, consider the system shown in Figure 3.3, where each $P_i(s)$ is in the form

$$P_i(s) = e^{-h_i s} \frac{Q_i(s)}{R_i(s)}, \qquad h_i > 0,$$

with some polynomials $Q_i$ and $R_i$ satisfying $\deg(Q_i) \leq \deg(R_i)$.

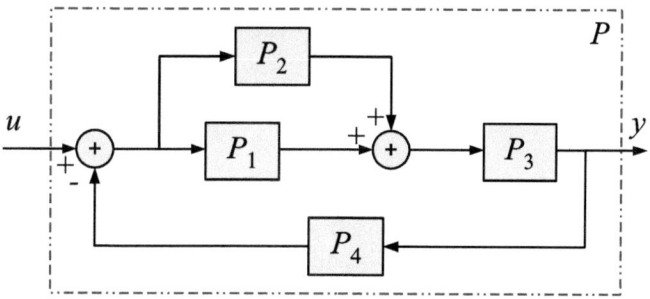

**Figure 3.3.** *An interconnected system with internal time delays.*

The transfer function of the plant, $P$, shown in Figure 3.3 is

$$P(s) = \frac{q_{n,1}(s)e^{-h_{n,1}s} + q_{n,2}(s)e^{-h_{n,2}s}}{q_{d,1}(s) + q_{d,2}(s)e^{-h_{d,2}s} + q_{d,3}(s)e^{-h_{d,3}s}}, \qquad (3.2)$$

where $q_{n,1} = Q_1 R_2 Q_3 R_4$, $q_{n,2} = R_1 Q_2 Q_3 R_4$, $q_{d,1} = R_1 R_2 R_3 R_4$, $q_{d,2} = Q_1 R_2 Q_3 Q_4$, $q_{d,3} = R_1 Q_2 Q_3 Q_4$, and $h_{n,1} = (h_3 + h_1)$, $h_{n,2} = (h_3 + h_2)$, $h_{d,2} = (h_4 + h_3 + h_1)$, $h_{d,3} = (h_4 + h_3 + h_2)$.

Generalization of the above example leads to the general structure of the retarded- and neutral-type delay systems:

$$P(s) = \frac{\sum_{i=1}^{v_n} q_{n,i}(s)e^{-h_{n,i}s}}{\sum_{k=1}^{v_d} q_{d,k}(s)e^{-h_{d,k}s}}, \qquad (3.3)$$

where $q_{n,i}$ and $q_{d,k}$ are polynomials for all $i = 1, \ldots, v_n$ and $k = 1, \ldots, v_d$, and the delays are enumerated to satisfy $0 \leq h_{n,1} < \cdots < h_{n,v_n}$ and $0 \leq h_{d,1} < \cdots < h_{d,v_d}$. The polynomials are assumed to satisfy $\deg q_{n,1} \geq \deg q_{n,i}$ for all $i = 2, \ldots, v_n$ and $\deg q_{d,1} \geq \deg q_{d,k}$ for all $k = 2, \ldots, v_d$. Furthermore, the plant (3.3) is assumed to be causal; that imposes the following restrictions: $\deg q_{n,1} \leq \deg q_{d,1}$ and $h_{d,1} \leq h_{n,1}$.

Note that, in this case, $e^{-h_{d,1}s}$ can be factored out from the denominator of (3.3), and hence it can be assumed that $h_{d,1} = 0$ holds, as in (3.2), without loss of generality.

**Definition 3.1.** *We say that the plant* (3.3) *is a time delay system of* retarded type *whenever deg $q_{d,1}$ > deg $q_{d,k}$ for all $k = 2, \ldots, v_d$. Otherwise, there are two possibilities. If deg $q_{d,1}$ = deg $q_{d,k}$ for some $k = 2, \ldots, v_d$, then the plant* (3.3) *is said to be a time delay system of* neutral type. *On the other hand, if deg $q_{d,1}$ < deg $q_{d,k}$ for some $k = 2, \ldots, v_d$, then the plant* (3.3) *is said to be a time delay system of* advance type.

By various connections of subsystems, as in Figure 3.3, it is possible to obtain a delay system of advance type. However, such systems rarely occur in nature and they are difficult to deal with in general, whereas retarded delay systems are more natural and easier to handle, because they have finitely many poles to the right of any vertical axis in the complex plane, whereas neutral-type systems may have infinitely many poles in the right half plane; see, e.g., [205].

### 3.1.1 ▪ A model of the investment decisions and delivery of the capital goods

In 1935, Kalecki proposed a mathematical model for investment decisions and delivery of capital goods, which can be represented by the dynamical system

$$\dot{y}(t) = \frac{\alpha}{h}y(t) - \left(\frac{\alpha}{h} + \beta\right) y(t - h) + u(t), \tag{3.4}$$

where $\alpha$ and $\beta$ are constants, $u(t) \equiv 0$, $y(t)$ represents the deviation of investment from a desired constant value, and $h$ is the "gestation period" (time delay) in getting new capital goods delivered and installed so that they can be used for production [117, 272]. The existence of an oscillatory response in $y(t)$, due to arbitrary initial conditions, is investigated in [73, 117]. Clearly, under the assumption that $u(t) = 0$ for all $t \geq 0$, (3.4) is a closed-market system where there are no external investments (inputs) entering into the system. One can also assume zero initial conditions for $y(t)$ and introduce a time function $u(t)$ to capture the effect of the initial conditions. Then, the transfer function from $u$ to $y$ is

$$P(s) = \left(s - \frac{\alpha}{h} + \left(\frac{\alpha}{h} + \beta\right) e^{-hs}\right)^{-1},$$

which is in the general form (3.3). The locations of the poles of $P(s)$ determine whether periodic oscillations are possible or not in $y(t)$ due to arbitrary initial conditions.

There are now many other time delay system models built on Kalecki's model: these are derived by including several other factors interfering with the investment decisions and by nonlinear extensions; see [272] for a review.

### 3.1.2 ▪ AQM scheme supporting TCP flows

A mathematical model of AQM (active queue management) schemes supporting TCP (transmission control protocol) flows in communication networks is proposed in [110, 163]. Consider the simplified network model shown in Figure 3.4. It is assumed that data is sent in blocks of packets, each of them containing several control bits assigned for flow control. One of the control bits is reserved for packet marking for congestion notification: as the packet goes through the router, the AQM mechanism writes the control bit: if it is 1, the packet is marked, and if it is 0, the packet is not marked. When the receiver gets the transmitted packet, it sends an acknowledgment (Ack) signal back to the sender. Since it is used for control purposes only, the size of the Ack packet is much smaller than the original data packet; so, typically, it goes from the receiver to the sender much faster than the travel time of the data packets on the forward path. One of the critical pieces of information in the Ack signal is the marked status of the packet received. In TCP, the sender transmits a block of $W$ packets at a time until an Ack signal is received. If the Ack signal says the packet block was not marked, then in the next round the sender transmits a $W + 1$ packet block; if the Ack indicates that the packet block was marked, then in the next round the sender transmits a $W/2$ packet block. The return trip time, or round trip time (RTT), $R$, is defined as the time elapsed between a packet block being sent and the corresponding Ack signal being received. Consider the case where a large number (say $N$) of senders and receivers use the same router, where packets are buffered and retransmitted based on a first-in-first-out queuing rule. The objective is to regulate the queue, denoted by $Q$. The AQM deals with the question of how to design a controller for packet markings (denoted by $P$) as a function of $Q$ at the router.

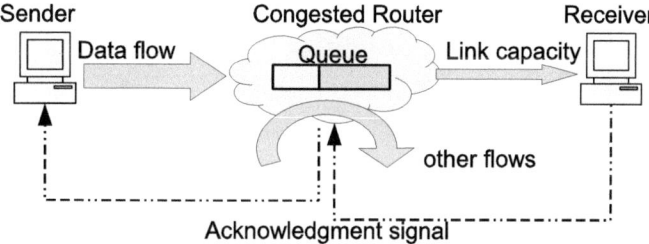

**Figure 3.4.** *AQM of TCP flows at a router.*

Assuming all senders and receivers obey the same TCP rules, for the case where $Q > 0$, an aggregate fluid flow model of the underlying system is represented by the following differential equations [110]:

$$\dot{Q}(t) = N \, \frac{W(t)}{R(t)} - C,$$

$$\dot{W}(t) = \frac{1}{R(t)} - \frac{W(t)}{2} \, M(t - R(t)),$$

$$M(t) = \frac{W(t)}{R(t)} P(t),$$

$$R(t) = \frac{Q(t)}{C(t)} + T_p,$$

where $C$ is the link capacity, $T_p$ is the propagation delay, and $M$ is the number of packets marked per RTT. Linearization of the above system around a positive equilibrium point $(Q_o, P_o)$ is done by defining $Q(t) = Q_o + q(t)$, $P(t) = P_o + p(t)$, $W(t) = W_o + w(t)$, and $R(t) = R_o + r(t)$ and solving for the resulting unique $W_o$ and $R_o$ [110, 213]. The feedback system corresponding to the AQM can be represented as shown in Figure 3.5. The linear plant transfer function seen by the controller from $p(t)$ to $q(t)$ is

$$P_{\text{AQM}}(s) = \frac{0.5 \, N \, W_o^3 \, e^{-R_o s}}{W_o \, R_o^2 \, s^2 + (W_o + 1) \, R_o \, s + 2 + R_o \, s \, e^{-R_o s}}, \tag{3.5}$$

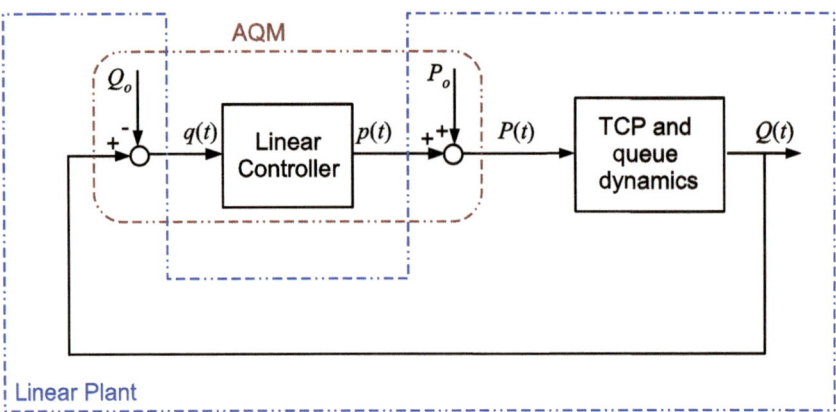

**Figure 3.5.** *Feedback system representation of the AQM of TCP flows.*

where the number of homogeneous TCP sources is $N$, the round trip delay is $R_o$, and the congestion window size is $W_o = R_o C/N$ at the operating point [110, 213].

## 3.2 ▪ Pseudorational transfer functions

Let us examine the transfer function given in (3.3) further. It takes the form of the ratio of two polynomials in $s$, $e^{-h_{d,i}s}$ and $e^{-h_{n,i}s}$, the latter exponentials denoting various delays. In terms of the output $y$ and input $u$, this means that the output is the function of its delayed values and also delayed values of the input.

This naturally leads to the generalization that

*finite-time past values and their derivatives $\mapsto$ output.*

The idea is to generalize this to the action of *distributions* of *compact support*. To see this more clearly, consider the following system where $u$ is the input and $y$ is the corresponding output:[1]

$$Y(s) = \frac{e^{-s}}{s - e^{-s}} U(s).$$

Multiplying both sides by the denominator, we obtain

$$(s - e^{-s})Y(s) = e^{-s}U(s),$$

whose time domain representation is

$$\dot{y}(t) - y(t-1) = u(t-1). \tag{3.6}$$

Recall the following identities from [223, 224] (see also [261] for a brief introduction):

$$\dot{\delta} \star y = \dot{y}, \qquad (\delta_1 \star y)(t) = y(t-1), \tag{3.7}$$

where $\star$ represents the usual convolution operator, $\delta_a(t)$ denotes the delta distribution placed at point $a$, namely $\delta(t-a)$, and $\dot{f}$ denotes $\frac{d}{dt}f(t)$; $\delta_0$ is denoted simply by $\delta$. Then, (3.6) may be rewritten as

$$\dot{\delta} \star y - \delta_1 \star y = \delta_1 \star u.$$

---

[1]Since we go back and forth between the time domain and frequency domain representations in this section, to avoid any confusion we use the notation $F(s) = \mathcal{L}\{f(t)\}$ for the Laplace transform of a signal $f(t)$.

That is, the action of a distribution $\dot{\delta} - \delta_1$, having compact support, acting on $y$ on the left-hand side equals the right-hand side $\delta_1 \star u$, which is also a result of a distribution with compact support acting on $u$.

The compact support property means that the processing of a finite time yields a pertinent quantity such as $(\dot{\delta} - \delta_1) \star y$. Generalizing this idea, systems in the form (3.3) can be represented by an equation in the form

*a distribution with compact support convolved with the output is equal to a distribution with compact support convolved with the input.*

Roughly speaking, when an I/O relation satisfies such a relation, we say that the underlying system is *pseudorational*.

We need some notations and preliminary concepts from distribution theory [224, 261]. Let $\mathcal{D}$ be the space of infinitely differentiable functions on the real line $(-\infty, \infty)$ having compact support. In another terminology, it is also written as $C_0^\infty(-\infty, \infty)$. This space is called the space of *test functions*. It is equipped with the so-called inductive limit topology of all subspaces with support bounded on a fixed finite interval. The description of this topology is beyond the scope of this book, and the reader is referred to [224, 261] for details. Once space $\mathcal{D}$ is equipped with a topology, it is possible to consider its *dual space*, i.e., the space of continuous-linear functionals defined on $\mathcal{D}$. This is the space of *distributions* defined on $(-\infty, \infty)$ and is denoted by $\mathcal{D}'(\mathbb{R})$ or $\mathcal{D}'(-\infty, \infty)$. For a distribution $\alpha \in \mathcal{D}'(\mathbb{R})$, its *support* is the largest closed set outside of which $\alpha$ is zero, and it is denoted by supp $\alpha$. Two subspaces of $\mathcal{D}'(\mathbb{R})$ are important to the subsequent developments. $\mathcal{E}'(-\infty, 0]$ denotes the subspace of $\mathcal{D}'(\mathbb{R})$ consisting of those having compact support contained in $(-\infty, 0]$. $\mathcal{D}'_+(\mathbb{R})$ consists of those having support bounded on the left.

Let us now give a more precise definition following [258].

**Definition 3.2.** *An impulse response function $g(t)$ is called* pseudorational *if there exist distributions $d$ and $n$ having compact support in $(-\infty, 0]$ such that*

1. *$g = d^{-1} \star n$ and*

2. *ord $d^{-1} = -$ ord $d$,*

*where $d^{-1}$ denotes the inverse with respect to convolution and* ord $d$ *is the order of distribution $d$ [224].*

Some remarks are in order. The order ord $d$ roughly denotes the number of differentiation contained in the action of a distribution. To be more precise, if a distribution $\alpha$ is expressible as

$$\alpha = (d/dt)^r \psi, \tag{3.8}$$

with measure $\psi$ and integer $r \geq 0$, then it is said to be of order at least $r$. If $\alpha$ is of order at least $r$ but not $r + 1$, then it is said to be of order $r$. If, on the other hand, $(d/dt)^r \alpha$ $(r \geq 0)$ is a measure, then it is of order no more than $-r$. If it is of order no more than $-r$ but not $-r - 1$, then it is of order $-r$. The differentiation here is taken in the sense of distributions [223, 224].

Here are some examples. The Dirac delta distribution $\delta_a$ is of order 0; its derivative $\dot{\delta}_a$ is of order 1. On the other hand, the Heaviside unit step function, denoted by $u(\cdot)$, is of order $-1$ since it is once differentiable in the sense of distributions as

$$\frac{d}{dt}u = \delta.$$

Note that the differentiation here is in the sense of distributions over the whole line $(-\infty, \infty)$. Because of the jump from 0 to 1 at the origin, it induces $\delta$ when differentiated. Since $(d/dt)\mathrm{u}$ is of order 0, $\mathrm{u}$ is of order $-1$.

The above technical condition on the order of $d$ does not concern us very much in what follows, but it is crucial for constructing a standard realization from the factorization $g = d^{-1} \star n$; see [258] for details. Definition 3.2 can be naturally extended to the multivariable case with matrix factorization; see [258].

Also observe that we have adopted the convention of expressing $g$ as the ratio of distributions with (compact) support contained in the *past* $(-\infty, 0]$. This is different from the convention employed in (3.3). Consider, for example,

$$\dot{y}(t) = y(t-1) + u(t-1).$$

This is expressed as

$$(\dot{\delta} - \delta_1) \star y = \delta_1 \star u. \tag{3.9}$$

But this can also be expressed as

$$(\dot{\delta} \star \delta_{-1} - \delta) \star y = u, \tag{3.10}$$

where $d := \dot{\delta} \star \delta_{-1} - \delta$, $n := \delta$ both have support contained in $[-1, 0]$. Noting the Laplace transforms $\mathcal{L}\{\dot{\delta}\} = s$, $\mathcal{L}\{\delta_{-1}\} = e^s$, and $\mathcal{L}\{\delta\} = 1$, (3.10) yields

$$Y(s) = \frac{1}{se^s - 1} U(s). \tag{3.11}$$

Hence, in place of the Laplace transform of (3.9), leading to

$$Y(s) = \frac{e^{-s}}{s - e^{-s}} U(s), \tag{3.12}$$

where we used $\mathcal{L}\{\delta_1\} = e^{-s}$, the representation (3.11) may be preferred. Equation (3.12) is in the form (3.3), but we may as well employ the expression (3.11), which is of type

$$P(s) = \frac{\sum_{i=1}^{v_n} q_{n,i}(s) e^{\ell_{n,i}s}}{\sum_{k=1}^{v_d} q_{d,k}(s) e^{\ell_{d,k}s}}, \tag{3.13}$$

where these $\ell_{n,i}$ and $\ell_{d,i}$ are different from those given in (3.3).

There is an advantage in expressing the impulse response as the convolutional ratio of distributions with compact support in the past. For example, for classical finite dimensional systems, the transfer functions are expressed as a ratio of polynomials in $s$, not in $s^{-1}$. Likewise, it is more in conformity with the convention of expressing transfer functions of delay systems as polynomials in $s$ and $e^s$, and not $e^{-s}$. This allows us a more unified treatment for dealing with a generalized class of pseudorational impulse responses (or transfer functions).

Let us give a characterization of the transfer function of a pseudorational impulse response.

**Theorem 3.3.** *Let* $g = d^{-1} \star n$ *be pseudorational as defined by Definition 3.2. Then its transfer function*

$$P(s) = \mathcal{L}\{g\}(s) = \frac{N(s)}{D(s)} \tag{3.14}$$

*is the ratio of entire functions of exponential type. Moreover, the denominator and numerator admit only a polynomial growth rate along the imaginary axis.*

***Proof.*** Suppose that a distribution $f$ has compact support contained in the interval $[-a, 0]$. We quote from the Paley–Wiener theorem [224] that the Laplace transform $\mathcal{L}\{f\}$ is the entire function in $s$ and satisfies

$$|F(s)| \leq \begin{cases} (1 + |s|)^m e^{a \operatorname{Re} s}, & \operatorname{Re} s \geq 0, \\ (1 + |s|)^m, & \operatorname{Re} s < 0. \end{cases} \tag{3.15}$$

Substituting $d$ and $n$ for $f$ readily yields the result.   □

Beyond the class of polynomials, entire functions of exponential type are the next simplest type of analytic functions, whence the name *pseudorational*. Certainly the numerator and denominator of delay-differential systems (3.13) satisfy this condition.

The following system provides an example that is not necessarily a lumped delay system.

**Example 3.4.** Let $g$ be an impulse response function that has bounded support; i.e., there exists $T > 0$ such that $\operatorname{supp} g \subset [0, T]$, where $\operatorname{supp} g$ denotes the support of $g$. Then, $g$ is pseudorational.   ∎

The above claim is obvious because

$$g = (\delta_{-T})^{-1} \star ((\delta_{-T}) \star g).$$

Note that the transfer function of the system defined in Example 3.4 is obtained by

$$\mathcal{L}\{g\} = G(s) = \int_0^T g(t) e^{-st} dt,$$

which is a distributed delay system. See Section 3.3.4 for a specific application example. In order to give an idea of how such a transfer function looks like, let us consider

$$g(t) = \begin{cases} k e^{at} & \text{for } 0 \leq t \leq \tau, \\ 0 & \text{for } t > \tau, \end{cases} \tag{3.16}$$

where $k, a \in \mathbb{R}$. The transfer function of this system is

$$G(s) = k \, \frac{1 - e^{-\tau(s-a)}}{s - a},$$

which is an entire function, and $G \in \mathcal{H}_\infty$. Clearly, this transfer function is a special case of lumped delay systems (3.3). On the other hand, if $g(t)$ is not in the generic form (3.16), i.e., not a linear combination of functions in the form $t^m e^{at}$, where $m$ is a nonnegative integer and $a \in \mathbb{R}$, then we have an infinite dimensional system whose transfer function cannot be written in the form (3.3). Therefore, pseudorational transfer functions is a larger class than those represented by lumped delay systems (retarded-, neutral-, or advance-type systems).

Using the fractional representation (3.14) in Definition 3.2, one can construct a state space and a standard observable realization out of these data, but in this book we will not work with the state space realizations of such systems. For further details, see [258, 259].

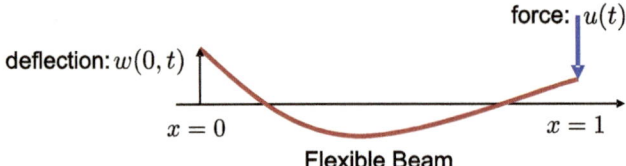

**Figure 3.6.** *A flexible beam.*

## 3.3 ▪ Other DPS examples

### 3.3.1 ▪ A flexible beam model

The dynamical behavior of the flexible beam system shown in Figure 3.6 is determined by the partial differential equation

$$\frac{\partial^2 w}{\partial t^2} + \varepsilon \frac{\partial^5 w}{\partial x^4 \partial t} + EI \frac{\partial^4 w}{\partial x^4} = 0,$$

where the material constants are normalized, $EI = 1$, and $\varepsilon > 0$ represents the Kelvin–Voigt damping constant. For the beam with free end points, and a transverse force input $u(t)$ applied at $x = 1$, the boundary conditions are

$$\frac{\partial^2 w}{\partial x^2}(0, t) + \varepsilon \frac{\partial^3 w}{\partial x^2 \partial t}(0, t) = 0, \quad \frac{\partial^2 w}{\partial x^2}(1, t) + \varepsilon \frac{\partial^3 w}{\partial x^2 \partial t}(1, t) = 0,$$

$$\frac{\partial^3 w}{\partial x^3}(0, t) + \varepsilon \frac{\partial^4 w}{\partial x^3 \partial t}(0, t) = 0, \quad \frac{\partial^3 w}{\partial x^3}(1, t) + \varepsilon \frac{\partial^4 w}{\partial x^3 \partial t}(1, t) = u(t).$$

Let the output of the system be the deflection at $x = 0$, i.e., $y_o(t) := w(0, t)$, and consider low-pass characteristics for the combined actuator and sensor dynamics: $H(s) = e^{-hs}/(1 + \tau s)$, with $h \geq 0$ and $\tau > 0$. Define the output available for feedback as $Y(s) = H(s)Y_o(s)$. Note that $w(0, t)$ is the deflection at the opposite end of the beam from the applied force $-u(t)$; i.e., even if the sensing delay is zero, the plant is in its nonminimum phase due to a noncollocated actuator and sensor. The transfer function of the plant (including the actuator and sensor dynamics) can be derived as in [137]:

$$\frac{Y(s)}{U(s)} = P(s) = \frac{(\sinh \beta - \sin \beta)\, e^{-hs}}{\beta^3(\cos \beta \cosh \beta - 1)(1 + \varepsilon s)(1 + \tau s)}, \quad \beta^4 = \frac{-s^2}{(1 + \varepsilon s)}. \quad (3.17)$$

One can show that $P(s)$ can be expressed as an infinite product of second order terms. This product representation displays the plant's poles and zeros clearly. It also facilitates inner/outer factorizations, to be discussed in Chapter 4:

$$P(s) = \frac{2\, e^{-hs}}{s^2\,(\tau s + 1)} \prod_{n=1}^{\infty} g_n(s), \quad (3.18)$$

where

$$g_n(s) = \frac{\left(1 + \varepsilon s - \frac{s^2}{4\alpha_n^4}\right)}{\left(1 + \varepsilon s + \frac{s^2}{\phi_n^4}\right)}$$

for values of $s$ where this infinite product converges and the coefficients $\alpha_n$ and $\phi_n$ are determined as the solutions of

$$\cos(\alpha_n)\sinh(\alpha_n) - \sin(\alpha_n)\cosh(\alpha_n) = 0 \ \text{ for } \alpha_n > 0,$$

$$\cos(\phi_n)\cosh(\phi_n) = 1 \ \text{ for } \phi_n > 0.$$

In [137, 138] it is shown that (3.18) converges everywhere in the open right half plane and can be written as a quotient of $\mathcal{H}_\infty$ functions.

Transfer functions of many other beam systems are given in [38]; see also [137].

### 3.3.2 ▪ The heat equation

The one dimensional heat flow on a conducting rod is described by the following partial differential equation:

$$\frac{\partial T(x,t)}{\partial t} = c^2 \frac{\partial^2 T(x,t)}{\partial x^2}, \tag{3.19}$$

where $T(x,t)$ is the temperature of the rod on the point $x \in [0,L]$ at time $t \geq 0$, and $c > 0$ is the thermal diffusivity constant. Typically, there are two types of boundary control problems associated with the heat equation (3.19). The first one is the Neumann boundary control:

$$\frac{\partial T(0,t)}{\partial x} = 0, \qquad c^2 \frac{\partial T(L,t)}{\partial x} = u(t), \tag{3.20}$$

where $u(t)$ is the heat flux input applied at $x = L$. The second possibility is to define the temperatures at both ends, i.e., Dirichlet boundary control:

$$T(0,t) = 0, \qquad T(L,t) = u(t). \tag{3.21}$$

In both cases we can think of $u(t)$ as the input. For the Neumann boundary control, define $y(t) = T(0,t)$ as the output. For the Dirichlet boundary control, we consider the temperature of a point $x_o \in (0,L)$ on the rod as the output, $y(t) = T(x_o,t)$. In the first case the transfer function of the underlying plant is

$$\frac{Y(s)}{U(s)} = P(s) = \frac{2e^{-L\xi}}{\xi(1 - e^{-2L\xi})}, \qquad \xi := \frac{\sqrt{s}}{c}; \tag{3.22}$$

see [143]. It can be shown that the plant (3.22) has a single pole at $s = 0$, and all other poles are in $\mathbb{C}_-$; see [38]. For the Dirichlet boundary control, the transfer function is

$$\frac{Y(s)}{U(s)} = P(s) = \frac{\sinh(x_o\xi)}{\sinh(L\xi)}, \qquad \xi := \frac{\sqrt{s}}{c}. \tag{3.23}$$

It can be shown that all the poles and zeros of the plant (3.23) are in $\mathbb{C}_-$; see [38].

### 3.3.3 ▪ Fractional order systems

In the last two decades, many papers and books have been published on the application of the fractional order calculus to modeling of dynamical systems; see [164] and its references. According to Caputo's definition, the Laplace transform of the *fractional order integral* of a signal $f(t)$ is

$$\int_0^\infty \left( \frac{1}{\Gamma(\alpha)} \int_0^t f(y)(t-y)^{\alpha-1} dy \right) e^{-st} dt = s^{-\alpha} F(s),$$

where $\Gamma$ is the Gamma function and $\alpha \in (0,1)$ is the fractional order. In particular, this allows more accurate modeling of certain physical systems where the Bode magnitude plot does not roll off at an integer multiple of $-20$ dB per decade. Note that (3.22) and (3.23) are functions of $\sqrt{s}$, so they can be thought of as special types of fractional order systems. However, in the literature, typically, a fractional order system refers to a rational function of $s^\alpha$, where $\alpha \in (0,1)$; these are commensurate systems. When the transfer function involves two or more polynomials of $s^{\alpha_1}$ and $s^{\alpha_2}$, with $\alpha_1/\alpha_2$ being an irrational number, the system is said to be a noncommensurate fractional order system. The literature on the stability analysis of fractional order systems (with or without time delays) is quite large; see, e.g., [61] for a method and further references.

Here we give an unstable commensurate fractional order system example. Recently, a fractional order plant model was derived in [131] for a nonlaminated magnetic suspension system:

$$P(s) = \frac{e^{-hs}}{(s^\alpha)^5 + (s^\alpha)^4 - c} = \frac{e^{-hs}}{s^2(\sqrt{s}+1) - c}, \qquad (3.24)$$

where $h > 0$ is the I/O delay, $\alpha = 0.5$ is the fractional order, and $c > 0$ is obtained from the physical properties of the system after a normalization.

From the robust control design perspective, it is critical to determine the right half plane poles and zeros of the plant (3.24); this information is used in order to perform coprime and inner-outer factorizations (see Chapter 4). In a recent work [204], it has been shown that such factorizations can be done for a large class of fractional order systems.

In [131], it has been shown that for all $c > 0$, the transfer function (3.24) can be written as

$$P(s) = \frac{e^{-hs}}{(s^\alpha - p)(s^\alpha - p_1)(s^\alpha - p_2)(s^\alpha - p_3)(s^\alpha - p_4)}, \qquad (3.25)$$

where $p > 0$ and $p_k \in \mathbb{C} \setminus \{re^{j\theta} : r \in \mathbb{R}_+, |\theta| < \alpha\pi/2\}$ for $k = 1, \ldots, 4$. Thus, by Matignon's theorem [149], the only unstable pole of this plant is at $s = p^2$.

### 3.3.4 ▪ Systems with distributed delays

In this section, as an example of a system with distributed delays, we consider a compartmental model representing the cell population dynamics in acute myeloid leukemia (AML). The origins of this model go back to [146], where a mathematical is introduced for cell dynamics in periodic hematopoiesis. Recently, the authors of [3] extended the model of [146] to AML. See also [12, 13, 47] for further refinements and extensions of this model.

In a very simplified form, the compartmental block diagram shown in Figure 3.7 illustrates how different types of cells lead to mature cells circulating in the blood; see, for example, [3, 215]. A particular type of AML is a result of blasts in progenitors; i.e., the progenitor cell population grows in an undesirable fashion, so that the percentage of cancer cells in the blood becomes higher than a certain threshold set by the World Health Organization or other medical associations. Let $x(t)$ denote the progenitor cell population at the resting phase at time $t$. Then, $w(t) := x(t)\beta(x(t))$ is the rate of progenitor cell population entering into the growth phase; typically, $\beta(\cdot)$ is a Hill function. The cells entering into the growth phase eventually die at a rate $\gamma > 0$ or divide before a maximal time instant $\tau > 0$. Let $u(t)$ denote the rate of differentiation of stem

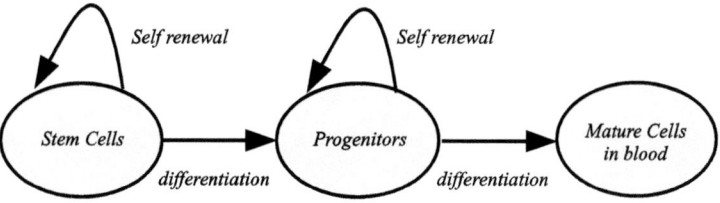

**Figure 3.7.** *Compartmental model of cell population dynamics in AML.*

cells contributing to progenitor cell population. Then, according to [3] the equation representing the progenitor cell population dynamics is given by

$$\dot{x}(t) = -\delta x(t) - w(t) + 2L \int_0^\tau e^{-\gamma \theta} f(\theta) w(t - \theta) d\theta + u(t), \qquad (3.26)$$

where $\delta > 0$ is the cell death rate at the resting phase, $0 < L < 1$ represents the fraction of divided cells that are self-renewing, and $f(\theta)$, $\theta \in [0, \tau]$, is the cell division probability density so that $\int_0^\tau f(\theta) d\theta = 1$. The system (3.26) is nonlinear; the first problem is to determine the existence condition of a unique positive equilibrium $x_e > 0$. Then, we can analyze the linearized system around $x_e$ to check whether we have local stability. Assume that $u(t) = u_e > 0$ is constant. Define

$$\alpha := 2L \int_0^\tau e^{-\gamma \theta} f(\theta) d\theta - 1,$$

and further assume that $\alpha > 0$ and $\beta(x) = b(a + x^n)^{-1}$ for some constants $a, b > 0$ and that $n$ is an integer greater than or equal to 2. Then, the equilibrium point $x_e > 0$ is the unique solution of

$$\beta(x_e) = \frac{1}{\alpha} \left( \delta - \frac{u_e}{x_e} \right).$$

Now define $u(t) = u_e + \tilde{u}(t)$, $x(t) = x_e + \tilde{x}(t)$, and

$$\mu := \frac{dw}{dx}\Big|_{x=x_e}.$$

Assuming that $|\tilde{u}| \ll u_e$ and $|\tilde{x}| \ll x_e$, the local behavior of (3.26) around the equilibrium is captured by the transfer function from $\tilde{u}$ to $\tilde{x}$, that is,

$$P(s) = \frac{1}{s + \delta + \mu - 2L\mu G(s)}, \quad \text{where} \quad G(s) = \int_0^\tau e^{-(s+\gamma)\theta} f(\theta) d\theta. \qquad (3.27)$$

Note that $G(s)$ is the Laplace transform of the signal $e^{-\gamma \theta} f(\theta)$, where $\theta$ is restricted to the finite interval $[0, \tau]$. If $f$ consists of finitely many Dirac delta terms only, then $P$ becomes a lumped delay system, as in Section 3.1. It is also worth noting that $G \in \mathcal{H}_\infty$ and $\|G\|_\infty = G(0)$; furthermore, $\alpha = 2LG(0) - 1$. When $\mu > 0$, by using the Nyquist stability criterion, it is a simple exercise to show that $P \in \mathcal{H}_\infty$ if and only if $\alpha < \delta/\mu$. For further discussion on the stability of $P$ under $\mu < 0$ and the global stability of (3.26), see [196].

# Chapter 4

# Factorizations

## 4.1 ▪ Feedback stabilization by coprime factorization

In this book we consider the standard feedback system shown in Figure 4.1 and discuss issues centered around robust controller design for LTI SISO distributed parameter plants. There can be several possible design objectives, but the most important constraint is stability of the feedback system.

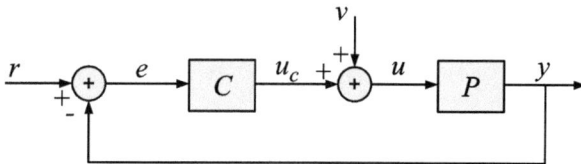

**Figure 4.1.** *Standard feedback system* $(C, P)$, *with controller $C$ and plant $P$.*

We say that the *feedback system is stable* if all transfer functions from external signals $(r, v)$ to internal signals $(e, u)$ are in $\mathcal{H}_\infty$. This is equivalent to having

$$S(s) = (1 + P(s)C(s))^{-1} \in \mathcal{H}_\infty, \quad C(s)S(s) \in \mathcal{H}_\infty, \quad P(s)S(s) \in \mathcal{H}_\infty. \quad (4.1)$$

Note that $S$ is the transfer function from $r$ to $e$ and also from $v$ to $u$; $CS$ is the transfer function from $r$ to $u$; and $PS$ is the transfer function from $v$ to $-e$. The transfer functions from $v$ to $y$ and from $r$ to $y$ are $PS$ and $T = 1 - S$, respectively. Clearly, $S \in \mathcal{H}_\infty$ if and only if $T \in \mathcal{H}_\infty$. The sensitivity of $T$ to variations in $P$ is formally defined as

$$\frac{P}{T} \frac{\partial T}{\partial P} = (1 + P C)^{-1} = S,$$

so $S$ is called the *sensitivity function*. Because of the algebraic relation $T = 1 - S$, the transfer function $T$ is called the *complementary sensitivity*. For a given plant $P$, all controllers $C$ satisfying (4.1) are called stabilizing controllers for the plant.

### 4.1.1 ▪ Small gain theorem

Let us consider the feedback system shown in Figure 4.2, where $G_1$ and $G_2$ are in $\mathcal{H}_\infty$. Note that the standard feedback system of Figure 4.1 is in this form, with $C = G_1$ and $P = G_2$.

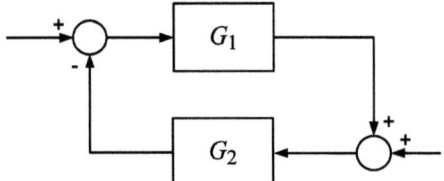

**Figure 4.2.** *Feedback system* $(G_1, G_2)$.

Since $G_1, G_2 \in \mathcal{H}_\infty$, the feedback system is stable if and only if the sensitivity function $S = (1 + G_1 G_2)^{-1}$ is in $\mathcal{H}_\infty$, which holds if and only if $U^{-1} \in \mathcal{H}_\infty$, where $U$ is the *characteristic function* defined by

$$U(s) = 1 + G_1(s)G_2(s). \tag{4.2}$$

Clearly, for $S = U^{-1}$ to be in $\mathcal{H}_\infty$, the characteristic function $U(s)$ should have no zeros in $\mathbb{C}_+$. The necessary and sufficient conditions for this are obtained from the Nyquist stability criterion (see, e.g., [50]), which requires plotting the graph of the *open-loop gain* $G(j\omega) = G_1(j\omega)G_2(j\omega)$ and counting the number of encirclements of the critical point $-1$ in the complex plane. In this special case, the feedback system $(G_1, G_2)$ is stable if and only if $-1$ is not encircled by the graph of $G(j\omega)$. There is a very simple sufficient condition for this to hold, formally known as the *small gain theorem*.

**Theorem 4.1 (small gain theorem).** *Let $G_1$ and $G_2$ be stable. Then, the feedback system $(G_1, G_2)$ is stable if*

$$\|G_1 G_2\|_\infty < 1. \tag{4.3}$$

***Proof.*** Clearly, when $\|G_1 G_2\|_\infty < 1$ we have that $|G(j\omega)| < 1$ for all $\omega$, and hence the graph of $G(j\omega)$ remains within the unit circle for all $\omega$, and hence it cannot encircle $-1$. Thus, the result follows from the Nyquist stability criterion. $\quad\square$

The small gain theorem can also be extended to nonlinear systems; see, e.g., a recent book [115, Chapter 8] and also classical books such as [126, 246] for detailed treatments. We should emphasize that the small gain is only a sufficient condition. That is, the feedback system may be stable even if $\|G_1 G_2\|_\infty \geq 1$, which happens in many practical applications. Another obvious, and useful, sufficient condition for feedback system stability is $\mathrm{Re}\big(G_1(j\omega)G_2(j\omega)\big) > -1$ for all $\omega$. See the above mentioned books for more details and connections to various other stability conditions.

Theorem 4.1 is very useful in establishing the robust stability of feedback systems in many applications. For example, let us consider the following special case: $G_1(s) = Q(s)$, where $Q \in \mathcal{H}_\infty$ represents a nominal closed-loop system transfer function, which is stable, and $G_2(s) = \Delta(s)$, where $\Delta \in \mathcal{H}_\infty$ is unknown, but its magnitude is known to be bounded, say $\|\Delta\| \leq \rho$. Recall that the feedback system is stable if and only if $U^{-1} \in \mathcal{H}_\infty$, where

$$U(s) = 1 + \Delta(s)Q(s).$$

Clearly, a sufficient condition for this comes from Theorem 4.1, that is,

$$\|Q\|_\infty < 1/\rho. \tag{4.4}$$

In fact, (4.4) is a necessary condition as well, because there is no restriction on the phase of $\Delta$. To see this, first suppose that there exists a finite $\omega_o$ such that $|Q(j\omega_o)| = 1/\rho$; then, a stable $\Delta(s)$ can be constructed such that

$$|\Delta(j\omega_o)| = \rho \quad \text{and} \quad \angle\Delta(j\omega_o) = -\pi - \angle Q(j\omega_o),$$

in which case $U(j\omega_o) = 0$. When $\lim_{\omega\to\infty} |Q(j\omega)| = 1/\rho$, then, choosing $\Delta(\infty) = \pm\rho$ and adjusting the sign, we can have $\lim_{s\to\infty} U(s) = 0$, which means that $U^{-1}$ is not in $\mathcal{H}_\infty$. Various robust stability conditions discussed in Chapter 5 are derived from similar arguments.

**Example 4.2.** Consider the feedback system shown in Figure 4.3, where $P_1 \in \mathcal{H}_\infty$, and there is no unstable pole-zero cancellation in the product $P := P_1 P_2$; $C$ is such that $(C, P)$ is a stable feedback system, and $\Delta_b \in \mathcal{H}_\infty$ is an uncertain transfer function in the internal feedback loop, with the norm bound

$$\|P_1\Delta_b\|_\infty \le \rho < 1. \tag{4.5}$$

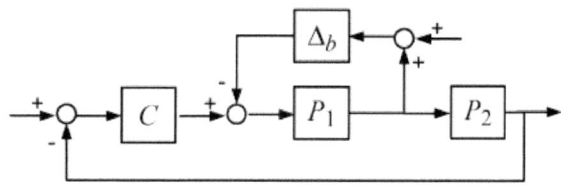

**Figure 4.3.** *Feedback system* $(C, P_{\Delta_b})$, *where* $P_{\Delta_b} = \frac{P_1 P_2}{1 + P_1\Delta_b}$.

Let us define $\Delta := P_1\Delta_b$. Since $\|\Delta\|_\infty < 1$, the feedback system is stable if and only if $U_b^{-1} \in \mathcal{H}_\infty$, where

$$U_b(s) = 1 + PC + \Delta = (1 + PC)\left(1 + (1 + PC)^{-1}\Delta\right).$$

By the assumption that $(C, P)$ is stable, we can conclude that the feedback system is stable if and only if

$$\|S\|_\infty = \|(1 + PC)^{-1}\|_\infty < 1/\rho.$$

Thus, $\rho_{\max}$, the largest allowable uncertainty level $\rho$ in (4.5), is $\|S\|_\infty^{-1}$.  ∎

**Exercise 4.3.**
As a numerical example, let us take

$$P_1(s) = \frac{1}{\tau s + 1}, \quad P_2(s) = \frac{e^{-hs}}{s},$$

where $\tau = \sqrt{2}$ sec, and $h = 0.5$ sec.
1. Show that the controller $C(s) = K$ stabilizes the feedback system $(C, P)$, with $K = 1$.
2. Compute $\rho_{\max}$ for $K = 1$.
3. Determine the largest $K > 0$ for which we have $\rho_{\max}(K) \ge 0.6$.

### 4.1.2 ▪ Parameterization of stabilizing controllers

We consider plant transfer functions which can be written in the form $P(s)=N(s)/D(s)$, where $N, D \in \mathcal{H}_\infty$ satisfy

$$\inf_{\mathrm{Re}(s)>0} (|N(s)| + |D(s)|) \geq \delta \quad \text{for some } \delta > 0. \tag{4.6}$$

Such a pair $(N, D)$ is called *strongly coprime* in $\mathcal{H}_\infty$ [227]. If the plant has finitely many poles (or zeros) in $\mathbb{C}_+$, then a strongly coprime factorization can be found easily. Let us consider a simple example, a first order unstable system with a time delay:

$$P(s) = \frac{e^{-hs}}{s-a}, \quad \text{where } h > 0, \ a > 0. \tag{4.7}$$

Clearly, $N(s) = e^{-hs}/(s+a)$ and $D(s) = (s-a)/(s+a)$ is a strongly coprime pair. Of course, such a factorization is not unique; we could have taken $N(s) = e^{-hs}/(s+b)$ and $D(s) = (s-a)/(s+b)$ with any $b > 0$.

Now suppose that we can find a controller in the form $C(s) = \widetilde{X}(s)/\widetilde{Y}(s)$, where $(\widetilde{X}, \widetilde{Y})$ is a strongly coprime pair in $\mathcal{H}_\infty$. Then, from the conditions (4.1) it is easy to see that the feedback system formed by the controller $C = \widetilde{X}/\widetilde{Y}$ and plant $P = N/D$ is stable if and only if $U^{-1} \in \mathcal{H}_\infty$, where

$$U(s) = N(s)\widetilde{X}(s) + D(s)\widetilde{Y}(s). \tag{4.8}$$

A function $U \in \mathcal{H}_\infty$ satisfying the condition $U^{-1} \in \mathcal{H}_\infty$ is called *unimodular in $\mathcal{H}_\infty$*. It is clear that whenever $U^{-1} \in \mathcal{H}_\infty$ the equality (4.8) can be rewritten as a Bezout equation, which is in the form

$$1 = N(s)\widetilde{X}(s)U^{-1}(s) + D(s)\widetilde{Y}(s)U^{-1}(s) = N(s)X(s) + D(s)Y(s), \tag{4.9}$$

where $X = \widetilde{X}U^{-1}$, $Y = \widetilde{Y}U^{-1}$. With these new variables the controller can be expressed as $C(s) = X(s)/Y(s)$. Note that if there exist $X, Y \in \mathcal{H}_\infty$ satisfying (4.9), then for any $Q \in \mathcal{H}_\infty$ we also have

$$1 = N(s)\left(X(s) + D(s)Q(s)\right) + D(s)\left(Y(s) - N(s)Q(s)\right), \tag{4.10}$$

which means that if the controller $C = \widetilde{X}/\widetilde{Y}$ is a stabilizing controller for the plant $P = N/D$, then $C = (X + DQ)/(Y - NQ)$ is also a stabilizing controller for the same plant. Moreover, the converse is also true; that is, if $C$ is a stabilizing controller for $P = N/D$, then there are $X, Y \in \mathcal{H}_\infty$ such that $X$ and $Y$ satisfy (4.9) and $C$ is in the form $C = (X + DQ)/(Y - NQ)$ for some $Q \in \mathcal{H}_\infty$. Thus, for a given plant $P = N/D$ with $X, Y \in \mathcal{H}_\infty$ satisfying (4.9), the set

$$\mathscr{C}(P) := \left\{ C = \frac{X + DQ}{Y - NQ} \ : \ Q \in \mathcal{H}_\infty, \ Q(\infty) \neq Y(\infty)N^{-1}(\infty) \right\} \tag{4.11}$$

can be defined as the *set of all stabilizing controllers* for the plant $P$. The result given by (4.11) is an extension of the well-known finite dimensional controller parameterization [6, 7, 135, 268] to infinite dimensional systems obtained by [227]. The condition that $Q(\infty) \neq Y(\infty)N^{-1}(\infty)$ is required to make sure that the resulting controller does not have a pole at $s = \infty$; otherwise, the controller becomes improper. Another point to remark is that if the plant is strictly proper, then so is $N(s)$, and hence the condition $Q(\infty) \neq Y(\infty)N^{-1}(\infty) = \infty$ is already captured by $Q \in \mathcal{H}_\infty$. Therefore, the condition on $Q(\infty)$ is dropped from (4.11) when the plant is strictly proper.

### 4.1.3 ▪ Systems with I/O delay (dead-time)

#### Smith predictor

Let us now return to the example (4.7) with $N(s) = e^{-hs}/(s+b)$ and $D(s) = (s-a)/(s+b)$, where $h > 0$, $a > 0$, $b > 0$. To find a stabilizing controller, we search for $X, Y \in \mathcal{H}_\infty$ satisfying (4.9), which can be rewritten as

$$Y(s) = \frac{s+b-e^{-hs}X(s)}{s-a}. \tag{4.12}$$

Clearly, any $X \in \mathcal{H}_\infty$ satisfying $X(a) = (a+b)e^{ha}$ puts the resulting $Y(s)$ in $\mathcal{H}_\infty$. The simplest choice $X(s) = (a+b)e^{ha}$ results in

$$Y(s) = 1 + (a+b)\frac{1-e^{-h(s-a)}}{s-a}. \tag{4.13}$$

All controllers stabilizing the plant (4.7) are characterized by

$$C(s) = \frac{(a+b)e^{ha} + \frac{(s-a)}{(s+b)}Q(s)}{1 + (a+b)\frac{1-e^{-h(s-a)}}{s-a} - \frac{e^{-hs}}{s+b}Q(s)}, \tag{4.14}$$

where $Q \in \mathcal{H}_\infty$ is the free parameter. In particular, $Q = 0$ is a feasible choice, but for different controller design objectives (see Chapter 5) we may have to choose different $Q$. Note that the second term in (4.13) represents a finite impulse response (FIR) transfer function (impulse response is restricted to a finite time interval) because

$$\mathscr{L}^{-1}\left\{\frac{1-e^{-h(s-a)}}{s-a}\right\} = \begin{cases} e^{at} & \text{for } t \in [0, h], \\ 0 & \text{otherwise.} \end{cases}$$

Also note that as $a \searrow 0$ the controller (4.14) for $Q = 0$ becomes

$$C(s) = b\left(1 + b\frac{1-e^{-hs}}{s}\right)^{-1}, \tag{4.15}$$

which is the classical Smith predictor (introduced by Otto J. M. Smith in the 1950s) for an integrating system with time delay; see, e.g., [191, 247]. For any $b > 0$, the controller (4.15) stabilizes the feedback system where the plant is $e^{-hs}/s$, and it leads to

$$T(s) = \frac{b}{s+b}\, e^{-hs}. \tag{4.16}$$

The parameter $b > 0$ is adjusted to place the closed-loop system pole and hence the bandwidth of the feedback system.

**Exercise 4.4.**
Consider the plant

$$P(s) = P_o(s)e^{-hs}, \quad P_o(s) = \frac{G_o(s)}{s}, \quad G_o(s) = \frac{4(s-2)}{(s+4)}, \quad h = 0.25$$

and a Smith predictor–based controller

$$C(s) = \frac{C_o(s)}{1 + G_o(s)C_o(s)H(s)}, \quad \text{where } H(s) := \left(\frac{1-e^{-hs}}{s}\right).$$

As before, $C_o(s)$ is a controller such that $(C_o, P_o)$ is stable. Define

$$N_o(s) = \frac{G_o(s)}{(s+b)} \quad \text{and} \quad D_o(s) = \frac{s}{(s+b)} \quad \text{for some } b > 0.$$

Then, $X_o(s) = -b/2$ and $Y_o(s) = (s+4+3b)/(s+4)$ satisfy $N_o X_o + D_o Y_o = 1$, and hence the controller

$$C_o(s) = -\frac{b}{2}\left(\frac{s+4}{s+4+3b}\right)$$

makes $(C_o, P_o)$ stable. If we define $N(s) = N_o(s)e^{-hs}$, then $P(s) = N(s)/D_o(s)$, and with

$$Y(s) = Y_o(s) + N_o(s)X_o(s)\frac{1-e^{-hs}}{D_o(s)} = Y_o(s) + G_o(s)X_o(s)\left(\frac{1-e^{-hs}}{s}\right)$$

we have $NX_o + D_o Y = 1$. Moreover,

$$T(s) = \frac{P(s)C(s)}{1 + P(s)C(s)} = T_o(s)e^{-hs},$$

where

$$T_o(s) = N_o(s)X_o(s) = \left(\frac{b}{s+b}\right)\left(\frac{2\,(2-s)}{(4+s)}\right).$$

Now, if we use a rational approximation $H_f(s)$ for the infinite dimensional FIR transfer function $H(s)$, so that

$$C_f(s) = \frac{C_o(s)}{1 + G_o(s)C_o(s)H_f(s)} = \frac{X_o(s)}{Y_o(s) + N_o(s)X_o(s)(s+b)H_f(s)}$$

is used instead of $C(s)$, then the stability of the resulting feedback system $(C_f, P)$ can be guaranteed by using the small gain theorem, as below.

1. Verify that the characteristic function of the feedback system $(C_f, P)$ is

$$U_f(s) = X_o(s)N_o(s)e^{-hs} + \Big(Y_o(s) + N_o(s)X_o(s)(s+b)H_f(s)\Big)D_o(s)$$

$$= 1 + N_o(s)X_o(s)\Big(s\,H_f(s) - (1 - e^{-hs})\Big)$$

$$= 1 + T_o(s)\Big(s\,(H_f(s) - H(s))\Big).$$

By the small gain theorem, if

$$\gamma := \left\| T_o(s)\Big(s\,(H_f(s) - H(s))\Big)\right\|_\infty < 1,$$

then the feedback system is stable. In fact, for the best results we would like to leave some extra margin and make $\gamma$ significantly smaller than 1. Consider the first order filter, which matches the characteristics of $H(s)$ at $s = 0$,

$$H_{f1}(s) = \frac{h}{1 + (hs/2)}.$$

Plot $\gamma$ versus $b$ for the values of $b$ in $[0.1, 10]$. Which values of $b$ do you prefer?

2. Implement the feedback system $(C_f, P)$ in Simulink with $H_{f1}$, and obtain the responses for a step reference input for the values of $b = 0.5$, $b = 2$, and $b = 5.3$; can we make a conclusion about the range of reasonable choices of $b$ with respect to the result of part 1?

**Extension of Smith predictor for plants with one unstable pole**

Let us now consider a plant whose transfer function is in the form

$$P(s) = \frac{e^{-hs}}{s-a}G_o(s), \quad h > 0, \; a > 0, \; G_o \in \mathcal{H}_\infty.$$

In the previous section we had $G_o(s) = 1$; now we extend the above ideas to a more general case where $G_o$ is an arbitrary stable transfer function. Let us define $P_o(s) = (s-a)^{-1}G_o(s)$ as the nondelayed system. The coprime factorizations

$$P(s) = N(s)/D(s), \quad P_o(s) = N_o(s)/D(s)$$

are obtained by defining

$$N(s) = e^{-hs}N_o(s), \; N_o(s) = G_o(s)/(s+a), \; D(s) = (s-a)/(s+a).$$

Then, solutions of the Bezout equations

$$1 = N(s)X(s) + D(s)Y(s) \quad \text{for } h > 0,$$
$$1 = N_o(s)X_o(s) + D(s)Y_o(s) \quad \text{for } h = 0,$$

where $X, Y, X_o, Y_o \in \mathcal{H}_\infty$, give stabilizing controllers for $P$ and $P_o$. In particular, the following relation must be satisfied between $X$ and $X_o$:

$$X(s) = e^{ha}X_o(s),$$

and $X_o(s)$ is determined from the nondelayed case. Note that $Y(s)$ must satisfy

$$
\begin{aligned}
Y(s) &= \frac{1 - e^{-hs}N_o(s)X(s)}{D(s)} = \frac{1 - e^{-h(s-a)}N_o(s)X_o(s)}{D(s)} \\
&= \frac{1 - N_o(s)X_o(s) + N_o(s)X_o(s)\big(1 - e^{-h(s-a)}\big)}{D(s)} \\
&= Y_o(s) + P_o(s)X_o(s)\big(1 - e^{-h(s-a)}\big).
\end{aligned}
\tag{4.17}
$$

The controllers

$$C_o(s) = \frac{X_o(s)}{Y_o(s)} \quad \text{and} \quad C(s) = \frac{X(s)}{Y(s)}$$

stabilize $P_o(s)$ and $P(s)$, respectively. From the relation (4.17) we get that

$$C(s) = \frac{X(s)}{Y(s)} = \frac{e^{ha}X_o(s)}{Y_o(s) + P_o(s)X_o(s)\big(1 - e^{-h(s-a)}\big)};$$

dividing the numerator and denominator of the right-hand side by $Y_o(s)$, a familiar structure is obtained for the controller stabilizing $P(s)$:

$$C(s) = \frac{e^{ha}\,C_o(s)}{1 + P_o(s)C_o(s)\big(1 - e^{-h(s-a)}\big)},
\tag{4.18}$$

where $C_o(s)$ is a stabilizing controller for the nondelayed system $P_o(s)$. The implementation of this controller can be done as shown in Figure 4.4.

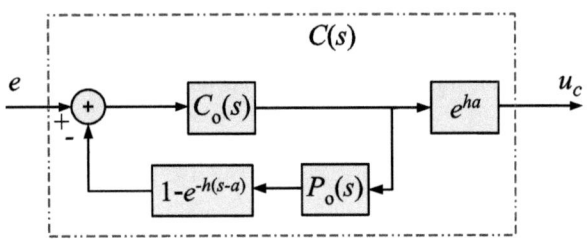

**Figure 4.4.** *Smith predictor for a plant with a single unstable pole.*

With this controller, the resulting closed-loop transfer function becomes

$$T(s) = \frac{P(s)C(s)}{1 + P(s)C(s)} = T_o(s)e^{-h(s-a)}, \quad \text{where} \quad T_o(s) = \frac{P_o(s)C_o(s)}{1 + P_o(s)C_o(s)}.$$

**Remark 4.5.**
1. It is clear that the above derivations are valid for plants in the form $P_o(s) = \left(\frac{s+b}{s-a}\right)G_o(s)$, where $a, b > 0$ and $G_o \in \mathcal{H}_\infty$. In other words, the plant does not have to be strictly proper.
2. The finite dimensional controller $C_o$ is freely chosen from the set $\mathscr{C}(P_o)$, i.e., it is in the form $C_o = \frac{X_o + DQ_o}{Y_o - N_o Q_o}$, where $Q_o \in \mathcal{H}_\infty$ and $Q_o(\infty) \neq Y_o(\infty)/N_o(\infty)$.
3. There are various design objectives which determine the controller $C_o$. For example, when $a > 0$, in order to have zero steady state error for step-like reference inputs we desire $T(0) = 1$, leading to $T_o(0) = e^{-ha}$. To satisfy this requirement, $C_o(0)$ must be

$$C_o(0) = \frac{a}{G_o(0)\left(1 - e^{ha}\right)},$$

and this places a pole at $s = 0$ in the controller $C(s)$.

An interesting question arises at this point: what if the value of the time delay used in the controller does not match its exact value: does the feedback system remain stable? How sensitive is the Smith predictor–based controller design? In order to analyze this situation, let us use $h_c$ in the controller, instead of $h$:

$$\widetilde{C}(s) = \frac{e^{h_c a}\, C_o(s)}{1 + P_o(s)C_o(s)\left(1 - e^{-h_c(s-a)}\right)} = \frac{e^{h_c a}\, C_o(s)S_o(s)}{1 - T_o(s)e^{-h_c(s-a)}},$$

where $S_o = (1 + P_o C_o)^{-1}$ and $T_o = 1 - S_o$. Then, the feedback system $(\widetilde{C}, P)$ is stable if and only if $\widetilde{S} = (1 + P\widetilde{C})^{-1}$, $\widetilde{C}\widetilde{S}$, and $P\widetilde{S}$ are in $\mathcal{H}_\infty$. A simple algebra shows that

$$\widetilde{S}(s) = \frac{S_o(s) + T_o(s)(1 - e^{-h_c(s-a)})}{1 + T_o(s)e^{h_c a}\left(e^{-hs} - e^{-h_c s}\right)} = \frac{1 - T_o(s)e^{-h_c(s-a)}}{1 + T_o(s)e^{h_c a}\left(e^{-hs} - e^{-h_c s}\right)},$$

$$\widetilde{C}(s)\widetilde{S}(s) = \frac{e^{h_c a}\, C_o(s)S_o(s)}{1 + T_o(s)e^{h_c a}\left(e^{-hs} - e^{-h_c s}\right)},$$

$$P(s)\widetilde{S}(s) = e^{-hs}\frac{P_o(s)S_o(s) - G_o(s)T_o(s)\frac{(1 - e^{-h_c(s-a)})}{s-a}}{1 + T_o(s)e^{h_c a}\left(e^{-hs} - e^{-h_c s}\right)}.$$

Recall that by design $S_o, C_o S_o, P_o S_o, T_o$ are in $\mathcal{H}_\infty$ and by assumption so is $G_o$. Moreover, $\frac{(1-e^{-h_c(s-a)})}{s-a}$ is a stable FIR transfer function. Therefore, the feedback system is stable if and only if the Nyquist graph of $T_o(j\omega)e^{h_c a}(e^{-jh\omega} - e^{-jh_c\omega})$ does not encircle $-1$. Hence, a sufficient condition for feedback system stability is

$$\|(1 - e^{-\delta_h s})T_o(s)\|_\infty < e^{-h_c a}, \tag{4.19}$$

where $\delta_h = |h_c - h|$. Note that

$$|1 - e^{-j\delta_h \omega}| \leq \begin{cases} \delta_h \omega & \text{for } \omega \leq 2/\delta_h, \\ 2 & \text{for } \omega > 2/\delta_h, \end{cases} \tag{4.20}$$

which implies that, typically, $T_o(s)$ should be a low-pass filter whose cut-off frequency is much less than $\delta_h^{-1}$. Thus, $C_o$ is designed accordingly. Conversely, once $C_o$ is designed, the highest $\delta_h$ satisfying (4.19) gives a lower bound on the allowable delay imprecision in the controller implementation. We will return to this discussion in Section 5.1.1. In fact, the bound (4.20) is conservative, and a tighter bound will be given in Section 5.1.2.

In closing this discussion, we would like to point out that for plants in the form $e^{-hs}P_o(s)$, where $P_o(s)$ is rational and possibly unstable, it has been shown [161] that every stabilizing controller has an observer–predictor-based structure.

**Exercise 4.6.**
Consider the plant

$$P(s) = P_o(s)e^{-hs}, \quad h \geq 0, \quad P_o(s) = \frac{G_o(s)}{(s-a)}, \quad a = 1, \quad G_o(s) = \frac{4(s-2)}{(s+4)}.$$

1. Let $h = 0$, and find a characterization of the set of all stabilizing controllers for this plant.
2. Now let $h = 0.25$ sec., and design a Smith predictor–based controller in the form

$$C(s) = \frac{e^{ha}\, C_o(s)}{1 + P_o(s)C_o(s)\left(1 - e^{-h(s-a)}\right)},$$

where $a = 1$ and $C_o$ is a stabilizing controller for $P_o$ such that the steady state error for a unit step reference input is zero.
3. Assume that we do not have exact knowledge of the delay in the plant and we are using $h_c$ in the controller, instead of the precise value $h = 0.25$ sec.,

$$\widetilde{C}(s) = \frac{e^{h_c a}\, C_o(s)}{1 + P_o(s)C_o(s)\left(1 - e^{-h_c(s-a)}\right)}.$$

Using (4.19), estimate the smallest $h_c \in [0, h)$ for which the feedback system $(\widetilde{C}, P)$ remains stable by using at least three different designs for $C_o(s)$ satisfying the constraints of part 2.

## 4.1.4 ▪ Stabilizing controllers for retarded and neutral systems

The idea developed above can be generalized to find stabilizing controllers for *retarded time delay systems*. In this case, the plant is in the form (3.3), i.e.,

$$P(s) = \frac{\sum_{i=1}^{v_n} q_{n,i}(s)e^{-h_{n,i}s}}{\sum_{k=1}^{v_d} q_{d,k}(s)e^{-h_{d,k}s}},$$

with $h_{d,1} = 0$ and $\deg q_{d,1} > \deg q_{n,1}$, and it has finitely many poles in $\mathbb{C}_+$, including the imaginary axis: these are the right half plane roots of the denominator quasi-polynomial. Let us label these unstable poles as $\alpha_1, \ldots, \alpha_\ell \in \mathbb{C}_+$ and for simplicity of the exposition assume they are distinct. Then, we can define

$$D(s) := \frac{\prod_{k=1}^{\ell}(s - \alpha_k)}{d_d(s)} \quad \text{and} \quad N(s) = P(s)D(s), \tag{4.21}$$

where $d_d(s)$ is an arbitrary $\ell$th order stable polynomial. When $\mathrm{Re}(\alpha_k) > 0$ for all $k = 1, \ldots, \ell$, choosing

$$d_d(s) = \prod_{k=1}^{\ell}(s + \overline{\alpha}_k)$$

makes $D(s)$ inner, which may simplify some of the robust control problems discussed in Chapters 5 and 6.

Note that $N, D \in \mathcal{H}_\infty$ and that $P = N/D$ is a strongly coprime factorization. Finding a stabilizing controller for this plant amounts to finding $X, Y \in \mathcal{H}_\infty$ satisfying

$$Y(s) = \frac{1 - N(s)X(s)}{D(s)}.$$

Now it is clear that the problem reduces to finding $X \in \mathcal{H}_\infty$ satisfying the interpolation conditions $X(\alpha_k) = 1/N(\alpha_k)$ for all $k = 1, \ldots, \ell$. There are many ways to solve this Lagrange interpolation problem. One of them is to define

$$X(s) = \frac{c_0 + c_1 s + \cdots + c_{\ell-1} s^{\ell-1}}{d_x(s)},$$

where $c_0, c_1, \ldots, c_{\ell-1}$ are constant coefficients to be determined, and $d_x(s)$ is an arbitrary stable polynomial whose degree is $\ell - 1$. Then, the solution is unique and given by

$$\begin{bmatrix} c_0 \\ \vdots \\ c_{\ell-1} \end{bmatrix} = \begin{bmatrix} 1 & \alpha_1 & \cdots & \alpha_1^{\ell-1} \\ \vdots & \vdots & & \vdots \\ 1 & \alpha_\ell & \cdots & \alpha_\ell^{\ell-1} \end{bmatrix}^{-1} \begin{bmatrix} d_x(\alpha_1)/N(\alpha_1) \\ \vdots \\ d_x(\alpha_\ell)/N(\alpha_\ell) \end{bmatrix}. \tag{4.22}$$

Extension of the above techniques to *neutral time delay systems* is a bit more delicate because such systems may have infinitely many poles in $\mathbb{C}_+$. However, if the plant (3.3) has finitely many zeros in $\mathbb{C}_+$ (including at $s = +\infty$), then the extension is still possible by reversing the roles of $D$ with $N$ and $X$ with $Y$. Technical details of such factorizations are given below with numerical examples.

**Example 4.7.** A plant obtained by a feedback connection of an integrator with time delay is in the form

$$P(s) = \frac{e^{-hs}/s}{1 + e^{-hs}/s} = \frac{e^{-hs}}{s + e^{-hs}}.$$

Let us take $h = \frac{\pi}{2}$, leading to poles on the imaginary axis. More precisely, this plant has two unstable poles at $\alpha_1 = +j$, $\alpha_2 = -j$. Now, we can define

$$D(s) = \frac{s^2 + 1}{(s+1)^2}, \quad N(s) = \frac{e^{-hs}(s^2 + 1)}{(s+1)^2 (s + e^{-hs})}$$

so that $P = N/D$ with strongly coprime $N, D \in \mathcal{H}_\infty$; in particular, $N(s)$ does not have any poles or zeros at $\pm j$; see below for the precise computation of $N(\pm j)$. We search for a first order stable $X(s)$ in the form (the stable denominator polynomial is arbitrarily chosen)

$$X(s) = \frac{c_1 s + c_0}{s + 1}$$

such that

$$Y(s) = \frac{1 - N(s)X(s)}{D(s)} \in \mathcal{H}_\infty.$$

This is accomplished by solving the linear system of equations (4.22); in this case,

$$\begin{bmatrix} c_0 \\ c_1 \end{bmatrix} = \begin{bmatrix} 1 & +j \\ 1 & -j \end{bmatrix}^{-1} \begin{bmatrix} (1+j)/N(j) \\ (1-j)/N(-j) \end{bmatrix}.$$

It is a simple exercise to show that

$$N(j) = \frac{1}{j - h}, \quad N(-j) = -\frac{1}{j + h}, \quad \text{where} \quad h = \frac{\pi}{2}.$$

This leads to $c_0 = -(1 + \frac{\pi}{2})$ and $c_1 = 1 - \frac{\pi}{2}$, and thus

$$X(s) = \frac{(1 - \frac{\pi}{2}) s - (1 + \frac{\pi}{2})}{s + 1},$$

and the resulting $Y(s)$ can be written as

$$Y(s) = 1 + H_Y(s), \quad \text{where} \quad H_Y(s) = \frac{2s}{s^2 + 1} - \frac{(c_1 s + c_0)e^{-hs}}{(s + 1)(s + e^{-hs})}.$$

An important point to note is that $H_Y(s)$ is a stable transfer function whose Bode plots are as shown in Figure 4.5. Therefore, all stabilizing controllers are parameterized by

$$C(s) = \frac{X(s) + D(s)Q(s)}{1 + H_Y(s) - N(s)Q(s)}, \quad Q \in \mathcal{H}_\infty,$$

and a particular stabilizing controller is $X(s)/(1 + H_Y(s))$. ∎

As one can observe from the above example, this approach requires computation of an $(\ell - 1)$st order stable transfer function $X(s)$ where $\ell$ is the number of unstable poles of the plant. The numerical errors in the solution of (4.22) may lead to imprecise $X$ and $Y$. If the plant is finite dimensional, the resulting error in $Y(s)$ can be eliminated by approximate pole-zero cancellations (we know that $Y$ does not have poles at the $\alpha_i$'s). In the infinite dimensional case, we typically try to write $Y(s) = D(\infty)^{-1} + H_Y(s)$ and find a stable approximation to $H_Y$ (since $d_d(s)$ is an arbitrary stable polynomial, we can always choose $D(s)$ such that $D(\infty) = 1$). Stability robustness to numerical imprecisions in $X$ and $Y$, and approximations of transfer functions in the form of $H_Y$, are further discussed in Chapters 5 and 6.

## 4.1.5 ▪ The heat equation with Neumann boundary control

Recall that the plant (3.22) has a single pole at $s = 0$ and all other poles are in $\mathbb{C}_-$. In this case, as before, we can write

$$D(s) = \frac{s}{s + b}, \quad N(s) = D(s)P(s) = \left(\frac{s}{s + b}\right) \frac{2e^{-L\xi}}{\xi (1 - e^{-2L\xi})}, \quad \text{where} \quad \xi := \frac{\sqrt{s}}{c}.$$

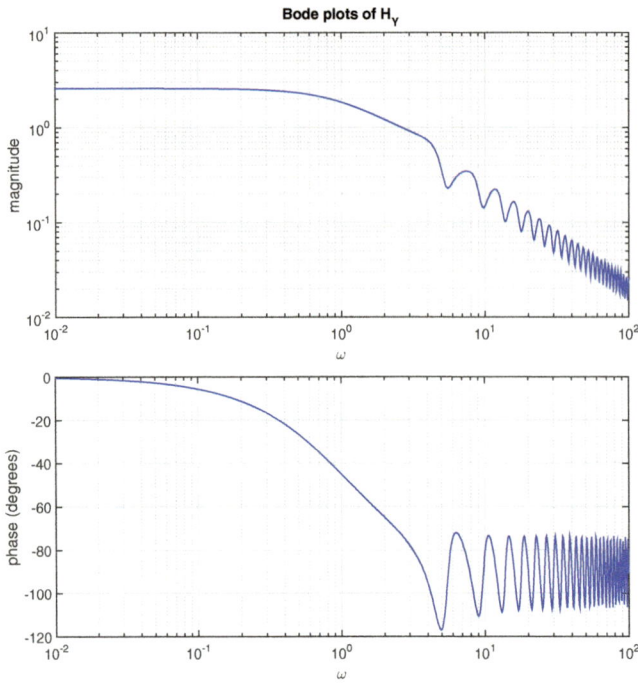

**Figure 4.5.** *Bode plots of $H_Y(s)$.*

Note that $N(0) = c^2/bL$, so a feasible solution for $Y = (1 - NX)/D$ to be in $\mathcal{H}_\infty$ is $X(s) = 1/N(0) = bL/c^2$. Recall that a stabilizing controller for the plant $P = N/D$ is obtained by putting $Q(s) = 0$ in (4.11):

$$ C(s) = \frac{X(s)}{Y(s)} = (s+b)^{-1}\left(\frac{N(0) - N(s)}{s}\right)^{-1}. $$

Note that this controller is stable; its Bode plots are shown in Figure 4.6 for different values of $b$, when $L = 1$ and $c = 10^{-3}$. Of course, a different stabilizing controller can be determined by a different choice of $Q \in \mathcal{H}_\infty$ in (4.11).

### 4.1.6 ▪ Unstable fractional order system

The fractional order plant modeling a nonlaminated magnetic suspension system is (3.24), and it can be written as (3.25). Therefore, a strongly coprime factorization is $P = N/D$, with

$$ D(s) = \frac{s - p^2}{s + p^2}, \quad N(s) = \frac{e^{-hs}(\sqrt{s} + p)}{(s + p^2)(\sqrt{s} - p_1)(\sqrt{s} - p_2)(\sqrt{s} - p_3)(\sqrt{s} - p_4)}, $$
$$ \tag{4.23} $$

where $p, p_1, \ldots, p_4$ are the roots of the polynomial $(z^5 + z^4 - c)$, with $p \in \mathbb{R}_+$. Therefore, by choosing

$$ X(s) = 1/N(p^2) = pe^{hp^2}\prod_{k=1}^{4}(p - p_k), $$

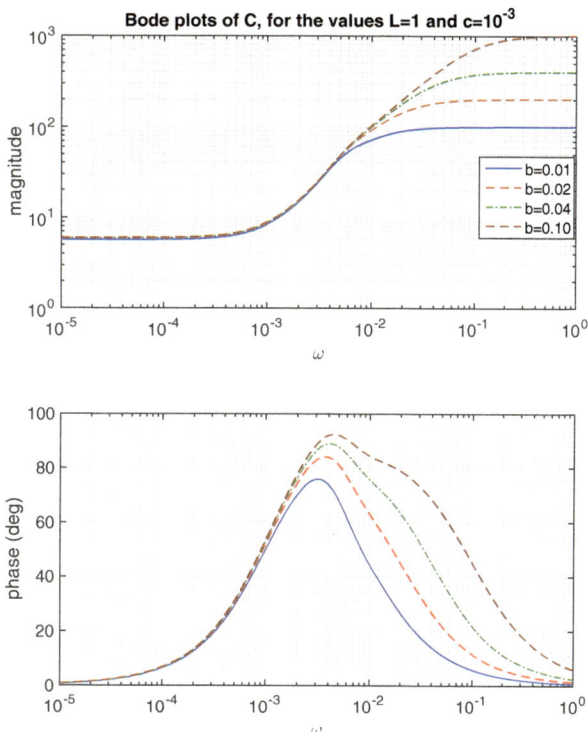

**Figure 4.6.** *Bode plots of C for L = 1 and c = $10^{-3}$.*

we obtain $Y \in \mathcal{H}_\infty$ satisfying (4.9), i.e., $Y = (1 - NX)/D \in \mathcal{H}_\infty$. In particular, $C = X/Y$ is a stabilizing controller for the plant (3.24).

As an example, let us take $c = 2$. Then, we have $p = 1$ and

$$p_{1,2} = -1.1898 \pm j\,0.6028, \quad p_{3,4} = 0.1898 \pm j\,1.0432.$$

For $h = 0.25$, we compute $X(s) = 11.5562$, and then $Y(s)$ can be written as

$$Y(s) = 1 + H_Y(s), \quad \text{where } H_Y(s) = \frac{2p^2}{s - p^2} - \frac{p e^{-h(s-p^2)} \prod_{k=1}^{4}(p - p_k)}{s^2(\sqrt{s}+1) - c}.$$

Note that $H_Y \in \mathcal{H}_\infty$, and in particular it can be shown that it does not have a pole at $p^2$; furthermore,

$$H_Y(p^2) = 2hp^2.$$

The Bode plots are shown in Figure 4.7. Since the phase of $H_Y$ stays above $-180°$, the transfer function $(1 + H_Y)^{-1}$ is stable, and hence the controller $C = X/Y$ is stable for these values of $h$ and $c$.

## 4.2 ▪ Inner-outer factorizations

In order to solve the $\mathcal{H}_\infty$ control problems to be defined in Chapter 5, coprime factorizations are taken one step further and *inner-outer factorizations* are performed. Recall

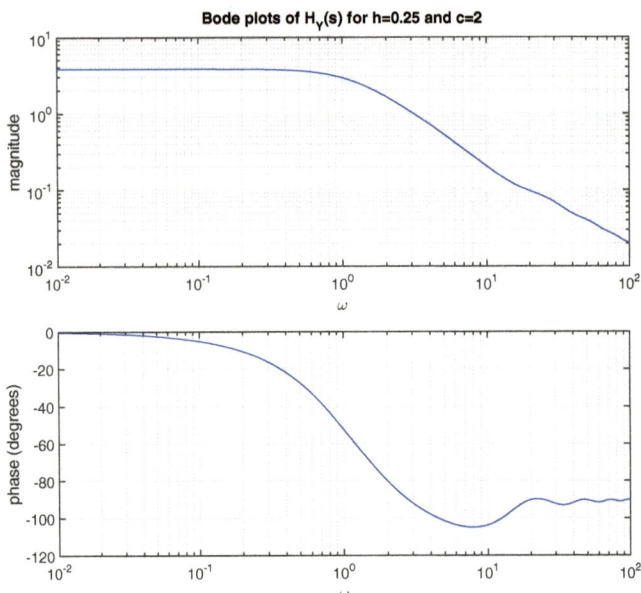

**Figure 4.7.** *Bode plots of $H_Y$ for $h = 0.25$ and $c = 2$.*

that whenever the plant has finitely many poles in $\mathbb{C}_+$, it can be factored as $P = N/D$, where $N, D$ are strongly coprime in $\mathcal{H}_\infty$. The examples given in Section 4.1 illustrate that plants with finitely many poles in $\mathbb{C}_+$, e.g., (4.21) and (4.23), admit coprime factorizations where $D$ can be chosen as a rational function. For solving the $\mathcal{H}_\infty$ control problems, we perform a factorization of the form

$$N(s) = M_n(s)N_o(s),$$

where $M_n \in \mathcal{H}_\infty$ is inner and $N_o \in \mathcal{H}_\infty$ is outer. Formally, every function $N \in \mathcal{H}_\infty$ admits such an inner-outer factorization; see, e.g., [109, 261] for technical details.

In the finite dimensional case, the inner-outer factorization is typically obtained from a spectral factorization which gives the outer part:

$$N(s)N(-s) = N_o(s)N_o(-s).$$

The spectral factor $N_o$ is obtained from an ARE; see, e.g., [281] for the state-space formulae that give $N_o$ and $M_n$ from a minimal state-space realization of $N(s) = C(sI - A)^{-1}B + D$.

For DPSs, we try to avoid infinite dimensional spectral factorization based on state-space methods (which is still possible, but numerically it can be difficult to do). Instead, we exploit the following observations:

(i) If $N(s)$ contains an I/O delay (i.e., a term $e^{-hs}$, with the largest possible $h \geq 0$ such that $e^{hs}N(s) \in \mathcal{H}_\infty$), then this is a factor of the inner part.

(ii) If $z \in \mathbb{C}_+$ is a zero of $N(s)$, then $M_n(s)$ contains a factor $\dfrac{s - z}{s + \overline{z}}$.

Once the product of all the terms in (i) and (ii) is formed to define $M_n(s)$, then the outer factor is obtained automatically from $N_o(s) = N(s)/M_n(s)$. This is a key step in the approach used in this book. For time delay systems in the form (3.3), inner-outer factorization can be done by finding the roots of the numerator quasi-polynomial in $\mathbb{C}_+$. See illustrative examples in the next sections of this chapter. Other types of DPSs can also be factored out by computing the zeros in $\mathbb{C}_+$. If there are infinitely many zeros in the right half plane, convergence of the infinite Blaschke products (as inner parts) constructed this way must be proven according to the condition (2.17). See [255] for more detailed discussion and further references on this topic.

Now consider the SISO system with the following coprime factorization:

$$P(s) = \frac{N(s)}{D(s)}. \tag{4.24}$$

Numerical methods used in Chapters 6 and 7 for robust controller designs require rewriting the plant as

$$P(s) = \frac{M_n(s)N_o(s)}{M_d(s)}, \tag{4.25}$$

where $M_n$ and $M_d$ are inner functions and $N_o$ is an outer function. One point is clear from (4.25); that is, we do not capture plants having poles on $\mathbb{I}$. However, such systems can still be handled by a shift of $\mathbb{I}$. For example, by using the mapping $s = \widehat{s} - \epsilon$ for sufficiently small $\epsilon > 0$, the poles on $\mathbb{I}$ appear in the right half plane of the $\widehat{s}$-plane. We design a stabilizing controller in the $\widehat{s}$-plane and then express it in terms of $\widehat{s} = s + \epsilon$. An example of this approach is given below, followed by motivation based on the $\mathcal{H}_\infty$ norm bounds.

**Example 4.8.** As we have seen in Chapter 3, an important type of plant is an integrator with I/O delay,

$$P(s) = \frac{K\,e^{-hs}}{s}, \quad K \in \mathbb{R}, \quad h > 0.$$

Clearly, any coprime factorization in the form $P = N/D$ has a $D(s)$ with a zero at $s = 0$, so the plant cannot be written in the form (4.25), where $M_d(s)$ is inner. Let us now make the transformation $s = \widehat{s} - \epsilon$ and define

$$\widehat{P}(\widehat{s}) = P(\widehat{s} - \epsilon) = \frac{Ke^{h\epsilon}e^{-h\widehat{s}}}{\widehat{s} - \epsilon} = \frac{\widehat{M}_n(\widehat{s})\widehat{N}_o(\widehat{s})}{\widehat{M}_d(\widehat{s})},$$

where

$$\widehat{M}_n(\widehat{s}) = e^{-h\widehat{s}}, \quad \widehat{N}_o(\widehat{s}) = \frac{Ke^{h\epsilon}}{\widehat{s} + \epsilon}, \quad \widehat{M}_d(\widehat{s}) = \frac{\widehat{s} - \epsilon}{\widehat{s} + \epsilon}.$$

Now a controller in the form (4.18) can be designed for $\widehat{P}(\widehat{s})$:

$$\widehat{C}(\widehat{s}) = \frac{e^{h\epsilon}\,\widehat{C}_o(\widehat{s})}{1 + \widehat{P}_o(\widehat{s})\widehat{C}_o(\widehat{s})\big(1 - e^{-h(\widehat{s}-\epsilon)}\big)},$$

where $\widehat{P}_o(\widehat{s}) = \frac{Ke^{h\epsilon}}{\widehat{s}-\epsilon}$ and $\widehat{C}_o(\widehat{s})$ is a stabilizing controller for $\widehat{P}_o(\widehat{s})$. Let us now choose $\widehat{C}_o(\widehat{s}) = K_o e^{-h\epsilon}/K$, with $K_o > \epsilon$. Then, $C(s)$ stabilizing $P(s)$ is obtained by $C(s) = \widehat{C}(s + \epsilon)$:

$$C(s) = \frac{K_o/K}{1 + K_o\big(\frac{1-e^{-hs}}{s}\big)}.$$

It can be verified that the choice of $K_o > \epsilon$ places the closed-loop system poles to the left of $\text{Re}(s) = -\epsilon$. Obviously, using Smith predictor design one could arrive at the same controller directly. The main point here was to show that axis shift allows us to transform the plant into a form which satisfies the structure of the inner-outer factorizations considered here. Another key point is that, with this technique, the poles of the feedback system $(C, P)$ are guaranteed to be in the half plane to the left of the vertical line $\text{Re}(s) = -\epsilon$. This is useful when such a design constraint is imposed on the controller. ∎

A word of caution is in order at this point: in the next chapters we will be dealing with control design problems that try to minimize the $\mathcal{H}_\infty$ norm of a certain closed-loop transfer function. If $F(s)$ denotes a stable closed-loop transfer function whose poles are in $\text{Re}(s) < -\epsilon$, then we can define a new transfer function by axis shift $\widehat{F}(\widehat{s}) = F(s) = F(\widehat{s} - \epsilon)$; note that $\widehat{F}$ is stable and its $\mathcal{H}_\infty$ norm satisfies

$$
\begin{aligned}
\|\widehat{F}\|_\infty &= \text{ess} \sup_{\text{Re}(\widehat{s})>0} |\widehat{F}(\widehat{s})| \\
&= \text{ess} \sup_{\text{Re}(s)>-\epsilon} |F(s)| \\
&\geq \text{ess} \sup_{\text{Re}(s)>0} |F(s)| = \|F\|_\infty.
\end{aligned}
$$

So, if $\|\widehat{F}\|_\infty \leq \gamma$ for some $\gamma > 0$, then we also have $\|F\|_\infty \leq \gamma$ automatically. But the converse is not necessarily true, so performing such an axis shift and doing the design in the $\widehat{s}$-domain introduces some conservatism. As $\epsilon$ decreases, so does the level of conservatism.

In order to complete the discussion on factorizations, we consider different classes of plants in the next sections. In Section 4.3, factorizations of pseudorational transfer functions are considered. In Sections 4.4 and 4.5, we focus on two different cases involving retarded or neutral time delay systems:

- In Section 4.4, we consider systems where $M_n$ is infinite dimensional and $M_d$ is rational (corresponding to systems with finitely many unstable poles).

- In Section 4.5, we consider systems where $M_n$ is rational and $M_d$ is infinite dimensional (corresponding to systems with infinitely many unstable poles and finitely many zeros in $\mathbb{C}_+$).

Note that the two cases differ due to the interchanging roles of finite and infinite dimensional terms, $M_n$ and $M_d$. Then, in Section 4.6 we discuss factorizations for other types of DPSs, such as flexible beams and fractional order systems. In each section, we give examples for these types of systems and show how to obtain their coprime and inner-outer factorization.

## 4.3 ▪ Factorization of pseudorational systems

Let us return to the pseudorational transfer functions introduced in Section 3.2. As we have seen there, a pseudorational transfer function $P$ assumes, by definition, a factorization $P(s) = N_p(s)/D_p(s)$, where $N_p$ and $D_p$ are entire functions of exponential type satisfying the Paley–Wiener-type estimate (3.15). A word of caution is in order at this point: the factorization in the previous section, Section 4.2, writes the plant as $P(s) = N(s)/D(s)$, where $N$ and $D$ are in $\mathcal{H}_\infty$. Here, a pseudorational transfer function is written as a ratio of $N_p$ and $D_p$ where these are entire functions.

In this section we further explore this factorization and indeed show that under some conditions on $N_p$ and $D_p$ such a factorization can yield an inner-outer factorization when $N_p/D_p$ is in $\mathcal{H}_\infty$. For an unstable system in the form $P = N/D$, where both $N$ and $D$ are pseudorational transfer functions in $\mathcal{H}_\infty$, we can write $N = N_n/D_n$ and $D = N_d/D_d$ (where $N_n, D_n, N_d, D_d$ are entire functions) and apply the inner-outer factorization given below for stable $P$ to $N$ and $D$ separately.

Before starting our discussion, we need some preliminary materials. Recall that for a given distribution $\alpha$, $\mathrm{supp}\,\alpha$ denotes its *support*. The following quantities are of importance; we allow them to be $\pm\infty$, but when $\mathrm{supp}\,\alpha$ is bounded, they are both finite:

$$
\begin{aligned}
\ell(\alpha) &:= \quad \inf\,\{t : t \in \mathrm{supp}\,\alpha\}, \\
r(\alpha) &:= \quad \sup\,\{t : t \in \mathrm{supp}\,\alpha\}.
\end{aligned}
\tag{4.26}
$$

The following lemma is a consequence of Titchmarsh's theorem on convolution; for details see [49].

**Lemma 4.9 (see [259]).** *For $\alpha, \beta \in \mathcal{D}'(\mathbb{R})$,*

$$
\ell(\alpha \star \beta) = \ell(\alpha) + \ell(\beta).
\tag{4.27}
$$

*Dually, for distributions $\alpha, \beta$ having support bounded on the right,*

$$
r(\alpha \star \beta) = r(\alpha) + r(\beta).
\tag{4.28}
$$

***Proof.*** Let $\alpha$ and $\beta$ have support bounded on the left. If one of them does not satisfy this condition, the conclusion follows easily from Titchmarsh's theorem. Without loss of generality, we assume that $\ell(\alpha) = \ell(\beta) = 0$. We claim $\ell(\alpha\star\beta) = 0$. By Titchmarsh's theorem [49], the convolution $\alpha \star \beta$ is nonzero at the origin. Hence $\ell(\alpha \star \beta) \leq 0$. On the other hand, since both $\alpha$ and $\beta$ are zero on $(-\infty, 0)$, $\ell(\alpha \star \beta) \geq 0$, and hence $\ell(\alpha \star \beta) = 0$. If $\ell(\alpha) \neq 0$, one can shift $\alpha$ suitably to make $\alpha = 0$. The same is true for $\beta$.

Finally, (4.28) follows similarly as above.    □

Let $\alpha^{-1}$ denote the inverse of $\alpha$, when it exists, with respect to convolution. We have the following corollary.

**Corollary 4.10.** *Suppose that $\alpha$ is a distribution with support bounded on the left such that $\alpha^{-1}$ exists and also has support bounded on the left. Then,*

$$
\ell(\alpha^{-1}) = -\ell(\alpha).
\tag{4.29}
$$

*Similarly, if both $\beta$ and $\beta^{-1}$ have support bounded on the right, then*

$$
r(\beta^{-1}) = -r(\beta).
\tag{4.30}
$$

***Proof.*** According to (4.27), $\ell(\alpha) + \ell(\alpha^{-1}) = \ell(\alpha \star \alpha^{-1}) = \ell(\delta) = 0$, when both $\alpha$ and $\alpha^{-1}$ have support that is bounded on the left. The second equality is obtained similarly.    □

We need two additional preliminary results to complete the factorization of pseudo-rational transfer functions.

**Lemma 4.11.** *Let $\tilde{\alpha}$ be the mirror image of distribution $\alpha$ defined by*

$$\langle \tilde{\alpha}, \phi(t) \rangle := \langle \alpha, \phi(-t) \rangle, \quad \phi \in \mathcal{D}. \tag{4.31}$$

*Then,*

$$\ell(\tilde{\alpha}) = -r(\alpha). \tag{4.32}$$

***Proof.*** The result is immediate from the definition (4.31).  $\square$

**Lemma 4.12 (see [121]).** *Let $f$ be a distribution with compact support contained in $(-\infty, 0]$, and assume $1/F(s)$ is stable, that is, there exists $\delta > 0$ such that all zeros of $F(s)$ belong to $\{s \in \mathbb{C} : \operatorname{Re} s < -\delta\}$. Then,*

$$M_f(s) := e^{-\ell(f)s} \frac{\tilde{F}(s)}{F(s)}$$

*is an inner transfer function.*

***Proof.*** Note that, by definition, $\tilde{F}(s) = F(-s)$ (recall that we only consider functions with real coefficients) and $F(-j\omega) = \overline{F(j\omega)}$ for all $\omega \in \mathbb{R}$. Clearly, $|M_f(j\omega)| = 1$, i.e., on the imaginary axis the magnitude condition is automatically satisfied, so it suffices to prove that $M_f$ belongs to $\mathcal{H}_\infty$. Recall that $M_f$ is in $\mathcal{H}_\infty$ if and only if it gives a bounded multiplication operator on $\mathcal{H}_2$ into itself, i.e., its inverse Laplace transform, namely $m_f(t)$, gives a bounded linear operator on $\mathcal{L}_2[0, \infty)$ into itself. Take an arbitrary $x \in \mathcal{L}_2[0, \infty)$, i.e., $X(s) \in \mathcal{H}_2$, and we show $m_f \star x \in \mathcal{L}_2[0, \infty)$. For this purpose, first note that $|M_f(j\omega)| = 1$ implies $M_f(j\omega)X(j\omega) \in \mathcal{L}_2(j\mathbb{R})$ and hence $m_f \star x \in \mathcal{L}_2(-\infty, \infty)$. Furthermore, noting that the inverse Laplace transform of $e^{-\ell(f)s}$ is $\delta_{\ell(f)}$, we have

$$\ell(m_f \star x) = \ell(\delta_{\ell(f)}) + \ell(\tilde{f}) + \ell(f^{-1}) + \ell(x) = \ell(f) - r(f) - \ell(f) + \ell(x) = -r(f) + \ell(x) \geq 0$$

by (4.27), (4.29), and (4.32) (note also that $r(f) \leq 0$). Thus, $m_f \star x \in \mathcal{L}_2[0, \infty)$, and hence $M_f$ is an inner function.  $\square$

**Example 4.13.** Take $F(s) = se^s + c$ for a small enough $c > 0$. Then, we have $\ell(f) = -1$, $\tilde{F}(s) = -se^{-s} + c$, and hence the inner function associated with $F$ is

$$M_f(s) = e^s \frac{-se^{-s} + c}{se^s + c} = \frac{ce^s - s}{se^s + c}.$$

Note that $M_f$ can be rewritten as the sum of two stable transfer functions

$$M_f(s) = \frac{c/s}{1 + (c/s)e^{-s}} - \frac{e^{-s}}{1 + (c/s)e^{-s}}.$$

From the Nyquist stability criterion, each of these two transfer functions is in $\mathcal{H}_\infty$ if and only if $c \in [0, \frac{\pi}{2})$. We can slightly generalize this example by taking $F(s) = se^{hs} + c$, which means $\ell(f) = -h$ and

$$M_f(s) = \frac{ce^{hs} - s}{se^{hs} + c} = \frac{c - se^{-hs}}{s + ce^{-hs}}, \tag{4.33}$$

which is in $\mathcal{H}_\infty$ if and only if $c \in [0, \frac{\pi}{2h})$.  ∎

In Section 6.1.2, we will consider a controller design example involving a pseudorational inner function in the form (4.33). Further, in Section 6.2.2, stable implementation of the $\mathcal{H}_\infty$-optimal controller is discussed for such a system.

We are now ready to give the main inner-outer factorization result of this section.

**Theorem 4.14.** *Suppose that a stable plant $P \in \mathcal{H}_\infty$ is factorized as*

$$P(s) = \frac{N_1(s)N_2(s)}{D(s)} \in \mathcal{H}_\infty, \tag{4.34}$$

*where $n_1$, $n_2$, $d$ are distributions with compact support, satisfying the requirements for the pseudorationality of $n_1 \star d^{-1}$ and $n_2 \star d^{-1}$ in Definition 3.2, and such that $D(s)^{-1}$, $N_1(s)^{-1}$, $e^{r(n_2)s}N_2(-s)^{-1} \in \mathcal{H}_\infty$. Define $L := -\ell(d) + \ell(n_1) - r(n_2)$, and let*

$$P_i(s) := e^{-Ls} \cdot \frac{N_2(s)}{N_2(-s)}, \tag{4.35}$$

$$P_o(s) := e^{Ls} \cdot \frac{N_1(s)N_2(-s)}{D(s)}. \tag{4.36}$$

*Then, $P(s) = P_i(s)P_o(s)$. Moreover, $P_i$ is inner and $P_o$ is outer.*

**Proof.** We start by showing that both $P_i$ and $P_o$ belong to $\mathcal{H}_\infty$. Indeed, by the assumption above, both functions have no unstable poles. Notice that, on the imaginary axis, $|P_i(j\omega)| = 1$ and $|P_o(j\omega)| = |P(j\omega)| \leq \|P\|_\infty < \infty$. It is thus sufficient to show that $p_i$ and $p_o$ have support in $[0, \infty)$, that is, $\ell(p_i) \geq 0$ and $\ell(p_o) \geq 0$. This readily follows from an argument similar to that in the proof of Lemma 4.12. More precisely, by (4.27), (4.29), and (4.32), we have

$$\ell(p_i) = L + \ell(n_2) - \ell(n_2\tilde{})$$
$$= \ell(d) + \ell(n_1) - r(n_2) + \ell(n_2) + r(n_2) \geq 0,$$

$$\ell(p_o) = -L + \ell(n_1) + \ell(n_2\tilde{}) - \ell(d)$$
$$= \ell(d) - \ell(n_1) + r(n_2) + \ell(n_1) - r(n_2) - \ell(d) = 0.$$

Thus, $n_i$ and $n_o$ have support in $[0, \infty)$ and hence belong to $\mathcal{H}_\infty$. This readily yields that $P_i$ is inner.

We need to show that $P_o$ is outer. Because of our assumptions, $P_o$ does not have any zeros in $\{s | \operatorname{Re} s \geq 0\}$. In order to show that $P_o$ is outer, it is sufficient to prove that $P_o\mathcal{H}_2$ is dense in $\mathcal{H}_2$ [109, Chapter 7, page 101, Corollary].

Since $\ell(p_o) = 0$ as above, there is no extra "delay element" induced by the multiplication by $P_o$. Since both the numerator and the denominator of $P_o$ are pseudorational, their orders are finite, and hence there exists $r \geq 0$ such that $(s + \alpha)^r P_o$ is invertible in $\mathcal{H}_\infty$. Here $\alpha > 0$ is any positive number such that $(s + \alpha)$ does not cancel with any zeros of $d(s)$. Then, $(s + \alpha)^r P_o\mathcal{H}_2 = \mathcal{H}_2$. This implies that $P_o\mathcal{H}_2$ consists of those elements in $\mathcal{H}_2$ that are expressible as $(s + \alpha)^{-r}\mathcal{H}_2$, that is, those elements integrated $r$ times with kernel $(s + \alpha)^{-1} = e^{-\alpha t}u(t)$. Such "smooth" functions clearly form a dense subspace of $\mathcal{H}_2$, and hence our claim is proved. $\square$

## 4.4 ▪ Factorization of time delay systems with finitely many unstable poles

In this section, we consider systems with finitely many unstable poles. By construction, the coprime factors $N$ and $D$ in (4.24) are stable and possibly infinite dimensional transfer functions. Since the system has finitely many unstable poles, the term $D$ has finitely many unstable zeros. Recall that our main objective is to write the plant transfer function in the form (4.25), and we assume that $P$ does not have any poles in $\mathbb{I}$. In this case, we can find an inner rational function $M_d(s)$ whose zeros coincide with the unstable zeros of the term $D$. Then, it is possible to rewrite the system (4.24) as

$$P(s) = \frac{N(s)}{D(s)} = \frac{N(s)}{M_d(s)\frac{D(s)}{M_d(s)}}.$$

Note that the term

$$N_{od} := \frac{D}{M_d}$$

is an outer function since all unstable zeros of $D$ are canceled by the zeros of $M_d$. Hence the plant is in the form

$$P(s) = \frac{N(s)}{M_d(s)}\, N_{od}^{-1}.$$

Now, there are two cases for inner-outer factorization of $N$ so that the plant $P$ is the form (4.25):

1. $N$ has finitely many unstable zeros.

2. $N$ has infinitely many unstable zeros, and there exists a *conjugate function* $\bar{N}$, a stable function such that $\bar{N}$ has finitely many unstable zeros; the transfer function $N/\bar{N}$ is causal and proper; and $|N(j\omega)| = |\bar{N}(j\omega)|$ for $\omega \in [0, \infty)$.

In the first case, the plant has finitely many poles and zeros in $\mathbb{C}_+$. In the second case, it has infinitely many unstable zeros and finitely many unstable poles.

**Example 4.15.** Consider the following plant:

$$P(s) = \frac{(s-1)(s+4)\, e^{-hs}}{(s^2 + 8s + 17)(s+1 - 3e^{-2hs})}, \quad h = \ln(2) \approx 0.693.$$

A coprime factorization is $P = N/D$, where

$$N(s) = \frac{(s-1)\, e^{-hs}}{(s^2 + 8s + 17)}, \quad D(s) = \frac{(s+1 - 3e^{-2hs})}{(s+4)}.$$

The function $D(s)$ has one zero in $\mathbb{C}_+$, at $s = 0.5$. Therefore, we define

$$M_d(s) = \frac{(s-0.5)}{(s+0.5)},$$

which makes

$$N_{od}(s) = \frac{D(s)}{M_d(s)} = \frac{(s+1 - 3e^{-2hs})\, (s+0.5)}{(s+4)\, (s-0.5)}$$

outer. Note that $N_{od}$ and $N_{od}^{-1}$ are in $\mathcal{H}_\infty$ since the zero of $D(s)$ at $s = 0.5$ is canceled by the zero of $M_d(s)$ at the same location. We observe that $N(s)$ defined here corresponds to case 1, i.e., it has finitely many zeros in $\mathbb{C}_+$. Its inner-outer factorization $N = N_i N_{on}$ is relatively easy: collect the I/O delay and the right half plane zeros into its inner part,

$$N_i(s) = e^{-hs} \frac{(s-1)}{(s+1)},$$

which defines the outer part automatically:

$$N_{on}(s) = \frac{N(s)}{N_i(s)} = \frac{(s+1)}{(s^2 + 8s + 17)}.$$

In summary, we have $P = M_n N_o / M_d$, where $M_n = N_i$ and $N_o = N_{on} N_{od}^{-1}$. ∎

**Example 4.16.** Consider the plant

$$P(s) = \frac{s + 3 + 2(s-1)e^{-0.4s}}{s^2 - 2s + 2},$$

with a factorization $P = N/D$, where

$$N(s) = \frac{s + 3 + 2(s-1)e^{-0.4s}}{s^2 + 2s + 2}, \quad D(s) = \frac{s^2 - 2s + 2}{s^2 + 2s + 2}.$$

It can be determined that the plant has infinitely many zeros in $\mathbb{C}_+$ (see below discussions on the roots of quasi-polynomials). Since $D$ is already inner, we define $M_d = D$ and $N_{od} = 1$. In order to proceed with the factorization of $N$, we define

$$\bar{N}(s) = \frac{2(s+1) + (s-3)e^{-0.4s}}{s^2 + 2s + 2},$$

which is stable and satisfies $|\bar{N}(j\omega)| = |N(j\omega)|$ for $\omega \in \mathbb{R}$. Furthermore, $\bar{N}$ has only one zero, at $s = \varrho = 0.247$, in $\mathbb{C}_+$. Moreover, it can computed that the quasi-polynomial $(s + 3 + 2(s-1)e^{-0.4s})$ has only one zero in $\mathbb{C}_-$, at $s = -\varrho$. Therefore, we can rewrite $N$ as

$$N(s) = \frac{s + 3 + 2(s-1)e^{-0.4s}}{2(s+1) + (s-3)e^{-0.4s}} \times \frac{2(s+1) + (s-3)e^{-0.4s}}{s^2 + 2s + 2} \times \frac{s - \varrho}{s + \varrho} \times \frac{s + \varrho}{s - \varrho}$$

$$= \frac{s + 3 + 2(s-1)e^{-0.4s}}{2(s+1) + (s-3)e^{-0.4s}} \times \frac{s - \varrho}{s + \varrho} \times \frac{2(s+1) + (s-3)e^{-0.4s}}{s^2 + 2s + 1} \times \frac{s + \varrho}{s - \varrho}.$$

Note that

$$N_i(s) = \frac{s + 3 + 2(s-1)e^{-0.4s}}{2(s+1) + (s-3)e^{-0.4s}} \times \frac{s - \varrho}{s + \varrho}$$

is inner (in particular, $N_i$ does not have a zero at $\varrho$ or a pole at $-\varrho$), and

$$N_{on}(s) = \frac{2(s+1) + (s-3)e^{-0.4s}}{s^2 + 2s + 1} \times \frac{s + \varrho}{s - \varrho}$$

is outer (note that $N_{on}$ does not have a pole at $\varrho$, but it does have a zero at $-\varrho$). Hence, $P = M_n N_o / M_d$, where $M_n = N_i$, $M_d = D$, $N_o = N_{on}$. ∎

The discussion below is a generalization based on the arguments used in these examples. The factorization when $N$ has finitely many unstable zeros is relatively easy.

Let $M_n$ be an inner function whose zeros are the zeros of $N$ in $\mathbb{C}_+$. Additionally, if there is an I/O delay in the system, it appears in $M_n$. Then, the plant (4.24) is written as

$$P(s) = \frac{N(s)}{D(s)} = \frac{M_n(s)\left(\frac{N(s)}{M_n(s)}\right)}{M_d(s)\left(\frac{D(s)}{M_d(s)}\right)} = \frac{M_n(s)}{M_d(s)}\left(\frac{N(s)}{M_n(s)}\right)\left(\frac{M_d(s)}{D(s)}\right), \qquad (4.37)$$

which admits the factorization in (4.25) since the terms in parentheses are outer functions.

The second case, where $N$ has infinitely many unstable zeros, requires the construction of an infinite dimensional inner function. Let $M_{\bar{n}}$ be a rational, inner function whose zeros are the unstable zeros of $\bar{N}$. Using the conjugate function $\bar{N}$ and the inner function $M_{\bar{n}}$, we write the plant (4.25) as

$$P(s) = \frac{N(s)}{D(s)} = \frac{\left(N(s)M_{\bar{n}}(s)/\bar{N}(s)\right)\left(\frac{\bar{N}(s)}{M_{\bar{n}}(s)}\right)}{M_d(s)\left(\frac{D(s)}{M_d(s)}\right)} \qquad (4.38)$$

$$= \frac{\left(N(s)M_{\bar{n}}(s)/\bar{N}(s)\right)}{M_d(s)}\left(\frac{\bar{N}(s)}{M_{\bar{n}}(s)}\right)\left(\frac{M_d(s)}{D(s)}\right). \qquad (4.39)$$

Note that the term $NM_{\bar{n}}/\bar{N}$ is a stable, infinite dimensional inner function. It is stable because $N$ and $M_{\bar{n}}$ are stable and unstable zeros of $\bar{N}$ are canceled by the zeros of $M_{\bar{n}}$. The inner function is the result of $\bar{N}$ having the same magnitude as $N$ on the imaginary axis. Thus, the plant (4.38) admits the coprime and inner-outer factorization in the form (4.25). The above discussion is summarized as follows.

**Corollary 4.17.** *The SISO system in* (4.24) *with finitely many unstable poles admits a coprime and inner-outer factorization in the form* (4.25) *if either $N$ has finitely many unstable zeros or has a conjugate function $\bar{N}$. The factorizations for each case are given in* (4.37) *and* (4.38), *respectively.*

Note that this result tells us when we can factorize the plant as in (4.25) but does not tell us how to do the factorization. For this, we need to compute the unstable zeros of $D$ and $N$. For the plant with finitely many unstable poles and infinitely many unstable zeros, we need to find the conditions when the conjugate function $\bar{N}$ exists and compute the unstable zeros of $D$ and $\bar{N}$. It is numerically difficult to do these computations for general plants, but the computations can be done for a large class of plants involving time delays, as illustrated in Example 4.16.

Now, we focus on time delay systems with finitely many unstable poles and show how to factorize them as in (4.25). The typical transfer function of a time delay system is a ratio of *quasi-polynomials*. A *quasi-polynomial* is a generalization of a polynomial and is frequently used in transfer function representation of time delay systems [156].

**Definition 4.18.** *Let $q_i(s)$, $i = 1, \ldots, v$, be polynomials with real coefficients and $h_i \in \mathbb{R}$ be such that $0 \le h_1 < \cdots < h_v$. A function of the form*

$$q(s) = \sum_{i=1}^{v} q_i(s)e^{-h_i s} \qquad (4.40)$$

*is called a* quasi-polynomial. *If deg $q_1$ > deg $q_i$ for all $i = 2, \ldots, v$, then $q$ is a* retarded *quasi-polynomial; it is a* neutral *quasi-polynomial if deg $q_1 \ge$ deg $q_i$ and there exists at* *least one $i \in \{2, \ldots, v\}$ such that deg $q_1 = $ deg $q_i$. If there exists $i \in \{2, \ldots, v\}$ such that $deg(q_i) > deg(q_1)$, then the quasi-polynomial is of* advanced *type. Time delays appearing in (4.40) are called* commensurate *if there exist $h_o \in \mathbb{R}_+$ and nonnegative integers $n_i$ such that $h_i = n_i h_o$ for all $i = 1, \ldots, v$.*

Advanced-type quasi-polynomials may appear in the factorization of pseudorational systems; for example, $M_f(s)$ defined by (4.33) contains an advanced-type quasi-polynomial in its numerator, whereas its denominator is a quasi-polynomial of retarded type. Note that in this case, the roots of the numerator, $c - se^{-hs}$, are infinitely many and their real parts converge to $+\infty$. Since $M_f$ is inner, under the assumption that $c \in [0, \frac{\pi}{2h})$, the "conjugate quasi-polynomial" (see below for a formal definition) which appears in the denominator, $(s + ce^{-hs})$, has infinitely many roots whose real parts converge to $-\infty$ (because of the property that $|M_f(j\omega)| = 1$, it has a pole-zero symmetry around $\mathbb{I}$).

Almost all of the numerical root finding tools assume retarded- or neutral-type quasi-polynomials. Therefore, in the rest of this section we will restrict our attention to these two classes. The root distributions of quasi-polynomials have been analyzed for more than half a century now; see [21] for early results. The retarded and neutral quasi-polynomials have infinitely many roots in the complex plane [101]. The asymptotic root chains of retarded quasi-polynomials extend to infinity in real and imaginary parts only inside the complex left half plane, whereas the asymptotic root chains of neutral quasi-polynomials contain at least one asymptotic chain of roots extending to infinity parallel to the imaginary axis. We define the roots of a quasi-polynomial $q(s)$ inside the complex left and right half planes as *stable* and *unstable* roots of $q(s)$, respectively.

We represent general SISO time delay systems by the transfer function

$$P(s) = \frac{N(s)}{D(s)} = \frac{q_n(s)}{q_d(s)} = \frac{\sum_{i=1}^{v_n} q_{n,i}(s)e^{-h_{n,i}s}}{\sum_{k=1}^{v_d} q_{d,k}(s)e^{-h_{d,k}s}}, \qquad (4.41)$$

where $q_n(s)$ and $q_d(s)$ are quasi-polynomials given in Definition 4.18. Any realizable proper and causal transfer function $P(s)$ in the form (4.41) also satisfies the necessary conditions, deg $q_{n,1} \leq$ deg $q_{d,1}$ and $h_{n,1} \geq h_{d,1}$. In order to perform coprime and inner-outer factorizations of the plant (4.41), we should find the roots of the quasi-polynomials $q_n(s)$ and $q_d(s)$ in $\mathbb{C}_+$. For this purpose we can use several alternative software tools, such as DDE-BIFTOOL [56, 218], QPmR.m [248, 249], and YALTA [11, 10]. We impose the following assumptions on the quasi-polynomials of the plant (4.41).

**Assumption 4.19.** *Time delays of $q_n(s)$ and $q_d(s)$ are commensurate. Furthermore, $h_{n,1} \geq h_{d,1}$, so that there is no time advance in the plant.*

**Assumption 4.20.** *The quasi-polynomials $q_n(s)$ and $q_d(s)$ are either of retarded or neutral type; in the latter case the asymptotic root chains do not coincide with the imaginary axis. Moreover, $deg(q_{n,1}) \leq deg(q_{d,1})$, so that the plant is proper.*

The first assumption is not restrictive for practical control applications; for example, when the delays are rational, they are commensurate. The second assumption excludes advanced-type systems and systems with poles or zeros asymptotically approaching the imaginary axis. The stabilization of these systems is an ongoing research area; see [176, 177] for an overview of recent developments and further references. As we shall see in the factorization of time delay systems, this case also requires the computation of infinitely many imaginary axis roots, which is a challenging task.

In this section, we consider systems with finitely many unstable poles. We use the conditions in Corollary 4.17 to obtain a coprime and inner-outer factorization of the plant (4.41). Therefore, we need to determine whether the SISO time delay system (4.41) has finitely many unstable zeros or poles. Infinitely many poles or zeros may occur at the asymptotic chains only, and we define an *asymptotic polynomial* to understand the behavior of quasi-polynomials at the asymptotes.

**Definition 4.21.** *Let $q(s)$ be a retarded or neutral quasi-polynomial with commensurate time delays as in Definition 4.18. The function $p(s) = \sum_{i=1}^{v} p_i s^{n_i - n_1}$ is called the asymptotic polynomial of $q(s)$, where $p_i = \lim_{s \to \infty} \frac{q_i(s)}{q_1(s)}$ for $i = 1, \ldots, v$.*

We determine whether a given quasi-polynomial has finitely many unstable roots from the roots of its asymptotic polynomial by the following lemma.

**Lemma 4.22.** *Let $q(s)$ be a quasi-polynomial given in Definition 4.18, satisfying Assumptions 4.19 and 4.20. Then, $q(s)$ has finitely many unstable roots if and only if it is a retarded quasi-polynomial or a neutral quasi-polynomial for which the asymptotic polynomial has all its roots outside the unit circle.*

*Proof.* By Definition 4.18, the quasi-polynomial $q(s)$ is either of retarded or neutral type. A retarded quasi-polynomial has finitely many unstable roots [101]. A neutral quasi-polynomial has finitely many unstable roots if and only if all its asymptotic root chains extending to infinity lie inside the complex left half plane. The asymptotic root chains of a neutral quasi-polynomial either lie inside the complex left half plane or extend to infinity in the imaginary part and to a constant value in the real part $\sigma_o$ [156]. Therefore, it is enough to show that the real part of any asymptotic root chain extending to infinity parallel to the imaginary axis is negative, i.e., $\sigma_o < 0$. For large values of the imaginary part $\omega$, the behavior of these root chains is characterized by the asymptotic polynomial $p(s)$; i.e., for any fixed $\sigma_o \in \mathbb{R}$ and $\epsilon > 0$, there exists $\omega_* > 0$ such that

$$\left| \frac{q(s)}{q_1(s) e^{-h_1 s}} - p\left(e^{-sh_o}\right) \right| = \left| \sum_{i=1}^{v} \left( \frac{q_i(s)}{q_1(s)} - p_i \right) e^{-(h_i - h_1)s} \right| < \epsilon \qquad (4.42)$$

for $s = \sigma_0 + j\omega$ and $\omega > w_*$. This establishes the connection between an asymptotic chain root of $q(s)$, $r_{q,i}$, and the corresponding root of its asymptotic polynomial $p(s)$, $r_{p,i}$, as

$$r_{q,i} = \sigma_0 + j\omega_i \iff r_{p,i} = e^{-(\sigma_0 + j\omega_i)h_o}.$$

The real part of $r_{q,i}$ is negative, $\sigma_o < 0$, if and only if the magnitude of the roots of the asymptotic polynomial $p(s)$ is greater than 1, i.e., $|r_{p,i}| > 1$. $\quad\square$

This lemma allows us to determine whether a quasi-polynomial has finitely many unstable roots. Another condition in Corollary 4.17 for the coprime and inner-outer factorization is the existence of a *conjugate function*, defined below.

**Definition 4.23.** *Let $q(s)$ be as given in Definition 4.18. Assume that $q(s)$ has infinitely many roots in $\mathbb{C}_+$, finitely many roots in $\mathbb{C}_-$, and no roots on $\mathbb{I}$. Then, the function defined by*

$$\bar{q}(s) = -q(-s)e^{-h_v s} \qquad (4.43)$$

*has finitely many roots in $\mathbb{C}_+$, infinitely many roots in $\mathbb{C}_-$, and no roots on $\mathbb{I}$, and it is called the* conjugate quasi-polynomial *of $q(s)$.*

An important point to keep in mind is that we use the conjugate function *only* when $q(s)$ has infinitely many unstable roots. Of course, one can still define $\bar{q}(s)$ as in (4.43) when $q(s)$ has infinitely many roots in $\mathbb{C}_+$ and in $\mathbb{C}_-$, but then $\bar{q}(s)$ will also have infinitely many roots in $\mathbb{C}_+$ and in $\mathbb{C}_-$. In such cases we may end up having both $M_n$ and $M_d$ infinite dimensional; since our design techniques depend on factorizations where one of them is finite dimensional, these situations fall beyond the scope of this book. Nevertheless, there are some cases where the quasi-polynomial in question has infinitely many roots in $\mathbb{C}_+$ and in $\mathbb{C}_-$, and the plant can still be factored as $P = M_n N_o / M_d$ with a finite dimensional $M_d$, as illustrated by the following example.

**Example 4.24.** Let us consider the plant

$$P(s) = \frac{q(s)}{(s-1)\,(s+2)}, \quad \text{with} \quad q(s) = c_0 s + (c_0 c_1 - s^2) e^{-hs} - c_1 s e^{-2hs},$$

where $h > 0$ and $c_0, c_1 \in [0, \frac{\pi}{2h})$. In this case, $q(s)$ has infinitely many roots in $\mathbb{C}_+$ and infinitely many roots in $\mathbb{C}_-$. To see this, note that $q(s)$ can be factored as

$$q(s) = (c_0 - s e^{-hs})\,(s + c_1 e^{-hs}),$$

where the first term has all its infinitely many roots in $\mathbb{C}_+$ and the second term has all its infinitely many roots in $\mathbb{C}_-$, as long as $c_0$ and $c_1$ are sufficiently small. In light of this observation, using the conjugate of the first term, a factorization in the form $P = M_n N_o M_d^{-1}$ is obtained as

$$P(s) = \left(\frac{c_0 - s e^{-hs}}{s + c_0 e^{-hs}}\right) \left(\frac{(s + c_0 e^{-hs})(s + c_1 e^{-hs})}{(s+1)\,(s+2)}\right) \left(\frac{(s-1)}{(s+1)}\right)^{-1}.$$

As we have seen here, if the quasi-polynomial in question can be factored out into products of quasi-polynomials having finitely many roots in $\mathbb{C}_+$ or finitely many roots in $\mathbb{C}_-$, then a desired plant factorization can be obtained. ∎

Let us now consider a quasi-polynomial $q(s)$ with finitely many roots in $\mathbb{C}_+$. Then, $Z(s) = q(s)/\bar{q}(s)$ is a causal and biproper transfer function. The magnitude property that $|Z(j\omega)| = 1$ holds on the imaginary axis by construction. But $Z$ need not be in $\mathcal{H}_\infty$. Since the roots of $\bar{q}$ in $\mathbb{C}_+$ are finitely many, one can construct an inner function $M(s) = Z(s)R_i(s)$, where $R_i(s)$ is a rational inner function whose zeros in $\mathbb{C}_+$ are those of $\bar{q}$ in $\mathbb{C}_+$. This is the main idea behind the inner-outer factorizations for neutral systems when the plant has infinitely many zeros in $\mathbb{C}_+$.

By combining Lemma 4.22 with the above discussion, we can explicitly give the conditions for the conjugate function of a quasi-polynomial to have finitely many roots in $\mathbb{C}_+$.

**Corollary 4.25.** *Let $q(s)$ be a neutral quasi-polynomial satisfying Assumptions* 4.19 *and* 4.20. *The conjugate quasi-polynomial $\bar{q}(s)$ has finitely many roots in $\mathbb{C}_+$ if and only if the roots of the asymptotic polynomial of $q(s)$ are inside the unit circle.*

**Proof.** The roots of quasi-polynomials $\bar{q}(s)$ and $q(-s)$ in the complex plane are the same. By following the steps of the proof in Lemma 4.22 for $q(-s)$, we obtain

$$r_{\bar{q},i} = \sigma_0 + j\omega_i \iff r_{p,i} = e^{(\sigma_0 + j\omega_i)h_o},$$

which is the reciprocal of the case in Lemma 4.22. □

Using Lemma 4.22, we can determine when a quasi-polynomial has finitely many unstable roots, and, by Corollary 4.25, we can compute its conjugate function if it exists without computing the rightmost roots of the quasi-polynomial, but only the roots of its asymptotic polynomial. The following example illustrates how we determine whether a quasi-polynomial has finitely many unstable roots and when to use a conjugate function.

**Example 4.26.** Consider the following quasi-polynomials:

$$q_1(s) = 3s + 0.5 + (2s+7)e^{-1.5s} + (s-1)e^{-2s} \quad \text{and} \quad q_2(s) = s + 3 + (2s-2)e^{-0.4s}.$$

The quasi-polynomial $q_1(s)$ is neutral, and its corresponding asymptotic polynomial is $p_1(s) = (1/3)s^4 + (2/3)s^3 + 1$. The magnitudes of its roots are $\{1.6796, 1.0312\}$, and they are greater than 1. By Lemma 4.22, $q_1(s)$ has finitely many unstable roots. Figure 4.8 shows that there are four roots of $q_1(s)$ in $\mathbb{C}_+$.

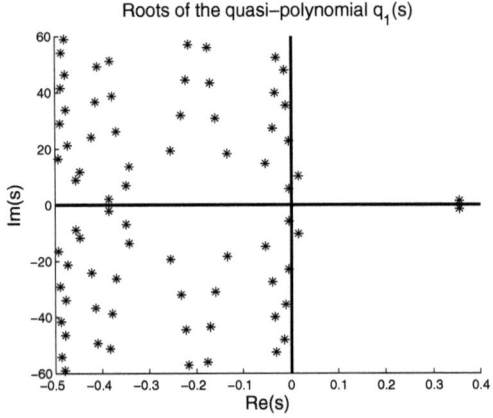

**Figure 4.8.** *The roots of the quasi-polynomial $q_1(s)$.*

The asymptotic polynomial of $q_2(s)$ is $p_2(s) = 2s + 1$. The magnitude of its root is smaller than 1, and by Lemma 4.22, $q_2(s)$ has infinitely many roots inside $\mathbb{C}_+$. This is illustrated in Figure 4.9 by computed roots of $q_2(s)$ shown with star markers. On the other hand, since the root is smaller than 1, the conjugate quasi-polynomial $\bar{q}_2(s)$ has finitely many roots inside $\mathbb{C}_+$ by Corollary 4.25. Figure 4.9 shows the roots of $\bar{q}_2(s)$ with circle markers, and there is one root in $\mathbb{C}_+$. ∎

Based on Lemma 4.22 and Corollary 4.25, we can rewrite Corollary 4.17 for SISO time delay systems.

**Corollary 4.27.** *The SISO time delay system in (4.41), satisfying Assumptions 4.19 and 4.20, admits a factorization of the form (4.25) if the quasi-polynomial $q_d(s)$ in the denominator is a retarded quasi-polynomial, or a neutral quasi-polynomial whose asymptotic polynomial has all its roots outside the unit circle, and one of the following conditions holds on $q_n(s)$:*

(i) *$q_n(s)$ is a retarded quasi-polynomial, or it is a neutral quasi-polynomial whose asymptotic polynomial has all its roots outside the unit circle, or*

(ii) *$q_n(s)$ is a neutral quasi-polynomial whose asymptotic polynomial has all its roots inside the unit circle.*

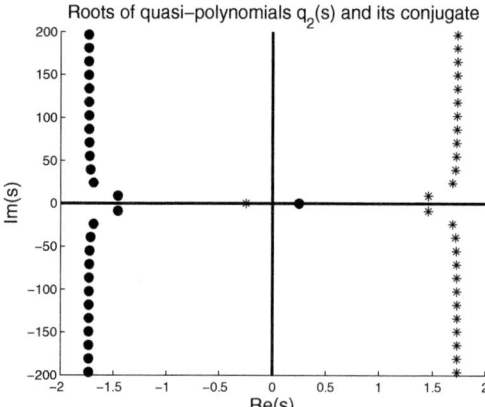

**Figure 4.9.** *The roots of quasi-polynomials $q_2(s)$ (with stars) and its conjugate function $\bar{q}_2(s)$ (with circles).*

In case (i), the quasi-polynomials $q_n(s)$ and $q_d(s)$ have finitely many unstable roots, and the coprime and inner-outer factorization is given in (4.37), where $N(s) = q_n(s)/(s+1)^k$ and $D(s) = q_d(s)/(s+1)^k$, where $k = \deg(q_{d,1})$. In case (ii), the quasi-polynomial $q_d(s)$ has finitely many unstable roots and $q_n(s)$ has infinitely many unstable roots, but it has a conjugate function with finitely many unstable roots, and hence the plant can be factorized as in (4.38), where $\bar{N}(s) = \bar{q}_n(s)/(s+1)^k$.

In summary, the coprime and inner-outer factorization requires the computation of all the unstable roots of the quasi-polynomials in question. In particular, the factorization in (4.37) requires the computation of unstable roots of the quasi-polynomials $q_n(s)$ and $q_d(s)$ and the factorization in (4.38) requires the location of unstable roots of quasi-polynomials $\bar{q}_n(s)$ and $q_d(s)$. There are available numerical methods and tools for the computation of unstable roots of quasi-polynomials; see [11, 29, 57, 58, 56, 142, 184, 248, 256].

## 4.5 ▪ Factorization of time delay systems with infinitely many unstable poles

In this section, we discuss factorization and stabilization of plants with infinitely many poles in $\mathbb{C}_+$. As before, let us assume that $P$ is given as a ratio of two quasi-polynomials, $P(s) = q_n(s)/q_d(s)$, where $q_d(s)$ does not have any roots on $\mathbb{I}$. Our goal is to obtain an inner-outer factorization in the form $P = M_n N_o / M_d$, where $M_n$ and $M_d$ are inner and $N_o$ is outer. Since $P$ has infinitely many poles in $\mathbb{C}_+$, in this case, $M_d$ has to be infinite dimensional, and its zeros correspond to the zeros of $q_d(s)$ in $\mathbb{C}_+$. We assume that $q_d$ has finitely many roots in $\mathbb{C}_-$. In fact, this is not absolutely necessary; as long as it can be factored as the product of two quasi-polynomials $q_d(s) = q_{d-}(s)q_{d+}(s)$, where $q_{d-}(s)$ has finitely many roots in $\mathbb{C}_+$ and $q_{d+}(s)$ has finitely many roots in $\mathbb{C}_-$, the desired factorizations can be done as illustrated in Example 4.24.

The controller design techniques presented in the following chapters require that either $M_n$ or $M_d$ is finite dimensional. Due to this restriction, in this section, we assume that the plant does not contain an I/O delay and that $q_n(s)$ has finitely many roots in $\mathbb{C}_+$.

Before diving into technical discussions, we should mention some relevant works from the literature. A neutral system transfer function $P(s) = q_n(s)/q_d(s)$ can be obtained from a "state-space" representation

$$\dot{x}(t) - E\dot{x}(t - h) = A_0 x(t) + A_1 x(t - h) + Bu(t), \qquad (4.44)$$

$$y(t) = C_0 x(t) + C_1 x(t - \tau) + Du(t), \qquad (4.45)$$

where $x(t) \in \mathbb{R}^{n_x}$, $h > 0$, $\tau > 0$, and $D$ is scalar, and the matrices $E, A_0, A_1$ are $n_x \times n_x$, $C_0, C_1$ are $1 \times n_x$, and $B$ is $n_x \times 1$. The stabilization of these types of systems by a state feedback has been studied extensively; see, e.g., [32, 141, 214] and their references. In particular, it has been proven that for the existence of a state feedback, $u(t) = Kx(t)$, stabilizing (4.44)–(4.45), a necessary condition is that

$$q_d(s) = \det \left( s(I - Ee^{-hs}) - A_0 - A_1 e^{-hs} \right)$$

has finitely many roots in the right half plane; see [141]. In other words, if $q_d(s)$ has infinitely many roots on $\mathbb{C}_+$, then it is not possible to stabilize this system by using a state-feedback controller. The difficulty lies in the fact that the plant is *strictly proper*; this point is discussed next.

### 4.5.1 ▪ A restriction on the relative degrees of $P$ and $C$

When dealing with systems with infinitely many poles in $\mathbb{C}_+$, we have to make one more assumption, that is, $P(s)$ is biproper. In other words, strictly proper plants with infinitely many unstable poles are ruled out, because they cannot be stabilized by a proper controller; see Proposition 4.28 below, which formally states this fact. Furthermore, the open-loop gain $PC$ cannot be strictly proper in this case, which means that the controller should be biproper as well. In the statement below we have neutral time delay systems with infinitely many unstable poles in mind. But the result is given for a general class of systems satisfying a certain set of conditions typically seen in neutral delay systems. See also [98] for an alternative statement, a historical overview, and further references.

**Proposition 4.28.** *Let* $(C, P)$ *be a feedback system formed by a proper controller* $C(s)$ *and a proper plant* $P(s)$. *Let the system satisfy the following assumptions:*

(a1) *There is no unstable pole-zero cancellation in the product* $G(s) = P(s)C(s)$, *and* $G$ *does not have any poles on* $\mathbb{I}$. *Moreover, there exists* $\epsilon > 0$ *such that* $G$ *has finitely many poles in the* $\epsilon$-neighborhood of $\mathbb{I}$, *i.e.,* $\{s \in \mathbb{C} : -\epsilon < \mathrm{Re}(s) < \epsilon\}$.

(a2) *There exist* $\sigma_2 > \sigma_1 > 0$ *such that* $G$ *has infinitely many poles in the strip* $\{s \in \mathbb{C} : \sigma_1 < \mathrm{Re}(s) < \sigma_2\}$ *and has no poles in* $\{s \in \mathbb{C} : \mathrm{Re}(s) \geq \sigma_2\}$.

(a3) *There exists* $\delta > 0$ *such that*

$$\inf_{k,j} |p_k - p_j| \geq \delta > 0, \qquad (4.46)$$

*where* $p_k$ *and* $p_j$ *denote poles of* $G$ *in* $\mathbb{C}_+$.

*Under* (a1)–(a3), *if* $G(s)$ *is strictly proper, then the feedback system is unstable.*

**Proof.** Assume that $G$ is strictly proper and satisfies (a1)–(a3). We deduce from (4.46) that for any finite $r > 0$, $G$ has finitely many poles in the closed semidisk $\Gamma_r := \{s = re^{j\theta} : |s| \leq r, \theta \in [-\frac{\pi}{2}, \frac{\pi}{2}]\}$. Given $r > 0$, let us define a positive contour $\overrightarrow{\Gamma}_r$ encircling $\Gamma_r$, by going over the path $s = j\omega$, $\omega$ increasing from $-r$ to $+r$, and then $s = re^{j\theta}$, $\theta$ decreasing from $\frac{\pi}{2}$ to $-\frac{\pi}{2}$. Now consider a sequence $r_k$, $k = 1, 2, \ldots$, such that

(i) $r_{k+1} > r_k$ and $r_k \to \infty$ as $k \to \infty$,

(ii) $G(s)$ does not have any poles on $\vec{\Gamma}_{r_k}$ for all $k$, and

(iii) there exist $\omega_*$ and $k_*$ such that $|G(j\omega)| < 1$ for all $\omega \geq \omega_*$, and $|G(r_k e^{j\theta})| < 1$ for all $k \geq k_*$, and all $\theta \in [-\frac{\pi}{2}, \frac{\pi}{2}]$, with $r_{k_*} > \omega_*$.

Note that for strictly proper neutral time delay systems satisfying (a1) and (a3) it is always possible to find such a sequence by choosing $\omega_*$ and $r_*$ large enough (see below for an illustrative example). For any $r = r_k$, according to Cauchy's argument principle [1], the number of zeros, $n_z$, of the function $1 + G(s)$ inside $\vec{\Gamma}_r$ is equal to $(n_p + n_{\vec{\partial}})$, where $n_p$ is the number of poles of $G$ inside $\vec{\Gamma}_r$ and $n_{\vec{\partial}}$ is the number of clockwise encirclements of $-1$ by the closed path $\vec{\Gamma}_G := G(\vec{\Gamma}_r)$. Clearly, a necessary condition for feedback system stability is to have $n_z = 0$ for any given $r > 0$ (as in the Nyquist stability criterion). That means the curve $\vec{\Gamma}_G$ must encircle $-1$ exactly $n_p$ times in the counterclockwise direction. On the other hand, by (iii), $n_{\vec{\partial}}$ is invariant for all $k \geq k_*$. By (a2), $G$ has infinitely many poles in $\mathbb{C}_+$, and so the number $n_p$ increases without a bound as $r_k \to \infty$. Thus, there is a sufficiently large $r_k \geq r_{k_*}$, for which $n_p$ is strictly greater than $-n_{\vec{\partial}}$, which means that $1 + G(s)$ has at least one zero inside $\Gamma_{r_k}$, and hence the feedback system is unstable.    $\square$

**Example 4.29.** For a strictly proper neutral time delay system $G$ we illustrate the construction of the sequence $r_k$ satisfying (i)–(iii) on the following transfer function:

$$G(s) = \frac{F(s)}{s - 2a}, \quad \text{where} \quad F(s) = \left( \frac{1}{1 - e^{-h(s-a)}} \right), \quad h > 0, \ a > 0.$$

This system has poles at $2a$ and at $p_k = a \pm j\frac{2\pi k}{h}$, $k = 0, 1, \dots$. Therefore, it satisfies (a1)–(a3) with $\sigma_1 = \frac{a}{2}$, $\sigma_2 = 2a + \epsilon$ for arbitrary finite $\epsilon > 0$, and $\delta = \min\{a, \frac{2\pi}{h}\}$. Define

$$r_k = \frac{(2\pi + 1)k}{h}.$$

Clearly, (i) and (ii) are satisfied for sufficiently large $r_k$; the arguments used below to show (iii) also apply to (ii). In order to find $\omega_*$, first note that

$$|G(j\omega)| = \frac{1}{|1 - e^{ha}e^{-jh\omega}|\sqrt{\omega^2 + 4a^2}}.$$

If $e^{ha} > 2$, then we can select $\omega_* = 1$ rad/sec; otherwise, we can choose $\omega_* = (e^{ha} - 1)^{-1}$ rad/sec to satisfy the condition that $|G(j\omega)| < 1$ for all $\omega \geq \omega_*$. Now, to determine $k_*$ required for (iii) we note that $|G(r_k e^{j\theta})| \leq \frac{|F(r_k e^{j\theta})|}{r_k - 2a}$. So, if $r_k \geq 2a + 3$, then $|G(r_k e^{j\theta})| \leq |F(r_k e^{j\theta})|/3$. Next we will show that $|F(r_k e^{j\theta})| \leq 2$ for all $\theta \in [0, \frac{\pi}{2}]$ and all $k$ sufficiently large. Note that $|G(r_k e^{j\theta})| = |G(r_k e^{-j\theta})|$, and so it is sufficient to analyze $|G(r_k e^{j\theta})|$ for $\theta \in [0, \frac{\pi}{2}]$. For this purpose we examine

$$|F(r_k e^{j\theta})| = \left| \frac{1}{1 - e^{ha - hr_k \cos(\theta)}(\cos(hr_k \sin(\theta)) - j\sin(hr_k \sin(\theta)))} \right|.$$

Let $k$ be sufficiently large so that

$$\varepsilon_k := \frac{ha + \ln(2)}{(2\pi + 1)k} < \frac{1}{2}. \tag{4.47}$$

*Case* 1: $\theta \leq \theta_k := \cos^{-1}(\varepsilon_k)$. In this case, $\cos(\theta) \geq \varepsilon_k$, and hence

$$e^{ha} e^{-hr_k \cos(\theta)} \leq \frac{1}{2}.$$

Note that

$$|F(r_k e^{j\theta})| \leq \frac{1}{1 - |e^{ha} e^{-hr_k \cos(\theta)}|} \leq 2.$$

*Case* 2: $\theta \geq \theta_k$. Let $k$ be such that (4.47) holds. In this case, we have

$$1 - \varepsilon_k^2 \leq \sqrt{1 - \varepsilon_k^2} = \sin(\theta_k) \leq \sin(\theta) \leq 1.$$

Now choose $k$ large enough to satisfy

$$\delta_k := hr_k \varepsilon_k^2 = \frac{(ha + \ln(2))^2}{(2\pi + 1)k} \leq \frac{\pi}{3}. \tag{4.48}$$

Then, we have

$$-1 \leq \cos(hr_k \sin(\theta)) \leq \cos((2\pi + 1)k - \delta_k) = -\cos(\delta_k) \leq -\frac{1}{2}.$$

Therefore, in this case we have

$$\mathrm{Re}\left(1 - e^{ha - hr_k \cos(\theta)} e^{-jhr_k \sin(\theta)}\right) \geq 1 + \frac{1}{2} e^{ha - hr_k \cos(\theta)}$$

and hence

$$|F(r_k e^{j\theta})| \leq \frac{1}{1 + 0.5\, e^{ha - hr_k \cos(\theta)}} \leq 1.$$

Thus, the property (iii) is shown to hold by selecting $k_*$ large enough so that inequalities (4.47)–(4.48) are satisfied for all $k \geq k_*$.

As this example illustrates, when $G$ is strictly proper and there is a minimum positive distance between its infinitely many poles, one can construct the sequence $r_k$ to satisfy the three properties stated in the proof of Proposition 4.28. In summary, the closed-loop system cannot be stable when $G = F(s)G_o(s)$ for any strictly proper $G_o$ whose poles and zeros are within a bounded set of $\mathbb{C}$.  ∎

Clearly, the condition on the imaginary axis poles was to simplify the notation in the proof; if $G(s)$ has finitely many poles on $\mathbb{I}$, then by a slight change in the contour $\overrightarrow{\Gamma}_r$, we can arrive at the same conclusion.

From the discussion in Section 4.4 we can determine whether a neutral time delay system has asymptotic root chains in $\mathbb{C}_\sigma$, for some $\sigma > 0$, by checking the roots of its asymptotic polynomial. Moreover, such systems have finitely many poles in compact subsets of $\mathbb{C}_+$.

In conclusion, in light of the above proposition, for controller design purposes we assume that if $P$ has infinitely many poles in $\mathbb{C}_+$, then it is proper but *not strictly proper*; otherwise, it is impossible to find a proper stabilizing controller.

**Example 4.30.** The following plant has infinitely many poles in $\mathbb{C}_+$, and it is biproper. Therefore, it satisfies the necessary condition derived above for the existence of a stabilizing controller, which has to be biproper. The plant admits a coprime and inner-outer factorization:

$$P(s) = \frac{4(s-1)}{q_d(s)}, \quad q_d(s) = (s + 3 + 2(s-1)e^{-0.4s}).$$

As we have seen in Examples 4.16 and 4.26, the quasi-polynomial $q_d(s)$ has infinitely many roots in $\mathbb{C}_+$ and a single root $-\varrho = -0.247$ in $\mathbb{C}_-$. Now, the conjugate quasi-polynomial $\bar{q}_d(s) = 2(s+1) + (s-3)e^{-0.4s}$ can be used to construct the factorization $P = M_n N_o / M_d$ as follows:

$$M_n(s) = \frac{s-1}{s+1},$$

$$M_d(s) = \frac{q_d(s)\,(s-\varrho)}{\bar{q}_d(s)\,(s+\varrho)},$$

$$N_o(s) = \frac{(s+1)\,(s-\varrho)}{(s+\varrho)\,\bar{q}(s)}.$$

We see that $M_n$ is finite dimensional and $M_d$ is infinite dimensional with infinitely many zeros in $\mathbb{C}_+$. The outer factor $N_o$ is invertible in $\mathcal{H}_\infty$; in particular, it is biproper and it does not have a zero at $s = \varrho$. The stabilizing controller can now be obtained from the above factorization as

$$C(s) = \frac{\widetilde{X} + M_d \widetilde{Q}}{\widetilde{Y} - M_n N_o \widetilde{Q}} = N_o^{-1}\left( \frac{X + M_d Q}{Y - M_n Q} \right),$$

where $Y(s) = 1/M_d(1)$, so that $X = (1 - M_d Y)/M_n \in \mathcal{H}_\infty$. Defining $Q_o(s) = M_d(1)Q(s)$, we obtain the controller parameterization

$$C(s) = \left( \frac{s+1}{s-1} \right) \left( \frac{M_d(1)}{1 - \left( \frac{s-1}{s+1} \right) Q_o(s)} - M_d(s) \right) N_o^{-1}(s),$$

where $Q_o \in \mathcal{H}_\infty$, with the restriction that $Q_o(\infty) \neq 1$. In particular, one can use $Q_o(s) = 0$ and obtain a stabilizing controller in the form

$$C_o(s) = \left( \frac{s+1}{3\;1} \right) \left( M_d(1) - M_d(s) \right) N_o^{-1}(s).$$

Clearly, $C_o$ does not have a pole at $s = 1$ and it is a stable transfer function.   ∎

Finally, we would like to point out that results similar to Proposition 4.28 can be found in [175, 207, 262]; see [98] for further references.

## 4.5.2 ▪ Controller design using duality

The following result provides a duality between the factorizations of Section 4.4 and the factorizations obtained in this section.

**Proposition 4.31.** *Let $C$ and $P$ be biproper transfer functions where $P = M_n N_o / M_d$ with a finite dimensional inner $M_n$ and an infinite dimensional inner $M_d$. The feedback system $(C, P)$ is stable if and only if the dual feedback system $(C^{-1}, P^{-1})$ is stable.*

**Proof.** Since $C$ and $P$ are biproper, their inverses $C_1 := C^{-1}$ and $P_1 := P^{-1}$ are also biproper transfer functions. Now, $(C, P)$ is stable if and only if $S = (1 + PC)^{-1}$, $PS$, $CS$ are in $\mathcal{H}_\infty$. Note that $S = T_1$, $CS = P_1 S_1$, $PS = C_1 S_1$, $T = S_1$, where $S_1 = (1 + P_1 C_1)^{-1}$ and $T_1 = P_1 C_1 (1 + P_1 C_1)^{-1}$. Thus, $(C, P)$ is stable if and only if $(C_1, P_1)$ is stable.   □

In particular, Proposition 4.31 tells us that controller design objectives involving $S$ and $T$ can be translated to design objectives involving $S_1$ and $T_1$. So, in order to design $C$ for a given $P = \frac{M_n}{M_d} N_o$, we can first design $C_1$ for $P_1 = \frac{M_d}{M_n} N_o^{-1}$, which is a system with finitely many poles in $\mathbb{C}_+$, and then set $C = C_1^{-1}$. An $\mathcal{H}_\infty$ controller design example is given in Section 6.3.2. For an extension to the MIMO case, see Section 7.5.2.

## 4.5.3 ▪ Repetitive controller

An interesting type of controller used in various applications, where tracking of a periodic reference signal is desired, is the *repetitive controller* [105]. The basic idea is to use a filter in the form $F_r(s) = (1 - e^{-Ls})^{-1}$ in the controller, i.e., $C(s) = F_r(s)C_a(s)$, in order to track periodic reference signals whose periods are $L > 0$. This is in the spirit of the internal model principle, where a copy of the reference signal generator appears as a factor of the controller. Since periodic signals of period $L$ have Fourier series expansions which include sinusoidal signals of the form $\sin(2\pi nt/L)$, with $n \geq 1$, the filter $F_r$ contains poles at $2\pi n/L$ for $n = 1, \ldots$. Also note that $s = 0$ is a pole of the filter; this is to make sure that the average value of the periodic reference and the resulting output match in the steady state.

For the sake of simplicity of the arguments to follow, we can assume that the underlying plant $P$ is stable and finite dimensional. In light of Proposition 4.28, we further assume that $P$ is biproper. In repetitive control, we can treat the system $P_a(s) = P(s)F_r(s)$ as the augmented plant to be controlled by the controller $C_a(s)$. Now, for a given $P_a$, designing a stabilizing controller $C_a$ is a challenging problem, because $P_a$ has infinitely many poles on $\mathbb{I}$, located at the roots of $1 - e^{-Ls} = 0$, i.e., at $s = \pm j\omega_n$, with $\omega_n = \frac{2\pi n}{L}$, for $n = 0, 1, 2, \ldots$. The trick is to do the axis shift with $\widehat{s} = s + \epsilon, \epsilon > 0$. Since $P$ is stable and finite dimensional, there exists an $\epsilon > 0$ such that $\widehat{P}(\widehat{s}) = P(\widehat{s} - \epsilon)$ does not have any poles for $\text{Re}(\widehat{s}) > 0$. So, $\widehat{P}_a(\widehat{s}) = \widehat{P}(\widehat{s})F_r(\widehat{s} - \epsilon)$ is in the form

$$\widehat{P}_a(\widehat{s}) = \frac{\widehat{M}_n(\widehat{s})\widehat{N}_o(\widehat{s})(e^{L\epsilon} - e^{-L\widehat{s}})^{-1}}{\widehat{M}_d(\widehat{s})},$$

where $\widehat{M}_n(\widehat{s})\widehat{N}_o(\widehat{s})$ is the inner-outer factorization of $\widehat{P}(\widehat{s})$ and

$$\widehat{M}_d(\widehat{s}) = \frac{(e^{L\epsilon} - e^{-L\widehat{s}})}{(1 - e^{L\epsilon}e^{-L\widehat{s}})}.$$

Note that $\widehat{M}_d(\widehat{s})$ has infinitely many poles on $\text{Re}(\widehat{s}) = \epsilon > 0$, and $\widehat{N}_o(\widehat{s})(e^{L\epsilon} - e^{-L\widehat{s}})^{-1}$ is invertible in $\mathcal{H}_\infty$ (relative to the half plane $\text{Re}(\widehat{s}) > 0$; by a slight abuse of notation, we simply use $\mathcal{H}_\infty$ whenever it is clear from the context that this is for the $\widehat{s}$-plane and it is not the same as the original $\mathcal{H}_\infty$ for the $s$-plane). Therefore, we can assume a controller structure

$$\widehat{C}_a(\widehat{s}) = \widehat{C}_i(\widehat{s})\widehat{N}_o^{-1}(\widehat{s})(e^{L\epsilon} - e^{-L\widehat{s}}),$$

where $\widehat{C}_i(\widehat{s})$ must be a stabilizing controller for $\widehat{M}_n(\widehat{s})/\widehat{M}_d(\widehat{s})$. From the controller parameterization we have that all such controllers are in the form

$$\widehat{C}_i = \frac{\widehat{X} + \widehat{M}_d\widehat{Q}}{\widehat{Y} - \widehat{M}_n\widehat{Q}},$$

where $\widehat{X}\widehat{M_n} + \widehat{Y}\widehat{M_d} = 1$, with $\widehat{X}, \widehat{Y} \in \mathcal{H}_\infty$, and the free parameter must be stable, with the restriction that the resulting controller is proper. Now, since $\widehat{M_n}$ is finite dimensional, as shown in Section 4.1, we can find a finite dimensional $\widehat{Y}$, from interpolation conditions at the zeros of $\widehat{M_n}$, so that

$$\widehat{X}(\widehat{s}) = \frac{1 - \widehat{Y}(\widehat{s})\widehat{M_d}(\widehat{s})}{\widehat{M_n}(\widehat{s})} \in \mathcal{H}_\infty.$$

Hence, $\widehat{C_i}$ can be obtained from a selected $\widehat{Q}$. Finally, the original controller, $C_a(s)$, can be obtained by the back transformation $\widehat{s} = s + \epsilon$.

**Example 4.32.** In order to illustrate the key points of the above design, we take

$$P(s) = \frac{0.001s + 1}{s + 1} = N_o(s).$$

Since the rightmost pole of $P$ is at $-1$, we choose $\epsilon < 1$; for example, $\epsilon = 0.1$ is admissible (however, the final controller does not depend on $\epsilon$; see below). Note that the plant is already outer, and so $\widehat{P}(\widehat{s}) = \widehat{N_o}(\widehat{s})$. Then, we have

$$\widehat{X}(\widehat{s}) = \left(1 - \widehat{Y}(\widehat{s})\widehat{M_d}(\widehat{s})\right);$$

for example, $\widehat{X}(\widehat{s}) = 1$ and $\widehat{Y}(\widehat{s}) = 0$ are admissible. Hence

$$\widehat{C_a}(\widehat{s}) = \left(\frac{1 + \widehat{M_d}(\widehat{s})\widehat{Q}(\widehat{s})}{-\widehat{Q}(\widehat{s})}\right)\widehat{N_o}^{-1}(\widehat{s})(e^{L\epsilon} - e^{-L\widehat{s}}),$$

with arbitrary $\widehat{Q} \in \mathcal{H}_\infty$ satisfying $\widehat{Q}(\infty) \neq 0$ is a stabilizing controller for $\widehat{P_a}$. Note that $(e^{L\epsilon} - e^{-L\widehat{s}})$ is outer; therefore, we can reparameterize the controller with $\widehat{Q_1} = (e^{L\epsilon} - e^{-L\widehat{s}})^{-1}\widehat{Q}$, which leads us to

$$\widehat{C_a}(\widehat{s}) = \left(\frac{1 + \widehat{M_d}(\widehat{s})(e^{L\epsilon} - e^{-L\widehat{s}})\widehat{Q_1}(\widehat{s})}{-\widehat{Q_1}(\widehat{s})}\right)\widehat{N_o}^{-1}(\widehat{s}).$$

Back transformation with $\widehat{s} = s + \epsilon$ yields

$$C_a(s) = \left(\frac{1 + (1 - e^{-Ls})Q_1(s)}{-Q_1(s)}\right)\left(\frac{s + 1}{0.001s + 1}\right), \quad Q_1 \in \mathcal{H}_\infty, \ Q_1(\infty) \neq 0.$$

Thus, the overall repetitive controller for the original plant $P(s)$ is

$$C(s) = -\left(\frac{1}{(1 - e^{-Ls})Q_1(s)} + 1\right)\left(\frac{s + 1}{0.001s + 1}\right).$$

The resulting sensitivity and complementary sensitivity functions are

$$S(s) = (1 + P_a(s)C_a(s))^{-1} = -(1 - e^{-Ls})Q_1(s), \quad T(s) = 1 + (1 - e^{-Ls})Q_1(s).$$

Clearly, $S(j\omega_n) = 0$ and $T(j\omega_n) = 1$ for all $n = 0, 1, 2, \ldots$, for all $Q_1 \in \mathcal{H}_\infty$. At this point, $Q_1$ can be chosen to shape the sensitivity in the frequency intervals $(\omega_n, \omega_{n+1})$, $n = 0, 1, \ldots$; for example, Figure 4.10 shows the plot of $|T(j\omega)|$ for $L = 2\pi$ sec and

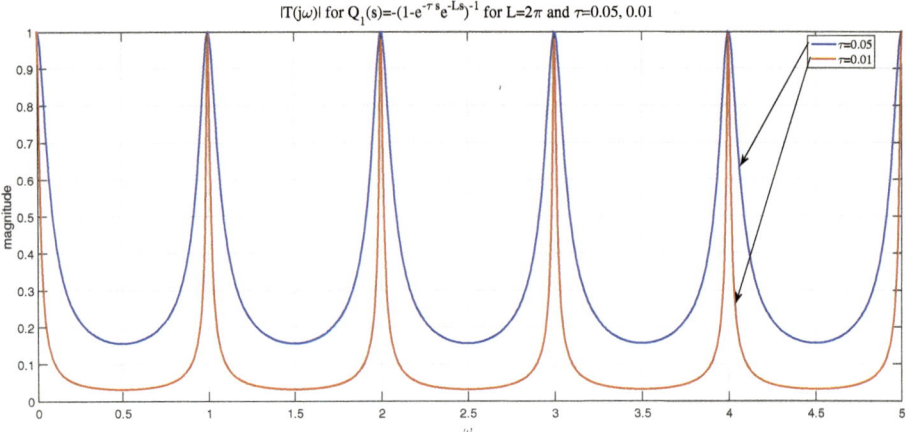

**Figure 4.10.** $|T(j\omega)|$ with $Q_1(s) = -(1 - e^{-\tau L}e^{-Ls})^{-1}$ for $L = 2\pi$ sec., $\tau = 0.05,\ 0.010$.

$Q_1(s) = -(1 - e^{-\tau L}e^{-Ls})^{-1}$ with two different values of $\tau$. The main idea behind this selection is to suppress $|T(j\omega)|$ at frequencies other than $\omega_n$, $n = 0,\ 1,\ldots$, so that unwanted modes are not excited. ∎

In many practical applications it is not possible to have infinite bandwidth; in this case, a modification can be made in the repetitive controller:

$$F_r(s) = \frac{1}{1 - w(s)e^{-Ls}},$$

where $w(s)$ is a strictly proper rational outer low-pass filter with $w(j\omega_n) \approx 1$ for $n = 0,\ 1,\ldots,N$ and $|w(j\omega)| \ll 1$ for all $\omega > \omega_N$. This also relaxes the condition that $P$ has to be biproper, because $F_r$ has finitely many unstable poles; see [105, 262, 263]. The modified filter $F_r$ is still useful in tracking periodic signals whose Fourier series expansions contain sinusoidal signals with periods $\frac{L}{n}$ for $n = 1,\ldots,N$. In this case, after the axis shift, $s = \hat{s} - \epsilon$, the augmented plant admits a factorization in the form

$$\widehat{P}_a(\hat{s}) = F_r(\hat{s} - \epsilon)P(\hat{s} - \epsilon) = \frac{\widehat{M}_n(\hat{s})\widehat{N}_o(\hat{s})}{\widehat{M}_d(\hat{s})},$$

where $\widehat{M}_d$ is a finite dimensional inner function. In order to see this, note that if $w(s) = w_n(s)/w_d(s)$, where $w_n$ and $w_d$ are coprime stable polynomials, then the zeros of $\widehat{M}_d$ are the roots of the retarded quasi-polynomial

$$\hat{q}_d(\hat{s}) := w_d(\hat{s} - \epsilon) - w_n(\hat{s} - \epsilon)e^{L\epsilon}e^{-L\hat{s}},$$

and there are finitely many in $\mathrm{Re}(\hat{s}) > 0$, because $\deg(w_d) > \deg(w_n)$. Thus, in this case, the factorization of $\widehat{P}_a$ falls within the framework of the method given in Section 4.4.

For different approaches to repetitive controller design and their industrial applications, see [34, 253, 172] and references therein.

## 4.6 • Factorizations of other DPS examples

### 4.6.1 • Flexible beam

Recall the beam transfer function (3.18). A coprime factorization is $P(s) = \frac{N(s)}{D(s)}$, where

$$D(s) = \frac{s^2}{(\tau s + 1)^2} \qquad (4.49)$$

and $N(s)$ can be factored as $N = N_o M_d$ with $M_n(s) = e^{-hs} B(s)$. This is done by separating the roots of $(1 + \varepsilon s - (s/2\alpha_n^2)^2$; those in $\mathbb{C}_+$ define the inner part. The zeros of the plant are at

$$s = 2\alpha_n^4 \left( \varepsilon \pm \sqrt{\varepsilon^2 + \frac{1}{\alpha_n^4}} \right) \qquad \text{for} \quad n = 1, 2, \ldots, \qquad (4.50)$$

where

$$\cos(\alpha_n) \sinh(\alpha_n) = \sin(\alpha_n) \cosh(\alpha_n) \text{ for } \alpha_n > 0.$$

The zeros (4.50) that are in $\mathbb{C}_+$ define the inner part,

$$B(s) = \prod_{n=1}^{\infty} \left( \frac{2\alpha_n^4 \left( \varepsilon + \sqrt{\varepsilon^2 + \frac{1}{\alpha_n^4}} \right) - s}{2\alpha_n^4 \left( \varepsilon + \sqrt{\varepsilon^2 + \frac{1}{\alpha_n^4}} \right) + s} \right), \qquad (4.51)$$

and, finally, $N_o = DP/M_n$ is determined as

$$N_o(s) = \frac{2}{(\tau s + 1)^3} \prod_{n=1}^{\infty} \frac{\left( 1 + s\sqrt{\varepsilon^2 + \frac{1}{\alpha_n^4}} + \frac{s^2}{4\alpha_n^4} \right)}{\left( 1 + \varepsilon s + \frac{s^2}{\phi_n^4} \right)}. \qquad (4.52)$$

It has been shown [137, 138] that $N_o(s) \in \mathcal{H}_\infty$ and $B(s) \in \mathcal{H}_\infty$ converge in the closed right half plane. This plant has a singularity at $-1/\varepsilon$, and except for the double pole at $s = 0$, all its poles are in $\mathbb{C}_-$, at

$$s = \frac{-\phi_n^4}{2} \left( \varepsilon \pm \sqrt{\varepsilon^2 - \frac{4}{\phi_n^4}} \right) \qquad \text{for} \quad n = 1, 2, \ldots, \qquad (4.53)$$

where $\cos(\phi_n) \cosh(\phi_n) = 1$ for $\phi_n > 0$.

Since the plant contains poles on $\mathbb{I}$, $D(s)$ cannot be written as an inner function whose zeros are the unstable poles. As described in Section 4.2, this problem can be circumvented by an axis shift, which puts the poles at $s = 0$ to a point $\hat{s} = \epsilon$ in the right half plane. Recall that $\epsilon$ should be sufficiently small not to introduce significant conservatism in the robust control design techniques discussed in this book.

### Exercise 4.33.
1. Let $\varepsilon = 0.001$, and compute the first 10 sets of zeros, (4.50), and poles, (4.53), of the plant. Draw the Bode plots of the plant.
2. Do the same exercise for $\varepsilon = 10^{-2}$, and observe the changes in the pole and zero locations as well as the Bode plots. Which plant is easier to control? Of course, we have not defined a level of difficulty for controlling a plant, but you may discuss how large the control effort (magnitude and bandwidth) should be in each case, to bring the closed-loop poles to the left of a specified vertical line in $\mathbb{C}_-$, say $\text{Re}(s) = -0.5$.

## 4.6.2 ▪ An unstable fractional order system

The plant model of a nonlaminated magnetic suspension system was given in Section 3.3.3 as in (3.25):

$$P(s) = \frac{e^{-hs}}{(s^\alpha - p)(s^\alpha - p_1)(s^\alpha - p_2)(s^\alpha - p_3)(s^\alpha - p_4)},$$

where $\alpha = 0.5$, $h > 0$, $p > 0$, and $p_k \in \mathbb{C} \setminus \{re^{j\theta} : r \in \mathbb{R}_+, |\theta| < \alpha\pi/2\}$ for $k = 1, \ldots, 4$. Recall that this plant has only one unstable pole at $s = p^2$. So, a coprime and inner-outer factorization can be obtained as follows:

$$P(s) = \frac{M_n(s)N_o(s)}{M_d(s)},$$

where

$$M_n(s) = e^{-hs}, \quad M_d(s) = \frac{s - p^2}{s + p^2},$$

and

$$N_o(s) = \frac{(s^\alpha + p)}{(s + p^2)(s^\alpha - p_1)(s^\alpha - p_2)(s^\alpha - p_3)(s^\alpha - p_4)}.$$

See [118] for a numerical example and a robust controller design for this plant.

### Exercise 4.34.
Let the parameter $c$ in (3.24) be given as $c = 2$, and take $h = 0.1$ sec.
1. Find $p, p_1, \ldots, p_4$.
2. Obtain a stabilizing controller using the structure given in (4.18), where $P_o = N_o/M_d$, and choose $C_o$ in the form

$$C_o(s) = \frac{K}{(\varepsilon s + 1)^3} N_o^{-1}(s), \quad K > 0, \quad \varepsilon > 0.$$

Select $\varepsilon$ as follows: draw $|N_o(j\omega)|$ versus $\omega$, and determine the smallest frequency $\omega_m$ for which we have $|N_o(j\omega)| \le 10^{-3}$ for all $\omega \ge \omega_m$; then let $\varepsilon = \min\{\omega_m^{-1}, 10^{-3}\}$. Then, select $K > 1$ so that the feedback system is stable and the poles of $T = PC(1 + PC)^{-1}$ are placed to the left of $\text{Re}(s) \le -0.5$ in $\mathbb{C}_-$.

In general, as we have seen in the previous sections, for a coprime and inner-outer factorization of a plant we need to find the unstable poles and zeros. For this purpose, in particular for fractional order systems, progress has been made recently; see, e.g., [61]. The software YALTA [11, 10] handles fractional order systems, as well as systems with time delays, for the computation of unstable poles and zeros.

There are of course various other controller design techniques for fractional order systems that may or may not require factorizations; see, e.g., [197] for a proportional-integral-derivative (PID) controller design [188] for approaches to implementation and synthesis, and [164] for several other analysis and controller design techniques and further references.

# Chapter 5

# $\mathcal{H}_\infty$ Control Problems

In this chapter we discuss $\mathcal{H}_\infty$ control problems associated with the feedback system shown in Figure 5.1, where the uncertain plant is represented by $P_\Delta$. We will define various robust control problems based on different uncertainty descriptions.

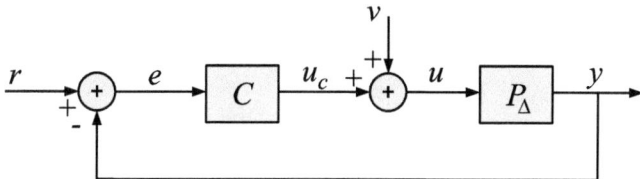

**Figure 5.1.** *Feedback system with controller $C$ and uncertain plant $P_\Delta$.*

## 5.1 ▪ Robust stability conditions

We define the *nominal feedback system* as the closed-loop system corresponding to the case where $P_\Delta = P$, with a given fixed *nominal plant* $P$ (no uncertainty). We assume that the nominal plant has finitely many unstable poles. To simplify the exposition, we define $n_u(P)$ as the number of poles of $P$ in $\overline{\mathbb{C}}_+$.

### 5.1.1 ▪ Additive and multiplicative uncertainty

One of the most common representations of an uncertain plant $P_\Delta$ is via additive uncertainty on the nominal plant:

$$\mathscr{P}_a = \{P_\Delta = P + \Delta_a \; : \; n_u(P_\Delta) = n_u(P), \; |\Delta_a(j\omega)| < |W_a(j\omega)| \, \forall \, \omega\}, \quad (5.1)$$

where $P(s)$ is the nominal plant and $W_a(s)$ is the uncertainty bound. In (5.1), $\mathscr{P}_a$ represents the set of all possible plants. A slightly modified representation of the set of all possible plants is by the multiplicative uncertainty

$$\mathscr{P}_m = \{P_\Delta = P(1 + \Delta_m) \; : \; n_u(P_\Delta) = n_u(P), \; |\Delta_m(j\omega)| < |W_m(j\omega)| \, \forall \, \omega\}, \quad (5.2)$$

where $W_m(s)$ is the multiplicative uncertainty bound. Typically, the following holds:

$$|W_a(j\omega)| = |W_m(j\omega)N_o(j\omega)| \quad \forall\, \omega, \tag{5.3}$$

where $N_o(s)$ is the outer part of the nominal plant (4.25). Hence, when the plant is given in the form (4.25), if the weights satisfy

$$W_a(s) = W_m(s)N_o(s),$$

with $W_a$ and $W_m$ being outer functions, we have $\mathscr{P}_m = \mathscr{P}_a$.

Let $C$ be a controller in $\mathscr{C}(P)$, i.e., the feedback system formed by the controller $C$ and the nominal plant $P$ is stable; then, $C \in \mathscr{C}(P_\Delta)$ for all $P_\Delta \in \mathscr{P}_a$ (respectively, $P_\Delta \in \mathscr{P}_m$), i.e., the feedback system is *robustly stable* if and only if

$$\|W_a CS\|_\infty \le 1 \quad (\text{respectively, } \|W_m T\|_\infty \le 1); \tag{5.4}$$

see, e.g., [51, 86, 87, 191] for the proof and discussions on how to construct the uncertainty bound $W_a$ (or $W_m$) from various modeling techniques.

**Example 5.1 (robustness analysis for the Smith predictor–based controller).** We have seen that for a nominal plant in the form $P(s) = \frac{e^{-hs}}{s-a}G_o(s)$, where $h > 0$, $a > 0$, and $G_o \in \mathcal{H}_\infty$, the controller $C$ given in (4.18) stabilizes the feedback system $(C, P)$, and the resulting complementary sensitivity function is in the form $T(s) = e^{-h(s-a)}T_o(s)$, where $T_o = P_oC_o(1 + P_oC_o)^{-1}$, with $C_o$ being a stabilizing controller for $P_o(s) = (s-a)^{-1}G_o(s)$. Now, one might ask whether this controller is a stabilizing controller for a plant $P_\Delta$ in the form

$$P_\Delta(s) = \frac{e^{-\tilde{h}s}}{s - \tilde{a}}G_o(s)(1 + \Delta_G(s)), \quad \tilde{h} > 0, \ \tilde{a} > 0, \ \Delta_G \in \mathcal{H}_\infty. \tag{5.5}$$

Since we have $n_u(P) = n_u(P_\Delta) = 1$, as long as $\tilde{a} > 0$, the controller given in (4.18) stabilizes all plants in the form (5.5) if it satisfies the robust stability inequality $\|W_m T\|_\infty \le 1$, where $W_m(s)$ is an upper bound for the multiplicative uncertainty

$$\left| 1 - e^{-j\Delta_h\omega}\left(1 - \frac{\Delta_a}{j\omega - \tilde{a}}\right)(1 + \Delta_G(j\omega)) \right| < |W_m(j\omega)| \quad \forall\, \omega,$$

where $\Delta_h = (\tilde{h} - h)$ and $\Delta_a = (\tilde{a} - a)$. Therefore, $C_o$ must be designed for $P_o$ in such a way that $\|W_m T_o\|_\infty \le e^{-ha}$.

In order to illustrate the key points in this discussion, let us consider the special case where $G_o(s) = 1$, $\Delta_G = 0$, $a = 1$, $\tilde{a} \in [0.85, 1.15]$, $h = 0.25$, $\tilde{h} \in [0.23, 0.27]$. For samples of $\tilde{a}$ and $\tilde{h}$ taken in the above intervals, the multiplicative uncertainty magnitudes are plotted, and an upper bound is determined as shown in Figure 5.2:

$$W_m(s) = 0.15 + \frac{2(s+1)}{(s+50)}.$$

So, $C_o$ should be designed to stabilize $P_o(s) = (s-1)^{-1}$ and satisfy the inequality $\|W_m C_o P_o(1 + C_o P_o)^{-1}\|_\infty \le e^{-0.25}$. This is a finite dimensional $\mathcal{H}_\infty$ control problem, which can be solved in an optimal way using well-established theory and associated MATLAB commands such as `mixsyn` and `hinfsyn`. Here, to complete the

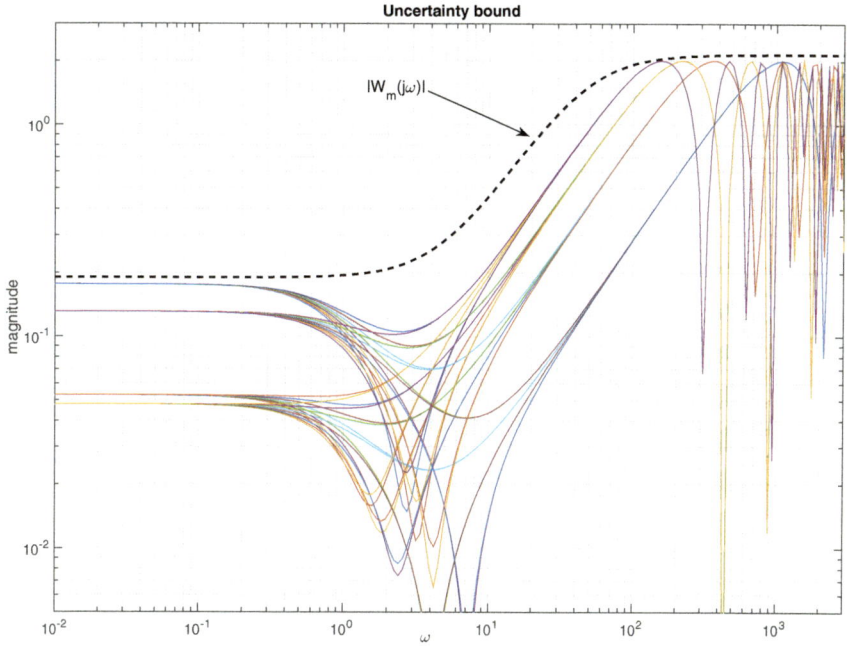

**Figure 5.2.** *Uncertainty bound* $|W_m(j\omega)|$.

design we can assume a special PI controller structure for $C_o(s) = K_p + \frac{K_i}{s}$ and investigate allowable parameters $K_p$ and $K_i$. It is clear that for all values of $K_p > 1$ and $K_i > 0$ the system $(C_o, P_o)$ is stable. For each fixed $K_p$ and $K_i$ we can now define the robustness margin (RM) as

$$\mathrm{RM} = \frac{e^{-0.25}}{\|W_m C_o P_o (1 + C_o P_o)^{-1}\|_\infty},$$

and if $\mathrm{RM} \geq 1$, then $C$ given by (4.18) robustly stabilizes all plants in the form (5.5) for $\Delta_G = 0$ and $\tilde{a} \in [0.85, 1.15]$, $\tilde{h} \in [0.23, 0.27]$. The RM is shown in Figure 5.3.

For each fixed $K_p > 0$ there is an optimal value of $K_i$ which maximizes the RM. Conversely, for each fixed $K_i \in (0, 160)$ there is an allowable interval for $K_p$ which makes $\mathrm{RM} \geq 1$, whose size shrinks as $K_i$ increases. For example, when $K_i = 15$, $\mathrm{RM} = 2.284$ is obtained by choosing $K_p = 7.45$, and if we define a minimum RM, say we want $\mathrm{RM} \geq 1.5$, then for $K_i = 15$ the allowable interval of $K_p$ values is $(3.57, 14.89)$. ∎

## 5.1.2 ▪ Delay margin optimization

One of the interesting problems in robust control is the *delay margin optimization* [158]. The goal is to find an optimal controller $C_o$ for a given finite dimensional plant $P_o$, such that the feedback system $(C_o, P)$ is stable for all $P(s) = P_o(s)e^{-\tau s}$, where $\tau$ is uncertain in $[0, h)$, with the maximal delay $h > 0$.

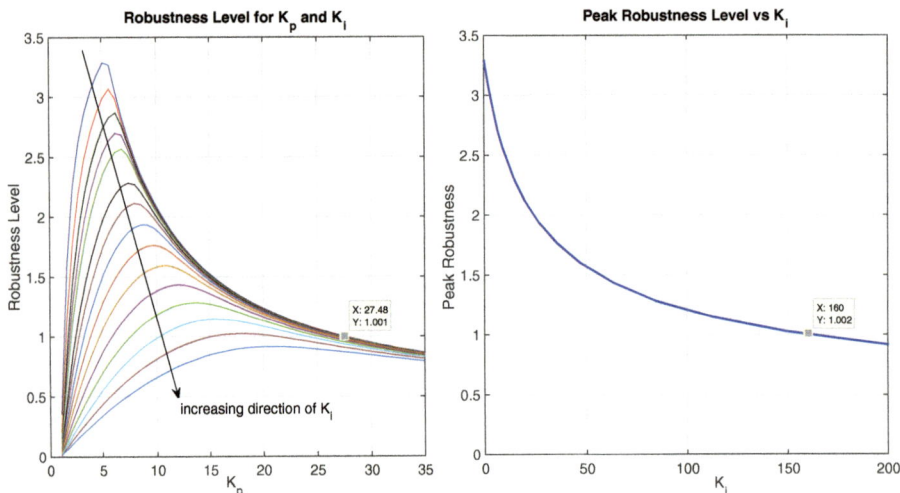

**Figure 5.3.** *RM with respect to $K_p$ and $K_i$.*

For a given finite dimensional controller $C_o$, let us examine the stability conditions for the feedback system $(C_o, P)$. The characteristic equation is

$$1 + P(s)C_o(s) = 1 + e^{-\tau s}P_o(s)C_o(s) = 1 + P_o(s)C_o(s) + \left(e^{-\tau s} - 1\right)P_o(s)C_o(s)$$
$$= \left(1 + P_o(s)C_o(s)\right)\left(1 + \Delta_\tau(s)T_o(s)\right),$$

where $T_o = P_o C_o (1 + P_o C_o)^{-1}$ and

$$\Delta_\tau(s) = e^{-\tau s} - 1, \quad \tau \in [0, h).$$

Therefore, $(C_o, P)$ is stable for all $\tau \in [0, h)$ if $(C_o, P_o)$ is stable and

$$\|W_h T_o\|_\infty < 1,$$

where $W_h \in \mathcal{H}_\infty$ is a tight bound on $|\Delta_\tau(j\omega)|$ for all $\omega \in \mathbb{R}$ and $\tau \in [0, h)$. For a fixed $h > 0$ we have

$$|W_h(j\omega)| = \begin{cases} 2\sin(\omega h/2), & \omega \in [0, \pi/h), \\ 2, & \omega \geq \pi/h. \end{cases}$$

Note that $W_h(s)$ is an infinite dimensional outer transfer function defined by its magnitude. Typically, by introducing some conservatism, a finite dimensional uncertainty bound $\widetilde{W}_h(s)$ is used: $|\widetilde{W}_h(j\omega)| \geq |W_h(j\omega)|$ for all $\omega$ [211]. By using the MATLAB built-in outer function construction tool, for $h = 1$ we obtain

$$\widetilde{W}_h(s) = \frac{(2+\delta)(s+\epsilon)(s+2.5)}{s^2 + (2+\sqrt{3})s + 5 + 2\delta}, \tag{5.6}$$

with $\delta = 0.028$ and $\epsilon = 2 \times 10^{-6}$. See Figure 5.4 for the magnitudes of these bounds. This gives an error bound satisfying

$$|W_h(j\omega)| \leq |\widetilde{W}_h(j\omega)| \leq 1.04\,|W_h(j\omega)|, \quad \omega \geq 10^{-5},$$
$$|W_h(j\omega)| \leq |\widetilde{W}_h(j\omega)| \leq 1.04 \times 10^{-5}, \quad \omega < 10^{-5}.$$

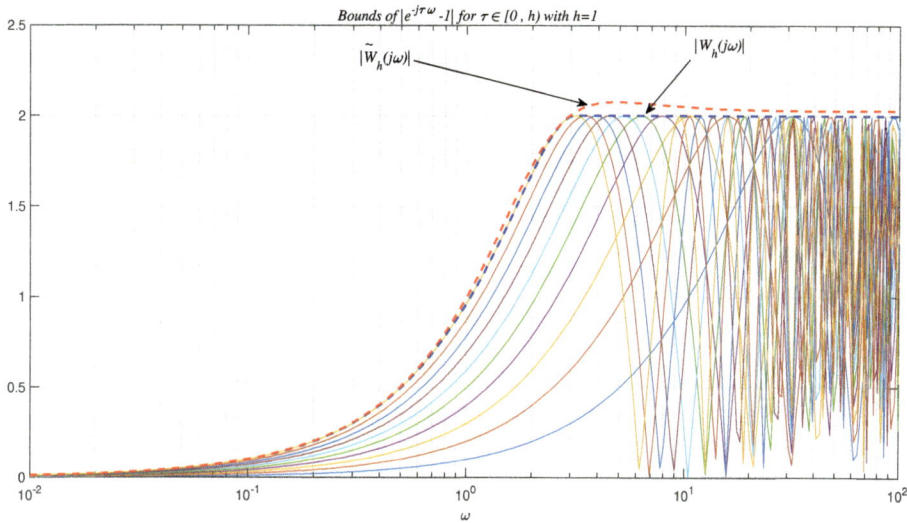

**Figure 5.4.** *Bounds of $|e^{-j\tau\omega} - 1|$ for $\tau \in [0, h)$ with $h = 1$.*

Let $p_i$ and $z_j$ be the poles and zeros of $P_o$ in $\mathbb{C}_+$. Then, for a given $h$, a feasible $C_o$ can be found if and only if one can find a transfer function $F \in \mathcal{H}_\infty$, such that $\|F\|_\infty < 1$ and $F(p_i) = \widetilde{W}_h(p_i)$ and $F(z_j) = 0$ for all $\mathbb{C}_+$ poles and zeros of $P_o$. Clearly, this problem can be solved using the optimal Nevanlinna–Pick interpolation. The corresponding controllers are computed using the optimal interpolation procedure of Section 2.4.1. More precisely, if $F_{opt}(s)$ is the optimal interpolant, then the optimal controller is

$$C_{opt}(s) = \frac{F_{opt}(s)}{P_o(s)\,(\,\widetilde{W}_h(s) - F_{opt}(s)\,)}.$$

Note that $\widetilde{W}_h$ and $F_{opt}$ are biproper, so if the plant is strictly proper, say with relative degree $n_o$, then the above controller is improper, which is not acceptable. In order to circumvent this problem, typically, we use an approximate controller $C_{app}(s) = \frac{1}{(\varepsilon s + 1)^{n_o}} C_{opt}(s)$, where $\varepsilon > 0$ is sufficiently small so that

$$\left\| \widetilde{W}_h \frac{P_o C_{app}}{1 + P_o C_{app}} \right\|_\infty < 1$$

for the largest delay value computed from the Nevanlinna–Pick interpolation.

Next, we illustrate the design procedure with a numerical example. Let us now consider the following family of plants, with poles in $\mathbb{C}_+$ as follows:

$$P_N(s) = G_o(s) \prod_{i=1}^{N} \frac{s + p_i}{s - p_i}, \quad \text{with } N = 3, 5, 7,$$

where $G_o, G_o^{-1} \in \mathcal{H}_\infty$, and

$$p_1 = 0.25\,\kappa,$$

$$p_{2,3} = (0.1 \pm j\,0.1)\,\kappa,$$

$$p_{4,5} = (0.2 \pm j\, 0.4)\, \kappa,$$

$$p_{6,7} = (0.05 \pm j\, 0.02)\, \kappa,$$

where $\kappa > 0$ is a scaling factor on the magnitude of the unstable poles.

We investigate the largest allowable $\kappa > 0$ such that the delay margin is greater than or equal to $h = 1$. From the Nevanlinna–Pick interpolation, it is computed that the largest allowable $\kappa$ which makes $\gamma_{opt} < 1$ is equal to 2.41, 1.13, and 1.03 for $P_3$, $P_5$, and $P_7$, respectively.

In this problem, $\kappa$ can be seen as a frequency scaling factor: in other words, for an arbitrary $h > 0$ and $h \neq 1$, we may define $\hat{s} = hs$, i.e., $s = \hat{s}/h$, and then $P_N$ has poles in the $\hat{s}$-plane at $p_i h$. So, with $\kappa = h$ we can solve the original delay margin optimization problem. Thus, the weight (5.6) can be used universally for any delay value $h$ by scaling the poles. In particular, when $P_o$ contains only one pole in $\mathbb{C}_+$, a lower bound for the largest allowable $(ph)$ can be calculated using the coefficients of $\widetilde{W}_h$ as the largest $(ph)$ which satisfies

$$f(ph) := \frac{(2+\delta)((ph)+\epsilon)((ph)+2.5)}{(ph)^2 + (2+\sqrt{3})(ph) + 5 + 2\delta} < 1.$$

The above inequality is satisfied for all $ph \in [0, 1.66)$; i.e., a lower bound of the maximum allowable $(ph)$ is found to be 1.66; see Figure 5.5. This result is consistent with [158], where the largest achievable delay margin is found as $(ph) = 2$. With a third or fifth order $\widetilde{W}_h$, it is possible to obtain (tighter) lower bounds closer to 2.0 [211, 283].

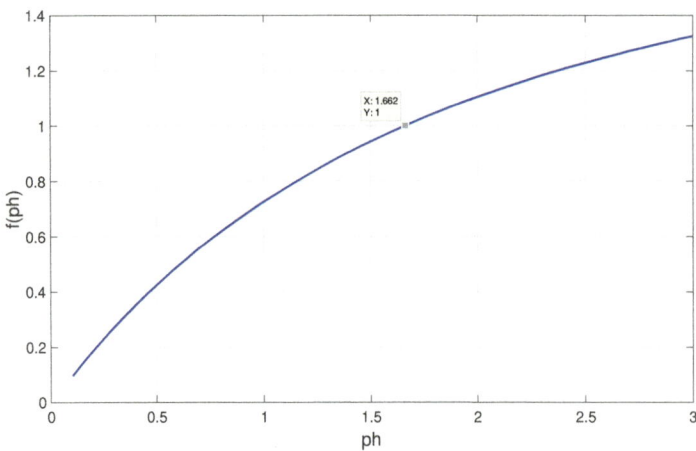

**Figure 5.5.** *Function $f(ph)$ versus ph.*

### 5.1.3 ▪ Robust stability condition for systems with infinitely many poles in $\mathbb{C}_+$

Recall that the set of all possible plants under additive (5.1) or multiplicative (5.2) uncertainty are defined with the assumption that the nominal plant $P$ and the uncertain plant $P_\Delta$ have the same number of poles in $\mathbb{C}_+$. The reason for this assumption is that the robust stability conditions given in (5.4) are based on the Nyquist stability criterion; see, e.g., [51]. However, when the nominal plant and all uncertain plants have infinitely many unstable poles, the Nyquist argument fails, because the number of encirclements of $-1$ by the graph of $PC$ is infinite (in the counterclockwise direction) when the nominal feedback system is stable.

Let us now define the following set of all possible plants:

$$\mathscr{P} = \{P_\Delta = P(1 + W_m \Delta_o) \ : \ \Delta_o \in \mathcal{H}_\infty \ \|\Delta_o\|_\infty \leq 1\}, \tag{5.7}$$

where $W_m \in \mathcal{H}_\infty$ is an outer transfer function characterizing the frequency distribution of the uncertainty, and $P$ is the nominal plant model which is allowed to have infinitely many poles in $\mathbb{C}_+$. A word of caution is in order here: in $\mathscr{P}$ the perturbations must be stable; in particular, this implies that the unstable pole locations in the nominal plant and the uncertain plant must be the same. However, in (5.1) or (5.2), an uncertain plant can have unstable poles that are different from those of the nominal plant.

**Theorem 5.2.** *Let $C$ be a controller such that $(C, P)$ is stable. Then, $(C, P_\Delta)$ is stable for all $P_\Delta \in \mathscr{P}$ if and only if*

$$\|W_m T\|_\infty < 1,$$

*where $T = PC(1 + PC)^{-1}$.*

**Proof.** *Sufficiency*: Assume that $\|W_m T\|_\infty < 1$. First we examine $S_\Delta = (1 + P_\Delta C)^{-1}$:

$$S_\Delta = (1 + P_\Delta C)^{-1} = (1 + PC + W_m PC\Delta_o)^{-1} = S (1 + W_m T \Delta_o)^{-1},$$

where $S = (1 + PC)^{-1} \in \mathcal{H}_\infty$, because $(C, P)$ is stable. Since by assumption $W_m$, $T$, and $\Delta_o$ are in $\mathcal{H}_\infty$, and $\|W_m T \Delta_o\|_\infty \leq \|W_m T\|_\infty \|\Delta_o\|_\infty < 1$, by the small gain theorem we have that $S_\Delta \in \mathcal{H}_\infty$. Next, note that $CS_\Delta = CS (1 + W_m T \Delta_o)^{-1}$, and since by assumption $(C, P)$ is stable, we have $CS \in \mathcal{H}_\infty$, and hence as before, $CS_\Delta \in \mathcal{H}_\infty$. Finally, we check $P_\Delta S_\Delta = PS(1 + W_m \Delta_o)(1 + W_m T \Delta_o)^{-1} \in \mathcal{H}_\infty$ as follows: since $(C, P)$ is stable, we have $PS \in \mathcal{H}_\infty$, and by assumption $W_m$ and $\Delta_o$ are in $\mathcal{H}_\infty$, so again, by the small gain theorem we have $P_\Delta S_\Delta \in \mathcal{H}_\infty$.

*Necessity*: Assume that $\|W_m T\|_\infty \geq 1$. In this case let us assume that there exists $\omega_o \in \mathbb{R}$ such that $|W_m(j\omega_o)T(j\omega_o)| = 1$; that is, if the norm is 1, then it is attained at $\omega_o$, and if the norm is strictly greater than 1, then the magnitude of the function $W_m T$ passes through 1 at $\omega_o$. If $\omega_o = 0$, then $\Delta_o(s) = \Delta_o(0) = -(W_m(0)T(0))^{-1}$ is an admissible destabilizing perturbation in the sense that $S_\Delta$ has a pole at $s = 0$. If $\omega_o > 0$, then we can still construct a stable transfer function $\Delta_o$ satisfying $\|\Delta_o\|_\infty = 1$ and

$$|\Delta_o(j\omega_o)| = 1, \quad \angle\Delta_o(j\omega_o) + \angle W_m(j\omega_o)T(j\omega_o) = -\pi.$$

That means once again that $W_m(j\omega_o)T(j\omega_o)\Delta(j\omega_o) = -1$, and hence in this case $S_\Delta$ has a pole at $s = j\omega_o$. To construct such a destabilizing $\Delta_o(s)$, first let the phase of $W_m T$ at $j\omega_o$ be between $-\pi$ and $\pi$. Then the phase of $\Delta_o$ at $j\omega_o$ must be between $0$ and $-2\pi$. One particular destabilizing $\Delta_o \in \mathcal{H}_\infty$ is

$$\Delta_o(s) = \frac{(s/\omega_o)\, e^{-\theta s}}{1 + (s/\omega_o) + (s/\omega_o)^2}.$$

Note that $\|\Delta_o\|_\infty = 1$, the norm is attained at $s = j\omega_o$, and the phase of $\Delta_o(j\omega_o) = -\theta\omega_o$. By adjusting $\theta > 0$, one can obtain any desired phase between $0$ and $-2\pi$. The above arguments assume that the norm is attained at a finite frequency. They can be extended to the case when the norm is the limit of a sequence of values taken by $W_m(j\omega)T(j\omega)$; technical details are omitted. $\square$

### 5.1.4 ▪ Coprime factor uncertainty

Another type of uncertain plant description is by coprime factor perturbations. Let us assume that the nominal plant admits a factorization in the form $P = N/D$, where $N, D \in \mathcal{H}_\infty$ are strongly coprime. Perturbations in these coprime factors lead to

$$\mathscr{P}_{cf} = \left\{ P_\Delta = \frac{N + \Delta_N}{D + \Delta_D} \ : \ \Delta_D, \Delta_N \in \mathcal{H}_\infty, \text{ with } \| [\Delta_D \ \ \Delta_N] \|_\infty < \delta \right\}, \quad (5.8)$$

where $\delta > 0$ represents the size of the uncertainty. Now consider a controller stabilizing the nominal plant, i.e., $C = (X + DQ)/(Y - NQ)$, with $X, Y, Q \in \mathcal{H}_\infty$ and $NX + DY = 1$. From the discussion of Section 4.1, we deduce that this controller is in $\mathscr{C}(P_\Delta)$ if and only if

$$U = 1 + \Delta_D(Y - NQ) + \Delta_N(X + DQ) \quad (5.9)$$

is unimodular in $\mathcal{H}_\infty$, i.e., $U^{-1} \in \mathcal{H}_\infty$. Since $\Delta_D$ and $\Delta_N$ are arbitrary in $\mathcal{H}_\infty$, subject to $\| [\Delta_D \ \ \Delta_N] \|_\infty < \delta$, we have that $U^{-1} \in \mathcal{H}_\infty$ if and only if (see [76, 81] and their references for more detailed discussions)

$$\left\| \begin{bmatrix} Y - NQ \\ X + DQ \end{bmatrix} \right\|_\infty \leq \delta^{-1}. \quad (5.10)$$

Note that $(Y - NQ) = D^{-1}S \in \mathcal{H}_\infty$ and $(X + DQ) = D^{-1}CS \in \mathcal{H}_\infty$; hence, the condition (5.10) can be rewritten as

$$\left\| D^{-1} \begin{bmatrix} S \\ CS \end{bmatrix} \right\|_\infty \leq \delta^{-1}. \quad (5.11)$$

Clearly, when $P = N/D$ with $D = M_d$, which is inner, (5.11) is equivalent to having

$$\left\| \begin{bmatrix} S \\ CS \end{bmatrix} \right\|_\infty \leq \delta^{-1}. \quad (5.12)$$

**Uncertainty in the controller**

In some practical situations, precise implementation of the controller is not possible. Let $P = N/D$ be given such that $N, D, X, Y \in \mathcal{H}_\infty$ satisfy $NX + DY = 1$. Recall that if $\alpha_i$ is a pole of the plant in $\mathbb{C}_+$, then $X$ must be such that $X(\alpha_i) = 1/N(\alpha_i)$. These interpolation conditions impose some restrictions that must be satisfied exactly. So, numerical errors in these computations lead to uncertainty in the "nominal controller," which is in the form $C = (X + DQ)(Y - NQ)^{-1}$. Therefore, an imprecise controller implementation may take the form

$$C_\Delta = \frac{X + DQ + \Delta_X}{Y - NQ + \Delta_Y}, \quad (5.13)$$

where $\Delta_X, \Delta_Y \in \mathcal{H}_\infty$. The feedback system formed by $C_\Delta$ and $P$ is stable if and only if

$$U = N(X + DQ + \Delta_X) + D(Y - NQ + \Delta_Y) = 1 + \begin{bmatrix} \Delta_X & \Delta_Y \end{bmatrix} \begin{bmatrix} N \\ D \end{bmatrix}$$

is unimodular. Now consider the spectral factorization

$$G_o(-s)G_o(s) = N(-s)N(s) + D(-s)D(-s), \quad (5.14)$$

where $G_o, G_o^{-1} \in \mathcal{H}_\infty$. Then, $U$ defined above is unimodular if

$$\left\| \begin{bmatrix} \Delta_X & \Delta_Y \end{bmatrix} G_o \begin{bmatrix} NG_o^{-1} \\ DG_o^{-1} \end{bmatrix} \right\|_\infty < 1,$$

which is holds if

$$\left\| \begin{bmatrix} \Delta_X & \Delta_Y \end{bmatrix} G_o \right\|_\infty < 1. \tag{5.15}$$

The inequality (5.15) illustrates how much imprecision can be tolerated in the implementation of $X$ and $Y$ as far as feedback stability is concerned. See [78, 147] for further discussions on this topic, and a stability condition under combined plant and controller uncertainty.

**Normalized coprime factor uncertainty**

A particular coprime factorization is $P = N_1/D_1$, where $N_1, D_1 \in \mathcal{H}_\infty$ satisfy

$$1 = N_1(-s)N_1(s) + D_1(-s)D_1(s). \tag{5.16}$$

Such a factorization is called a *normalized* coprime factorization. From any given coprime factorization $P = N/D$ we can obtain a normalized coprime factorization $N_1 = NG_o^{-1}$, $D_1 = DG_o^{-1}$, where $G_o, G_o^{-1} \in \mathcal{H}_\infty$ is determined from the spectral factorization (5.14). Now, considering $P = N_1/D_1$ and its perturbation $P_\Delta = (N_1 + \Delta_{N_1})/(D_1 + \Delta_{D_1})$, with $\|[\Delta_{D_1} \ \Delta_{N_1}]\|_\infty < \delta_1 < 1$, the condition (5.10) can be rewritten as

$$\|(N_1^* Y_1 - D_1^* X_1) - Q_1\|_\infty \leq \sqrt{\delta_1^{-2} - 1}, \tag{5.17}$$

where $X_1, Y_1 \in \mathcal{H}_\infty$ satisfy $N_1 X_1 + D_1 Y_1 = 1$ and $C = (X_1 + D_1 Q_1)/(Y_1 - N_1 Q_1)$ is in $\mathscr{C}(P)$, with $Q_1 \in \mathcal{H}_\infty$. In (5.17) we used the notation $N_1^* := N_1(-s)$, and similarly $D_1^* := D_1(-s)$. The problem of minimizing the left-hand side of (5.17) over $Q_1 \in \mathcal{H}_\infty$ is called *robustness optimization in the gap metric* [76, 77, 239].

# 5.2 ▪ Mixed sensitivity minimization problem

## 5.2.1 ▪ Stability margin: Combined gain and phase perturbations

Consider a feedback system with a given nominal plant $P$ and a stabilizing controller $C \in \mathscr{C}(P)$. Define $G = PC$, and assume that $n_u(P) + n_u(C) = n_u(G)$ is finite. Then, the Nyquist graph of $G(j\omega)$ encircles $-1$ in the counterclockwise direction $n_u(G)$ times. An important stability margin can be defined as the distance between $G(j\omega)$ and $-1$:

$$\beta := \inf_{\omega \in \mathbb{R}} |G(j\omega) - (-1)|. \tag{5.18}$$

In order to have good stability robustness to combined gain and phase perturbations in the form $ke^{-j\phi}G(j\omega)$, where $k > 1$ and $\phi > 0$, we should have a large $\beta$. From (5.18), it is clear that

$$\beta^{-1} = \|S\|_\infty, \quad \text{where} \quad S = (1 + PC)^{-1}. \tag{5.19}$$

Therefore, an important motivation to *minimize the peak sensitivity* is to have a large stability margin in the presence of combined gain and phase perturbations.

## 5.2.2 ▪ Reference tracking

Now consider the reference tracking problem associated with the feedback system shown in Figure 5.1, where $P_\Delta = P$. In this problem we try to minimize the worst error energy $\|e\|_2$ over the set of all possible reference inputs

$$\mathscr{R} = \{R(s) = W_r(s)R_o(s) \ : \ \|r_o\|_2 \le 1\}, \qquad (5.20)$$

where $W_r(s)$ is an outer function, which can be seen as a reference generator. Typically, if we are interested in tracking "step-like" inputs, then we choose

$$W_r(s) = \frac{1}{s + \varepsilon}, \quad \text{where } \varepsilon \searrow 0. \qquad (5.21)$$

Similarly, if the reference of interest is a sinusoidal signal of known frequency, say $\omega_o$, then we choose

$$W_r(s) = \frac{1}{s^2 + 2\varepsilon\omega_o s + \omega_o^2}, \quad \text{where } \varepsilon \searrow 0. \qquad (5.22)$$

In summary, the poles of $W_r(s)$ include the poles of $R(s)$.

The $\mathcal{H}_\infty$-based tracking problem can now be stated in precise terms as follows: given a plant $P$, determine a controller $C \in \mathscr{C}(P)$ such that the cost

$$\gamma(C) := \sup_{r \in \mathscr{R}} \|e\|_2 \qquad (5.23)$$

is minimized. Since the transfer function from $r_o$ to $e$ is $W_r S$, we have that the cost defined in (5.23) is $\gamma(C) = \|W_r(1 + PC)^{-1}\|_\infty$. Hence, the $\mathcal{H}_\infty$ sensitivity minimization problem is to find

$$\gamma_{\text{opt}} := \inf_{C \in \mathscr{C}(P)} \|W_r(1 + PC)^{-1}\|_\infty \qquad (5.24)$$

and the corresponding optimal controller achieving the performance level $\gamma_o$.

## 5.2.3 ▪ Disturbance attenuation

Consider the feedback system Figure 5.1, with $P_\Delta = P$. Similar to (5.20), we can define the class of disturbance signals of interest,

$$\mathscr{V} = \{V(s) = W_v(s)V_o(s) \ : \ \|v_o\|_2 \le 1\}, \qquad (5.25)$$

where $W_v(s)$ is an outer function, which is the disturbance generating filter. For a controller $C \in \mathscr{C}(P)$ define

$$\gamma(C) := \sup_{v \in \mathscr{V}} \|y\|_2 \qquad (5.26)$$

as the worst disturbance impact on the output. Recall that the transfer function from disturbance, $v$, to output, $y$, is $PS$. Therefore, to minimize the worst disturbance impact we try to find

$$\gamma_{\text{opt}} := \inf_{C \in \mathscr{C}(P)} \|W_v P(1 + PC)^{-1}\|_\infty \qquad (5.27)$$

and the corresponding optimal controller achieving the performance level $\gamma_{\text{opt}}$.

## 5.2.4 ▪ The mixed sensitivity minimization problem

The problems discussed above can be put into a single framework: find

$$\gamma_{\text{opt}} := \inf_{C \in \mathscr{C}(P)} \left\| \begin{bmatrix} W_1(1 + PC)^{-1} \\ W_2 PC(1 + PC)^{-1} \end{bmatrix} \right\|_\infty \tag{5.28}$$

and the corresponding optimal controller $C_{\text{opt}} \in \mathscr{C}(P)$ achieving the cost $\gamma_{\text{opt}}$. The $\mathcal{H}_\infty$ control problem defined by (5.28) is called the *mixed sensitivity minimization*. In particular, if $P$ is in the form (4.25), with $N_o$ being outer, then Table 5.1 represents the mapping from earlier defined problems to (5.28).

**Table 5.1.** *Weight selection in the mixed sensitivity minimization problem.*

| $\mathcal{H}_\infty$ control problem | $W_1(s)$ | $W_2(s)$ |
|:---:|:---:|:---:|
| (5.4) | $W_1(s) = \varepsilon \searrow 0$ | $W_2(s) = W_m(s)$ |
| (5.12) | $W_1(s) = 1$ | $W_2(s) = 1/N_o(s)$ |
| (5.24) | $W_1(s) = W_r(s)$ | $W_2(s) = \varepsilon \searrow 0$ |
| (5.27) | $W_1(s) = W_v(s)N_o(s)$ | $W_2(s) = \varepsilon \searrow 0$ |

The mixed sensitivity minimization problem is also useful in obtaining a suboptimal solution for the *robust performance problem*, which can be defined as follows: find a controller $C$ robustly stabilizing the feedback system such that

$$\|W_r(1 + P_\Delta C)^{-1}\|_\infty \leq \gamma \tag{5.29}$$

for all $P_\Delta \in \mathscr{P}_m$, for the smallest possible $\gamma > 0$.

It can be shown that (see [51]) if a controller $C \in \mathscr{C}(P)$ achieves

$$\left\| \begin{bmatrix} W_1(1 + PC)^{-1} \\ W_2 PC(1 + PC)^{-1} \end{bmatrix} \right\|_\infty \leq \frac{1}{\sqrt{2}}, \tag{5.30}$$

with $W_1(s) = \gamma^{-1}W_r(s)$ and $W_2(s) = W_m(s)$, then this controller satisfies (5.29).

**Exercise 5.3.**
1. Show that for a given controller $C$ stabilizing the feedback loop $(C, P)$ the robust performance condition (5.29) is equivalent to having

$$|W_1(j\omega)S(j\omega)| + |W_2(j\omega)T(j\omega)| \leq 1 \quad \forall \omega, \tag{5.31}$$

with $W_1(s) = \gamma^{-1}W_r(s)$ and $W_2(s) = W_m(s)$, where, as usual, $S = (1 + PC)^{-1}$ and $T = 1 - S$.
2. For a given $\gamma > 0$, prove that if the inequality (5.30) holds true, then the robust performance condition (5.31) is satisfied.
*Hint: a good reference on 1 and 2 is* [51]. □

In the Robust Control Toolbox of MATLAB, the command mixsyn solves the following problem for finite dimensional plants: find $C \in \mathscr{C}(P)$ minimizing

$$\widehat{\gamma}(C) = \left\| \begin{bmatrix} \widehat{W}_1(1 + PC)^{-1} \\ \widehat{W}_3 P(1 + PC)^{-1} \\ \widehat{W}_2 PC(1 + PC)^{-1} \end{bmatrix} \right\|_\infty . \tag{5.32}$$

For plants in the form (4.25), by defining $W_2 = \widehat{W}_2$ and $W_1 = \widehat{G}_1$, where $\widehat{G}_1, \widehat{G}_1^{-1} \in \mathcal{H}_\infty$ is determined from the spectral factorization

$$\widehat{W}_1(-s)\widehat{W}_1(s) + \widehat{W}_3(-s)N_o(-s)N_o(s)\widehat{W}_3(s) = \widehat{G}_1(-s)\widehat{G}_1(s),$$

we can show that (5.32) is equivalent to (5.28).

**Exercise 5.4.**
Prove the above statement by using the following arguments:

(i) *Fact*: for any $n \times m$ matrix $F$ whose elements are in $\mathcal{H}_\infty$ and any $n \times n$ unitary matrix $M$ whose entries are in $\mathcal{L}_\infty$ with the property that $M(j\omega)^* M(j\omega) = M(j\omega)M(j\omega)^* = I$, we have $\|F\|_\infty = \|MF\|_\infty$. Here the notation $M(j\omega)^*$ denotes the *transpose of* $M(-j\omega)$.

(ii) For plants in the form (4.25), we have

$$\widehat{\gamma}(C) = \left\| \begin{bmatrix} \widehat{W}_1 S \\ \widehat{W}_3 N_o \frac{M_n}{M_d} S \\ \widehat{W}_2 T \end{bmatrix} \right\|_\infty = \left\| \begin{bmatrix} \widehat{W}_1 \widehat{G}_1^{-1}\widehat{G}_1 S \\ \widehat{W}_3 N_o \widehat{G}_1^{-1}\widehat{G}_1 S \\ \widehat{W}_2 T \end{bmatrix} \right\|_\infty$$

$$= \left\| \begin{bmatrix} \widehat{W}_1^* \widehat{G}_1^{-*} & \widehat{W}_3^* N_o^* \widehat{G}_1^{-*} & 0 \\ -\widehat{W}_3 N_o \widehat{G}_1^{-1} & \widehat{W}_1 \widehat{G}_1^{-1} & 0 \\ 0 & 0 & 1 \end{bmatrix} \begin{bmatrix} \widehat{W}_1 \widehat{G}_1^{-1}\widehat{G}_1 S \\ \widehat{W}_3 N_o \widehat{G}_1^{-1}\widehat{G}_1 S \\ \widehat{W}_2 PC(1 + PC)^{-1} \end{bmatrix} \right\|_\infty,$$

where $S = (1 + PC)^{-1}$, $T = 1 - S$. In the first identity we used the property given in (i) with the diagonal unitary matrix

$$M(j\omega) = \begin{bmatrix} 1 & 0 & 0 \\ 0 & M_d(j\omega)/M_n(j\omega) & 0 \\ 0 & 0 & 1 \end{bmatrix}.$$

(iii) Check that $\widehat{\gamma}(C)$ obtained in step (ii) is the same as

$$\left\| \begin{bmatrix} W_1 S \\ W_2 T \end{bmatrix} \right\|_\infty,$$

with $W_1 = \widehat{G}_1$ and $W_2 = \widehat{W}_2$.

Recall that the controller parameterization (4.11) leads to $S = D(Y - NQ)$ and $T = N(X + DQ)$; hence, by using the identity $NX + DY = 1$, we see that the problem (5.28) can be transformed to

$$\gamma_{\text{opt}} = \inf_{Q \in \mathcal{H}_\infty} \left\| \begin{bmatrix} W_1 \\ 0 \end{bmatrix} - \begin{bmatrix} W_1 \\ -W_2 \end{bmatrix} N(X + DQ) \right\|_\infty, \qquad (5.33)$$

which is equivalent to

$$\gamma_{\text{opt}} = \inf_{Q \in \mathcal{H}_\infty} \left\| \begin{bmatrix} G^{-*}W_1^* W_1 - GN(X + DQ) \\ G^{-1}W_1 W_2 \end{bmatrix} \right\|_\infty, \qquad (5.34)$$

where $G^{-*} \equiv G^{-1}(-s)$ (similarly, $W_1^* \equiv W_1(-s)$) and the outer function $G$ is obtained from the spectral factorization

$$W_1(-s)W_1(s) + W_2(s)W_2(-s) = G(-s)G(s). \tag{5.35}$$

In the literature, (5.34) is called the *two-block problem*. It is interesting to note that, by (5.34), an immediate lower bound is determined for $\gamma_{\text{opt}}$:

$$\gamma_{\min} := \|G^{-1}W_1W_2\|_\infty \leq \gamma_{\text{opt}}, \tag{5.36}$$

which depends on the weights only; i.e., (5.36) is independent of the plant.

Similarly, an upper bound, $\gamma_{\max}$ can be determined by putting $Q = 0$ in (5.33):

$$\gamma_{\text{opt}} \leq \left\| \begin{bmatrix} W_1 DY \\ W_2 NX \end{bmatrix} \right\|_\infty =: \gamma_{\max}. \tag{5.37}$$

**Remark 5.5.** In most practical problems, the nominal plant is strictly proper, with a relative degree, say $n_{rel}$. Then, the multiplicative uncertainty bound $W_2$ is typically an improper function whose relative degree is $(-n_{rel})$. That makes $(W_2 N_o)$ a biproper outer function, and the optimal $Q$, the solution of (5.34), is proper. This leads to a proper optimal controller. In order to get a strictly proper optimal controller, the function $(W_2 N_o)^{-1}$ should be strictly proper, which means that the relative degree of $W_2^{-1}$ should be greater than $n_{rel}$. It should also be noted that $G$ defined by the spectral factorization (5.35) is improper, with a relative degree the same as that of $W_2$, provided that $W_1$ is biproper or strictly proper (which is the case in most practical situations).

## 5.3 ▪ $\mathcal{H}_\infty$-based estimation

Consider the system shown in Figure 5.6, where $z(t)$ represents the signal to be estimated from the observation $y(t)$; $Q \in \mathcal{H}_\infty$ is the estimator to be designed; $P_w \in \mathcal{H}_\infty$ is the underlying plant whose input is the process noise $w(t)$; $v(t)$ is the measurement noise, whose frequency content and strength are captured by the filter $W_v(s)$; and the signal $err(t)$ is the estimation error. We will assume that in this problem the sensor $H(s)$ gives a delayed version of $z$, i.e., $H(s) = e^{-hs}$, where $h$ is the measurement delay.

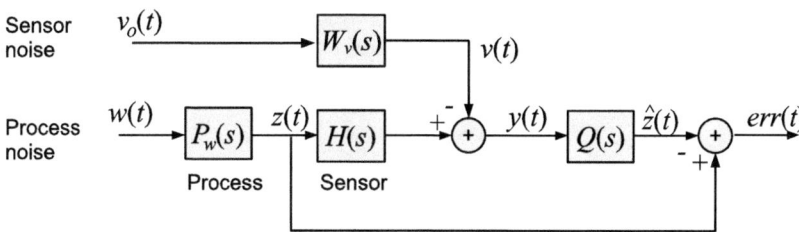

**Figure 5.6.** *Estimation problem.*

The goal in the estimation problem is to minimize the impact of process and measurement noise, $w(t)$ and $v(t)$, on the estimation error $err(t)$. More precisely, given $P_w$, $W_v$, and the time delay $h > 0$, our objective is to find $Q \in \mathcal{H}_\infty$ so that the $\mathcal{H}_\infty$ cost

$$\gamma(Q) := \sup_{\left\| \begin{bmatrix} w \\ v_o \end{bmatrix} \right\|_2 \leq 1} \|err\|_2 \tag{5.38}$$

is minimized. It is a simple exercise to show that

$$err = \left[\ (1 - Q(s)e^{-hs})P_w(s) \qquad - Q(s)W_v(s)\ \right] \left[\begin{array}{c} w \\ v_o \end{array}\right].$$

Therefore,

$$\gamma(Q) = \left\|\left[\begin{array}{c} P_w(s) \\ 0 \end{array}\right] - \left[\begin{array}{c} e^{-hs}P_w(s) \\ -W_v(s) \end{array}\right] Q(s)\right\|_\infty. \tag{5.39}$$

We will assume that $P_w$ and $W_v$ are outer; this is without loss of generality since their inner factors do not play a role in the computation of $\gamma(Q)$ by (5.39). Therefore, if we define $P(s) = e^{-hs}$, $W_1(s) = P_w(s)$, and $W_2(s) = W_v(s)$, we see that (5.39) is equivalent to (5.33): we use the observation that in this case we have $N(s) = M_n(s) = e^{-hs}$ and $D(s) = 1$, $X(s) = 0$; the equivalence is obtained by multiplying the $\{2, 1\}$ block of (5.39) by $M_n$, which is norm preserving. In this case the controller parameterization with $D = 1$, $X = 0$ gives $C = Q/(1 - M_nQ)$, and we solve for $Q$ as

$$Q = \frac{C}{1 + M_nC},$$

where $C$ is the solution of the associated mixed sensitivity minimization (two-block) problem.

In conclusion, the estimation problem defined above can be solved by solving a two-block problem, which comes from the mixed sensitivity minimization, (5.28).

## 5.4 • Strong stabilization by a small gain approach

For a given plant $P$ the problem of finding a stable stabilizing controller $C$, i.e., having $C \in \mathscr{C}(P) \cap \mathcal{H}_\infty$, is called *strong stabilization*. There are many practical reasons why one wants to use a stable controller in the feedback loop; see, e.g., [193, 275] and their references. Stable $\mathcal{H}_\infty$ controller design for infinite dimensional systems has been studied recently in [96, 97, 250]. Here we give a method which uses the mixed sensitivity minimization problem [193].

It is well known that such a controller can be found if and only if $P$ satisfies the *parity interlacing property* (PIP); that is, between every pair of zeros of $P$ in the extended positive real axis, $\overline{\mathbb{R}}_+ \cup \{\infty\}$, the number of poles of $P$ (including multiplicities), is even. For example, consider the plant

$$P(s) = e^{-hs}\frac{(s^2 - 2\zeta\omega_o s + \omega_o^2)}{(s+1)^2(s - \rho\omega_o)}, \qquad \rho > 0,\ \omega_o > 0,\ h \geq 0. \tag{5.40}$$

Clearly, when $\rho > 2\zeta > 2$, the plant does not satisfy the PIP; hence, $\mathscr{C}(P) \cap \mathcal{H}_\infty = \emptyset$. On the other hand, when $0 < \zeta < 1$, the PIP holds, so it is possible to find a stable stabilizing controller.

Let us now consider a stable controller $C = Q \in \mathcal{H}_\infty$; then, the feedback system with a plant $P = N/D$ is stable if and only if

$$S = \frac{D}{D + NQ} \in \mathcal{H}_\infty \quad \text{and} \quad PS = \frac{N}{D + NQ} \in \mathcal{H}_\infty,$$

i.e., $(D + NQ)^{-1} \in \mathcal{H}_\infty$. Note that

$$(D + NQ)^{-1} = \left(1 - (1 - D - NQ)\right)^{-1}.$$

Then, by the small gain theorem, a sufficient condition for a controller $C = Q \in \mathcal{H}_\infty$ to stabilize the feedback system is

$$\|(1 - D) - NQ\|_\infty < 1. \tag{5.41}$$

In fact, for any given outer function $W \in \mathcal{H}_\infty$ such that $W^{-1} \in \mathcal{H}_\infty$ we could have written

$$(D + NQ)^{-1} = W^{-1}\big(1 - (1 - W^{-1}(D + NQ))\big)^{-1}.$$

Therefore, a stable controller $Q \in \mathcal{H}_\infty$ stabilizes the feedback system if it satisfies

$$\|(1 - W^{-1}D) - NW^{-1}Q\|_\infty < 1. \tag{5.42}$$

In other words, for a given $P = N/D$ a strongly stabilizing controller can be found if there exists an outer function $W, W^{-1} \in \mathcal{H}_\infty$ such that

$$\gamma_{\text{opt},1} := \inf_{Q_1 \in \mathcal{H}_\infty} \|(1 - W^{-1}D) - NQ_1\|_\infty < 1. \tag{5.43}$$

If the above one-block problem is solvable, then the optimal $Q_{\text{opt},1} \in \mathcal{H}_\infty$ solving (5.43) gives a strongly stabilizing controller $Q = W \, Q_{\text{opt},1}$.

In order to keep the controller magnitude bounded by a prespecified function, we may add a restriction in the form

$$|Q(j\omega)| < |W_c(j\omega)| \quad \forall \, \omega, \tag{5.44}$$

where $W_c$ is an outer function. Then, we can try to find a strongly stabilizing controller by checking whether the following inequality is satisfied:

$$\gamma_{\text{opt}} = \inf_{Q_1 \in \mathcal{H}_\infty} \left\| \begin{bmatrix} 1 - W^{-1}D \\ 0 \end{bmatrix} - \begin{bmatrix} N \\ -W_c^{-1}W \end{bmatrix} Q_1 \right\|_\infty < 1. \tag{5.45}$$

Clearly, if (5.45) holds, then (5.43) and (5.44) are satisfied. The problem (5.45) is a two-block problem; its solution technique is essentially the same as the solution of the mixed sensitivity minimization problem, which is discussed in detail in Chapter 6. In order to make this connection, we will assume that the plant factorization is such that $P = M_n N_o/D$, where $M_n$ and $D$ are inner and $N_o$ is outer (in particular, this means that the plant does not have any poles on $\mathbb{I}$). Then, we take an outer $W$ such that $|W(j\omega)| > 1$ for all $\omega$. This implies that $W_1 := (1 - W^{-1}D)$ satisfies $W, W^{-1} \in \mathcal{H}_\infty$, because $|W^{-1}(j\omega)D(j\omega)| < 1$ for all $\omega$. With this set-up, it can be shown that (5.45) is equivalent to a mixed sensitivity minimization problem for the stable plant $N$ with weights $W_1$ and a properly defined $W_2$.

**Exercise 5.6.**
Take $W_1$ as above, and define $W_2 = W_1 W W_c^{-1} N_o^{-1}$; prove that the mixed sensitivity minimization problem (5.28) defined with these weights and the stable plant $P = N$ is equivalent to (5.45). For this, you may use a controller parameterization in the form $Q_2/(1 - NQ_2)$, where $Q_2 \in \mathcal{H}_\infty$ is the free parameter, and the connection between these two problems is now made by defining $Q_2 = W_1^{-1}Q_1$ and $Q_1 = W^{-1}Q$. The optimal $Q_{\text{opt},2}$ solving this mixed sensitivity minimization problem with $\gamma_{\text{opt}} < 1$ gives a strongly stabilizing controller for the original plant $P = N/D$ as $Q = W_1 W Q_{\text{opt},2}$.

In order to illustrate the above approach, let us consider the plant (5.40): it has a factorization in the form $P = M_n N_o/M_d$, where

$$M_n(s) = e^{-hs}\frac{s^2 - 2\zeta\omega_o s + \omega_o^2}{s^2 + 2\zeta\omega_o s + \omega_o^2}, \quad M_d(s) = \frac{s - \rho\omega_o}{s + \rho\omega_o}, \quad N_o(s) = \frac{s^2 + 2\zeta\omega_o s + \omega_o^2}{(s+1)^2(s + \rho\omega_o)}.$$

Suppose we take a bound in the form $W_c(s) = \sigma/(s + \delta)$, with $\sigma > 0$ and $\delta > 0$. The simplest outer function $W$ satisfying $|W(j\omega)| > 1$ for all $\omega$ is

$$W(s) = 1 + \epsilon, \quad \epsilon > 0.$$

This leads to

$$W_1(s) := \left( \frac{1}{1+\epsilon} \right) \left( \frac{\epsilon s + (2 + \epsilon)\rho\omega_o}{s + \rho\omega_o} \right).$$

Then, it can be shown that (5.45) is equivalent to

$$\gamma_{opt} = \inf_{Q_2 \in \mathcal{H}_\infty} \left\| \begin{bmatrix} W_1 \\ 0 \end{bmatrix} - \begin{bmatrix} W_1 \\ -W_2 \end{bmatrix} NQ_2 \right\|_\infty < 1, \tag{5.46}$$

where

$$W_2(s) := \left( \frac{\epsilon s + (2 + \epsilon)\rho\omega_o}{\sigma} \right) \left( \frac{(s + \delta)(s + 1)^2}{s^2 + 2\zeta\omega_o s + \omega_o^2} \right).$$

Now it should be clear that (5.46) is in the form (5.33) with $N = M_n N_o$, $D = 1$, and $X = 0$. Since (5.46) is just a sufficient condition for solvability of the strong stabilization problem, it can be solved only for a certain restricted set of parameters $\omega_o$, $\rho$, $\zeta$, $\sigma$, and $h$. Computational issues and the parameter set for which the problem (5.46) is solvable are discussed in Section 6.3.4.

# Chapter 6

# $\mathcal{H}_\infty$ Optimal Controller: Computational Issues

Given an infinite dimensional plant $P$ and two weighting functions $W_1$ and $W_2$, the mixed sensitivity minimization problem is to find

$$\gamma_{\text{opt}} := \inf_{C \in \mathscr{C}(P)} \left\| \begin{bmatrix} W_1(1+PC)^{-1} \\ W_2PC(1+PC)^{-1} \end{bmatrix} \right\|_\infty, \qquad (6.1)$$

where $\mathscr{C}(P)$ is the set of all controllers $C$ for which the feedback system formed by $C$ and $P$ is stable.

## 6.1 ▪ Computation of the $\mathcal{H}_\infty$ optimal controller

In this section, first, an explicit formula will be given from [238] for the optimal controller solving the mixed sensitivity minimization problem (6.1). Next, some computational simplifications, previously reported in [194], will be presented.

### 6.1.1 ▪ Toker–Özbay formula

In this section, the plant in (6.1) is assumed to have finitely many poles in $\mathbb{C}_+$ and no poles on $\mathbb{I}$. Recall that in such a case, $P(s)$ can be factored as

$$P(s) = \frac{M_n(s)N_o(s)}{M_d(s)}, \qquad (6.2)$$

where $M_n$ is an inner (all-pass) function, $N_o(s)$ is an outer (minimum phase) function, and $M_d(s)$ is a rational inner function. Let $\alpha_1, \ldots, \alpha_\ell \in \mathbb{C}_+$ be the zeros of $M_d(s)$, i.e., unstable poles of the plant. For simplicity of notation, it is assumed that $\alpha_1, \ldots, \alpha_\ell$ are distinct.

Since $W_1$ is rational, it can be written as $W_1(s) = nW_1(s)/dW_1(s)$ for two coprime polynomials $nW_1$ and $dW_1$; it is assumed that $\deg(nW_1) \leq \deg(dW_1) =: n_1 \geq 1$. Define

$$E_\gamma(s) := \left( \frac{W_1(-s)W_1(s)}{\gamma^2} - 1 \right), \qquad (6.3)$$

and let $\beta_1, \ldots, \beta_{2n_1}$ be the zeros of $E_\gamma(s)$, enumerated in such a way that $-\beta_{n_1+k} = \beta_k \in \overline{\mathbb{C}}_+$ for $k = 1, \ldots, n_1$. Note that each $\beta_k$ is dependent on $\gamma > 0$, which is a candidate for $\gamma_{\mathrm{opt}}$. We assume that for $\gamma = \gamma_{\mathrm{opt}}$ the zeros of $E_\gamma$ are distinct. Note that this condition is satisfied generically (if not, a small perturbation in the problem data changes $\gamma_{\mathrm{opt}}$, which moves the locations of $\beta_1, \ldots, \beta_{n_1}$).

Now define a rational function which depends on $\gamma > 0$ and the weights $W_1$ and $W_2$,

$$F_\gamma(s) := \gamma \, \frac{dW_1(-s)}{nW_1(s)} \, G_\gamma(s), \tag{6.4}$$

where $G_\gamma \in \mathcal{H}_\infty$ is an outer function determined from the spectral factorization

$$G_\gamma(-s)G_\gamma(s) = \left( 1 + \frac{W_2(-s)W_2(s)}{W_1(-s)W_1(s)} - \frac{W_2(-s)W_2(s)}{\gamma^2} \right)^{-1}. \tag{6.5}$$

With the above definitions, the $\mathcal{H}_\infty$ optimal controller can be expressed as

$$C_{\mathrm{opt}}(s) = E_\gamma(s)M_d(s)\frac{F_\gamma(s)L(s)}{1 + M_n(s)F_\gamma(s)L(s)}N_o^{-1}(s), \tag{6.6}$$

where $\gamma = \gamma_{\mathrm{opt}}$ and $L(s)$ is a transfer function of the form

$$L(s) = \frac{[1 \ s \ \ldots \ s^{n-1}] \, \Psi_2}{[1 \ s \ \ldots \ s^{n-1}] \, \Psi_1}, \qquad n := n_1 + \ell, \tag{6.7}$$

with the coefficient vectors

$$\Psi_1 = [\psi_{10} \ \ldots \ \psi_{1(n-1)}]^{\mathrm{T}} \quad \text{and} \quad \Psi_2 = [\psi_{20} \ \ldots \ \psi_{2(n-1)}]^{\mathrm{T}} \tag{6.8}$$

to be determined from the interpolation conditions given in [238]. These interpolation conditions can be expressed in matrix form as shown below. In order to prepare a basis for the simplifications of Section 6.1.3, the following notation is introduced first: let $\mathfrak{J}_k$ be the $k \times k$ diagonal matrix, $k \geq 1$, whose $i$th diagonal entry is $(-1)^{i+1}$. For a given vector $\mathbf{x} = [x_1, \ldots, x_k]^{\mathrm{T}} \in \mathbb{C}^k$, with $x_i \neq x_j$ for $i \neq j$, and a positive integer $m \geq 1$, we define the associated Vandermonde matrix of size $k \times m$ as

$$\mathcal{V}_{\mathbf{x}}^m := \begin{bmatrix} 1 & x_1 & \ldots & x_1^{m-1} \\ \vdots & \vdots & & \vdots \\ 1 & x_k & \ldots & x_k^{m-1} \end{bmatrix}. \tag{6.9}$$

Similarly, define $\mathcal{V}_\alpha^n$ and $\mathcal{V}_\beta^n$ for the vectors $\alpha = [\alpha_1, \ldots, \alpha_\ell]^{\mathrm{T}}$ and $\beta = [\beta_1, \ldots, \beta_{n_1}]^{\mathrm{T}}$, respectively, and form the square matrix

$$\mathcal{V}_n := \begin{bmatrix} \mathcal{V}_\alpha^n \\ \mathcal{V}_\beta^n \end{bmatrix}.$$

Define the diagonal matrices

$$\begin{aligned}
\mathcal{D}_\alpha &= \mathrm{diag}\{\alpha_1^\ell, \ldots, \alpha_\ell^\ell\}, \\
\mathcal{D}_\beta &= \mathrm{diag}\{\beta_1^\ell, \ldots, \beta_{n_1}^\ell\}, \\
\mathcal{D}_\ell &= \mathrm{diag}\{M_n(\alpha_1)F_\gamma(\alpha_1), \ldots, M_n(\alpha_\ell)F_\gamma(\alpha_\ell)\}, \\
\mathcal{D}_{n_1} &= \mathrm{diag}\{M_n(\beta_1)F_\gamma(\beta_1), \ldots, M_n(\beta_{n_1})F_\gamma(\beta_{n_1})\}, \\
\mathcal{D}_n &= \mathrm{block \ diag}\{\mathcal{D}_\ell, \mathcal{D}_{n_1}\}.
\end{aligned}$$

**Theorem 6.1 (see [238]).** *Consider the mixed sensitivity minimization problem with the weights $W_1$, $W_2$ and the plant $P$. Assume that the set $\{\alpha_1, \ldots, \alpha_\ell, \beta_1, \ldots, \beta_{2n_1}\}$ consists of distinct elements in the feasible interval $\gamma_{\mathrm{opt}} \in [\gamma_{\min}, \gamma_{\max}]$. Then, the smallest achievable $\mathcal{H}_\infty$ mixed sensitivity level $\gamma_{\mathrm{opt}}$, defined in (6.1), is the largest $\gamma \in [\gamma_{\min}, \gamma_{\max}]$ for which the set of linear equations*

$$\mathcal{V}_n \Psi_1 + \mathcal{D}_n \mathcal{V}_n \Psi_2 = 0, \tag{6.10}$$

$$\mathcal{D}_n \mathcal{V}_n \mathfrak{J}_n \Psi_1 + \mathcal{V}_n \mathfrak{J}_n \Psi_2 = 0 \tag{6.11}$$

*has a nontrivial solution $\{\Psi_1, \Psi_2\}$. Moreover, the corresponding optimal controller $C_{\mathrm{opt}}(s)$ is determined by (6.6), where $\gamma = \gamma_{\mathrm{opt}}$ and $L(s)$ is given by (6.7) with nonzero $\Psi_1$, $\Psi_2$ satisfying (6.10)–(6.11).*

**Corollary 6.2 (see [238]).** *Since (6.10)–(6.11) can be written in the compact form*

$$\mathcal{T}(\gamma)\Psi = 0, \tag{6.12}$$

*where*

$$\mathcal{T}(\gamma) := \begin{bmatrix} \mathcal{V}_n & \mathcal{D}_n \mathcal{V}_n \\ \mathcal{D}_n \mathcal{V}_n \mathfrak{J}_n & \mathcal{V}_n \mathfrak{J}_n \end{bmatrix}, \qquad \Psi := \begin{bmatrix} \Psi_1 \\ \Psi_2 \end{bmatrix}, \tag{6.13}$$

*the smallest mixed sensitivity level $\gamma_{\mathrm{opt}}$ is the largest $\gamma$ for which $\mathcal{T}(\gamma)$ becomes singular.*

There is a program based on MATLAB [241] published in 1996 for the computation of $\gamma_{\mathrm{opt}}$ from (6.13) and the corresponding optimal controller (6.6). A newer version of this program is currently being developed; its outline and beta-versions have been reported recently in [265, 266].

Note that the first set of equations, (6.10), leads to

$$\Psi_1 = -\mathcal{W}_n \, \Psi_2, \quad \text{where } \mathcal{W}_n := (\mathcal{V}_n)^{-1} \mathcal{D}_n \mathcal{V}_n. \tag{6.14}$$

Inserting (6.14) into (6.11) and multiplying the resulting set of equations by $\mathfrak{J}_n (\mathcal{V}_n)^{-1}$ from the left gives

$$\left( I - (\mathfrak{J}_n \mathcal{W}_n)^2 \right) \Psi_2 = 0.$$

Thus, $\gamma_{\mathrm{opt}}$ can be computed as the largest $\gamma$ for which 1 is an eigenvalue of $(\mathfrak{J}_n \mathcal{W}_n)^2$. This way, $2n$ linear equations in (6.13) can be reduced to a set of $n$ equations. Also, if we put

$$\Psi_1 = \pm \mathfrak{J}_n \Psi_2 \tag{6.15}$$

in (6.10), we obtain (6.11). Therefore, (6.14) and (6.15) can replace (6.10) and (6.11), provided that the sign in (6.15) is determined. Moreover, this observation leads to

$$(\mathfrak{J}_n \mathcal{W}_n \pm I) \Psi_2 = 0. \tag{6.16}$$

With (6.14) and (6.15) we have

$$L(s) = -\frac{[1 \ \ s \ \ \ldots \ \ s^{n-1}] \, \Psi_2}{[1 \ \ s \ \ \ldots \ \ s^{n-1}] \, \mathcal{W}_n \Psi_2} = \pm \frac{[1 \ \ s \ \ \ldots \ \ s^{n-1}] \, \Psi_2}{[1 \ \ s \ \ \ldots \ \ s^{n-1}] \, \mathfrak{J}_n \Psi_2}, \tag{6.17}$$

which leads to $L(0) = \pm 1$, and $|L(j\omega)| = 1$ for all $\omega \in \mathbb{R}$. The set of $n = n_1 + \ell$ equations in (6.16) can further be reduced to a set of $n_1$ equations as described in Section 6.1.3. For this purpose the following observations play an important role.

We further decompose $\Psi_2$ as

$$\mathfrak{J}_n \Psi_2 =: \Phi = [\Phi_1^\mathsf{T} \ \ \Phi_2^\mathsf{T}]^\mathsf{T}, \quad \text{with} \quad \Phi_1 = [\phi_0, \dots, \phi_{\ell-1}]^\mathsf{T} \text{ and } \Phi_2 = [\phi_\ell, \dots, \phi_{n-1}]^\mathsf{T}, \tag{6.18}$$

and transform (6.10) into the form

$$\mathcal{R}(\gamma) \, \Phi = 0, \tag{6.19}$$

where

$$\mathcal{R}(\gamma) := \left[ \begin{array}{cc} \mathcal{V}_\alpha^\ell & \mathcal{D}_\alpha \mathcal{V}_\alpha^{n_1} \\ \mathcal{V}_\beta^\ell & \mathcal{D}_\beta \mathcal{V}_\beta^{n_1} \end{array} \right] \pm \left[ \begin{array}{cc} \mathcal{D}_\ell & \mathbf{0} \\ \mathbf{0} & \mathcal{D}_{n_1} \end{array} \right] \left[ \begin{array}{cc} \mathcal{V}_\alpha^\ell & \mathcal{D}_\alpha \mathcal{V}_\alpha^{n_1} \\ \mathcal{V}_\beta^\ell & \mathcal{D}_\beta \mathcal{V}_\beta^{n_1} \end{array} \right] \mathfrak{J}_{n_1+\ell}. \tag{6.20}$$

Thus, $\gamma_{\text{opt}}$ is the largest $\gamma$ which makes the matrix $\mathcal{R}(\gamma)$ singular with the $+$ or $-$ sign in (6.20). The corresponding $\Phi$ determines $C_{\text{opt}}$ defined in (6.6) via (6.7) and (6.18). Equivalently, $\gamma_{\text{opt}}$ is the largest $\gamma$ for which $(\mathfrak{J}_n \mathcal{W}_n \pm I)$ is singular; the corresponding $\Psi_2$ determines $L_{\gamma_{\text{opt}}}(s)$.

In conclusion, when the plant $P$ has finitely many unstable poles admitting coprime and inner-outer factorizations as in (4.37) or (4.38), for rational weight functions $W_1$ and $W_2$, the $\mathcal{H}_\infty$ optimal controller for the problem (6.1) is computed as

$$C_{opt}(s) = \frac{M_d(s) E_{\gamma_{\text{opt}}}(s) F_{\gamma_{\text{opt}}}(s) L_{\gamma_{\text{opt}}}(s) N_o^{-1}(s)}{1 + M_n(s) F_{\gamma_{\text{opt}}}(s) L_{\gamma_{\text{opt}}}(s)}, \tag{6.21}$$

where $E_{\gamma_{\text{opt}}}(s)$, $F_{\gamma_{\text{opt}}}(s)$, and $L_{\gamma_{\text{opt}}}(s)$ are rational transfer functions depending on $\gamma_{\text{opt}}$ as defined above.

### 6.1.2 ▪ Example: Design of integral action controllers for a stable pseudorational plant

In this section we consider a stable pseudorational plant whose outer part $N_o(s)$ is invertible in $\mathcal{H}_\infty$ and whose inner part is given by

$$M_n(s) = \frac{0.2 - se^{-hs}}{s + 0.2e^{-hs}}, \quad h = 5.$$

Note that for all $h \in [0, \frac{5\pi}{2})$ the transfer function $M_n$ is in $\mathcal{H}_\infty$. We examine the controller structure for a specific choice of weights:

$$W_1(s) = \frac{1}{s}, \quad W_2(s) = k\,s, \tag{6.22}$$

where $k > 0$ represents the relative importance of the multiplicative uncertainty with respect to the tracking performance under step-like reference inputs. With (6.22) the functions $E_\gamma(s)$ and $F_\gamma(s)$ are computed as

$$E_\gamma(s) = \frac{1 + \gamma^2 s^2}{-\gamma^2 s^2}, \quad F_\gamma(s) = \frac{-\gamma s}{ks^2 + k_\gamma s + 1}, \quad \text{where} \quad k_\gamma = \sqrt{2k - \frac{k^2}{\gamma^2}}. \tag{6.23}$$

Recall the lower bound computed in (5.36); for the weights given in (6.22), it translates into $\gamma_{\text{opt}} > \sqrt{k/2}$. Therefore, the search for $\gamma_{\text{opt}}$ is conducted for the values of $\gamma$ which make $k_\gamma$ real and positive.

With the above $E_\gamma$ and $F_\gamma$, the optimal controller is in the form

$$C_{\text{opt}}(s) = \left( \frac{1}{\gamma s} \right) \left( \frac{(1 + \gamma^2 s^2)\, L(s)}{(k\, s^2 + k_\gamma s + 1) - \gamma\, s\, M_n(s)L(s)} \right) N_o^{-1}(s). \qquad (6.24)$$

Since $|L(j\omega)| = 1$ for all $\omega \in \mathbb{R}$, we have that $L(0) \neq 0$. Furthermore, when the plant $P(s)$ does not have a pole at the origin we have $N_o^{-1}(0) \neq 0$. Hence the controller (6.24) contains an integral action due to the term $1/(\gamma\, s)$.

Note that with (6.22) we have $n_1 = 1$ and from (6.23), $\beta_1 = j/\gamma$. In particular, since the plant is stable, we have $\ell = 0$. In this case, $L(s) = \pm 1$, and $\mathcal{R}(\gamma) = 0$ if and only if

$$\mathcal{X}(\gamma) := 1 \pm F_\gamma(j/\gamma)M_n(j/\gamma) = 0. \qquad (6.25)$$

For the numerical example with $k = 2$ and $M_n(s)$ given above, the function $\mathcal{X}(\gamma)$ versus $\gamma$ is shown in Figure 6.1 for $L(s) = +1$ and $L(s) = -1$; the largest $\gamma$ which satisfies $X(\gamma) = 0$ is $\gamma_{\text{opt}} = 5.6878$ for $L(s) = 1$.

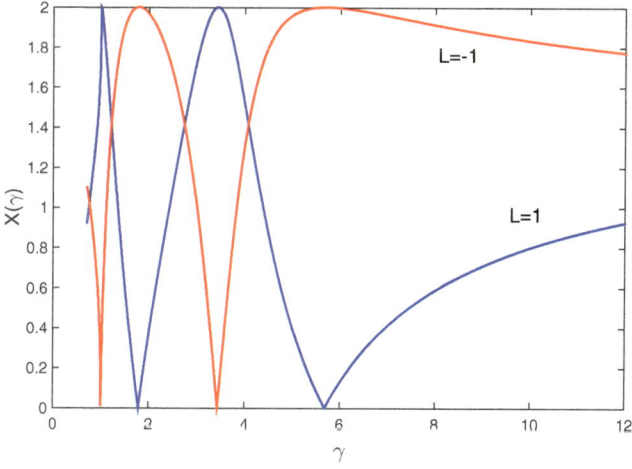

**Figure 6.1.** $\mathcal{X}(\gamma)$ *versus* $\gamma$.

## 6.1.3 ▪ Further simplifications in the set of equations for the computation of $\gamma_{\text{opt}}$

Note that there are $2n$ ($n = n_1 + \ell$) equations in the set (6.10)–(6.11), originally derived in [238]. Then, by a simple algebra, these are reduced to (6.19), where there are $n$ equations to find $\gamma_{\text{opt}}$. From a computational point of view, this is an improvement, since in the first case one needs to take the SVD of a $2n \times 2n$ matrix over a parameter $\gamma$, whereas in the latter case one computes the SVDs of two matrices (one corresponding to the $+$ sign and the other one to the $-$ sign) over the same parameter $\gamma$. In [194] it has been shown that under an assumption, which is generically satisfied, the equations (6.19) can be further reduced and $\gamma_{\text{opt}}$ can be determined by solvability conditions of a set of $n_1$ equations only. From the computational point of view, this approach may or may not be very attractive since it requires taking the inverse of an $\ell \times \ell$ matrix to construct $n_1$ equations. The details of the above mentioned simplifications are given below; they are based on [194].

First, note that (6.19) can be separated into two pieces:

$$(I \pm \mathcal{F}_\ell \mathfrak{J}_\ell)\Phi_1 + (\mathcal{V}_\alpha^\ell)^{-1}\mathcal{D}_\alpha(\mathcal{V}_\alpha^{n_1} \pm \mathcal{D}_\ell \mathcal{V}_\alpha^{n_1}(-1)^\ell \mathfrak{J}_{n_1})\Phi_2 = 0, \qquad (6.26)$$

$$(\mathcal{V}_\beta^{n_1})^{-1}\mathcal{D}_\beta^{-1}(\mathcal{V}_\beta^\ell \pm \mathcal{D}_{n_1}\mathcal{V}_\beta^\ell \mathfrak{J}_\ell)\Phi_1 + (I \pm \mathcal{F}_{n_1}(-1)^\ell \mathfrak{J}_{n_1})\Phi_2 = 0, \qquad (6.27)$$

where

$$\mathcal{F}_\ell = (\mathcal{V}_\alpha^\ell)^{-1}\mathcal{D}_\ell \mathcal{V}_\alpha^\ell, \qquad (6.28)$$

$$\mathcal{F}_{n_1} = (\mathcal{V}_\beta^{n_1})^{-1}\mathcal{D}_{n_1}\mathcal{V}_\beta^{n_1}. \qquad (6.29)$$

Now define

$$A_d = \begin{bmatrix} 0 & \cdots & 0 & -a_0 \\ 1 & & & -a_1 \\ & \ddots & & \vdots \\ & & 1 & -a_{\ell-1} \end{bmatrix}, \qquad (6.30)$$

where $a_0, \ldots, a_{\ell-1}$ are determined from the identity

$$\prod_{j=1}^\ell (s - \alpha_j) =: s^\ell + a_{\ell-1}s^{\ell-1} + \cdots + a_0.$$

Note that $A_d$ is the "A-matrix" of the observable canonical realization of $1/M_d(s)$. Its eigenvalues are $\alpha_1, \ldots, \alpha_\ell$, with the corresponding left eigenvectors being the rows of $\mathcal{V}_\alpha^\ell$. So,

$$\mathcal{F}_\ell = (\mathcal{V}_\alpha^\ell)^{-1}\mathcal{D}_\ell \mathcal{V}_\alpha^\ell = M_n(A_d)F_\gamma(A_d). \qquad (6.31)$$

Now assume that $(I \pm \mathcal{F}_\ell \mathfrak{J}_\ell)$ is nonsingular for $\gamma = \gamma_{\text{opt}}$. Then, from (6.26) we have

$$\Phi_1 = -(I \pm \mathcal{F}_\ell \mathfrak{J}_\ell)^{-1}(\mathcal{V}_\alpha^\ell)^{-1}\mathcal{D}_\alpha(\mathcal{V}_\alpha^{n_1} \pm \mathcal{D}_\ell \mathcal{V}_\alpha^{n_1}(-1)^\ell \mathfrak{J}_{n_1})\Phi_2. \qquad (6.32)$$

Substituting (6.32) into (6.27), the computation of $L(s)$, $\gamma_{\text{opt}}$, and $\Phi_2$ is reduced to the following set of $n_1$ equations:

$$\mathcal{P}(\gamma)\Phi_2 = 0, \qquad (6.33)$$

where

$$\mathcal{P}(\gamma) := -(\mathcal{V}_\beta^{n_1})^{-1}\mathcal{D}_\beta^{-1}(\mathcal{V}_\beta^\ell \pm \mathcal{D}_{n_1}\mathcal{V}_\beta^\ell \mathfrak{J}_\ell)(I \pm \mathcal{F}_\ell \mathfrak{J}_\ell)^{-1}(\mathcal{V}_\alpha^\ell)^{-1}\mathcal{D}_\alpha(\mathcal{V}_\alpha^{n_1} \pm \mathcal{D}_\ell \mathcal{V}_\alpha^{n_1}(-1)^\ell \mathfrak{J}_{n_1})$$
$$+ (I \pm \mathcal{F}_{n_1}(-1)^\ell \mathfrak{J}_{n_1}). \qquad (6.34)$$

The optimal mixed sensitivity level $\gamma_{\text{opt}}$ is the largest $\gamma$ for which there exists a nonsingular $\Phi_2$ satisfying (6.33).

In summary, $\gamma_{\text{opt}}$ can be determined from the minimum singular values of

- $\mathcal{T}(\gamma)$, which is $2(n_1 + \ell) \times 2(n_1 + \ell)$, or

- $\mathcal{R}(\gamma)$, or $(\mathfrak{J}_n W_n \pm I)$, which are $(n_1 + \ell) \times (n_1 + \ell)$, or

- $\mathcal{P}(\gamma)$, which is $n_1 \times n_1$,

where $n_1$ is the order of $W_1(s)$ and $\ell$ is the order of $M_d(s)$.

In particular, when $W_1$ is a first order rational transfer function, $\mathcal{P}(\gamma)$ given by (6.34) is a scalar function of $\gamma$. In this case,

$$(\mathcal{V}_\beta^{n_1})^{-1} = 1, \quad \mathcal{D}_\beta^{-1} = \beta_1^{-\ell}, \quad \mathcal{V}_\beta^\ell = [1, \beta_1, \ldots, \beta_1^{\ell-1}], \quad \mathfrak{J}_{n_1} = 1,$$

and
$$\mathcal{F}_{n_1} = \mathcal{D}_{n_1} = M_n(\beta_1)F_\gamma(\beta_1), \quad \mathcal{D}_\alpha \mathcal{V}_\alpha^{n_1} = [\alpha_1^\ell, \ldots, \alpha_\ell^\ell]^\mathsf{T}.$$

Moreover, for $n_1 = 1$,
$$(\mathcal{V}_\alpha^\ell)^{-1}\mathcal{D}_\alpha \mathcal{V}_\alpha^{n_1} = \mathbf{a},$$

where $\mathbf{a}$ is the last column of $A_d$, (6.30), i.e.,
$$\mathbf{a} := -[a_0, \ldots, a_{\ell-1}]^\mathsf{T}. \tag{6.35}$$

Let us define the vector
$$\mathbf{b} := -\beta_1^{-\ell} [1, \beta_1, \ldots, \beta_1^{\ell-1}]. \tag{6.36}$$

Then, for the case $n_1 = 1$ the function (6.34) becomes
$$\begin{aligned} \mathcal{P}(\gamma) = \mathbf{b} \, (I \pm M_n(\beta_1)F_\gamma(\beta_1)\mathfrak{J}_\ell) \, (I \pm \mathcal{F}_\ell \mathfrak{J}_\ell)^{-1} \, (I \pm (-1)^\ell \mathcal{F}_\ell) \, \mathbf{a} \\ + (1 \pm (-1)^\ell M_n(\beta_1)F_\gamma(\beta_1)). \end{aligned} \tag{6.37}$$

Since $\mathcal{P}(\gamma)$ in (6.37) is a scalar, in this case $\Phi_2$ in (6.33) can be chosen as any nonzero constant, say $\Phi_2 = 1$. Recall that $\mathcal{F}_\ell = M_n(A_d)F_\gamma(A_d)$; also, in (6.37) the terms $M_n(A_d)$, $\mathfrak{J}_\ell$, and $\mathbf{a}$ are independent of $\gamma$. The coefficients of $F_\gamma(A_d)$ depend on $\gamma$. When $n_1 = 1$, the roots of $E_\gamma$, i.e., $\beta_1$ and $\beta_2 = -\beta_1$, can be computed explicitly in terms of $\gamma$. So, the vector $\mathbf{b}$ and scalars $M_n(\beta_1)$ and $F_\gamma(\beta_1)$ can be evaluated numerically very easily.

Furthermore, when $W_1$ is a first order filter, and the plant has a single unstable pole $\alpha_1 > 0$, we have $n_1 = 1$ and $\ell = 1$; then, from (6.37), the condition $\mathcal{P}(\gamma) = 0$ is equivalent to having $\mathcal{Y}(\gamma) = 0$, where

$$\mathcal{Y}(\gamma) := |\, \beta_1(1 \mp Z_\gamma(\beta_1))(1 \pm Z_\gamma(\alpha_1)) - \alpha_1(1 \mp Z_\gamma(\alpha_1))(1 \pm Z_\gamma(\beta_1)) \,|, \tag{6.38}$$

and $Z_\gamma(s) := M_n(s)F_\gamma(s)$. In this case, when $\Phi_2 = 1$, the scalar $\Phi_1$ is determined from (6.32) as

$$\Phi_1 = -\alpha_1 \frac{(1 \mp Z_\gamma(\alpha_1))}{(1 \pm Z_\gamma(\alpha_1))}. \tag{6.39}$$

Then, by (6.17) and (6.18), $L(s)$ appearing in the optimal controller expression (6.6) is determined as
$$L(s) = \pm \frac{\Phi_1 - s}{\Phi_1 + s}, \tag{6.40}$$

where $\Phi_1$ is given by (6.39), and the corresponding $\gamma_{\mathrm{opt}}$ and sign to be used, $+$ or $-$, are determined from the largest $\gamma$ which satisfies $\mathcal{Y}(\gamma) = 0$, (6.38).

## 6.1.4 ▪ Example: Integral action $\mathcal{H}_\infty$ controllers for a first order unstable process with transport delay

Consider the mixed sensitivity minimization problem (5.28) with the weights (6.22). As before, let us take $k = 2$. In this case, $F_\gamma(s)$ is as determined in (6.23):

$$F_\gamma(s) = \frac{-\gamma s}{2s^2 + k_\gamma s + 1}, \quad \text{where} \quad k_\gamma = \frac{2}{\gamma}\sqrt{\gamma^2 - 1}.$$

The lower bound of $\gamma_{\mathrm{opt}}$ is computed as $1 \le \gamma_{\mathrm{opt}}$. These are independent of the plant considered.

Now, consider a first order unstable process with time delay:

$$P(s) = \frac{e^{-hs}}{s - \alpha}, \qquad h > 0, \quad \alpha > 0,$$

with the factorization $M_n(s) = e^{-hs}$, $N_o(s) = \frac{1}{s+\alpha}$, $M_d(s) = \frac{s-\alpha}{s+\alpha}$. For the numerical computations we take $k = 2$, $h = 0.2$, and $\alpha = 1$. Recall that

$$\mathcal{Y}(\gamma) := |\, \beta(1 \mp Z_\gamma(\beta))(1 \pm Z_\gamma(\alpha)) - \alpha(1 \mp Z_\gamma(\alpha))(1 \pm Z_\gamma(\beta)) \,|,$$

where $\beta = j/\gamma$, and

$$Z_\gamma(\beta) = \frac{-j\gamma^2\, e^{-j0.2/\gamma}}{\gamma^2 - 2 + j2\sqrt{\gamma^2 - 1}}, \qquad Z_\gamma(\alpha) = \frac{-\gamma^2\, e^{-0.2}}{3\gamma + 2\sqrt{\gamma^2 - 1}}.$$

The graphs of $\mathcal{Y}(\gamma)$ versus $\gamma$, for $\gamma > 1$, with the $+$ sign and the $-$ sign, are as shown in Figure 6.2. For comparison, we also construct $\mathcal{T}(\gamma)$, which is defined in (6.13):

$$\mathcal{T}(\gamma) = \begin{bmatrix} 1 & \alpha & Z(\alpha) & \alpha Z(\alpha) \\ 1 & \beta & Z(\beta) & \beta Z(\beta) \\ Z(\alpha) & -\alpha Z(\alpha) & 1 & -\alpha \\ Z(\beta) & -\beta Z(\beta) & 1 & -\beta \end{bmatrix}.$$

The graph of the minimum singular value of $\mathcal{T}(\gamma)$ versus $\gamma$ is also shown in Figure 6.2. As expected, the values of $\gamma$ which make $\mathcal{T}(\gamma)$ singular coincide with the zeros of $\mathcal{Y}(\gamma)$ for both signs. In this particular example, $\gamma_{\mathrm{opt}} = 4.648$, and it is obtained with the $-$ sign in $\mathcal{Y}(\gamma)$.

The following MATLAB code is used for computing $\gamma_{\mathrm{opt}}$ and $L_{\gamma_{\mathrm{opt}}}$ in this example. This template can be used for other mixed sensitivity minimization problems with appropriate modifications.

```
k=2; h=0.2; aa=1; Mna=exp(-h*aa);
%% Computation of gamma_opt from min(svd(Tgamma))
%gamma=linspace(1.01,10,1000);  %first see the big picture
gamma=linspace(4.6,4.7,1000);   %then zoom in
for j=1:length(gamma);
    g=gamma(j); bb=1i/g; Mnb=exp(-h*bb); kg=(k/g)*sqrt((2/k)*g^2-1);
    Fga=-g*aa/(2*aa^2+kg*aa+1);          Fgb=-g*bb/(2*bb^2+kg*bb+1);
    Dn=diag([Mna*Fga,Mnb*Fgb]);   Vp=[1,aa;1,bb]; Vm=[1,-aa;1,-bb];
    Tg=[Vp,Dn*Vp;Dn*Vm,Vm];       minsvdTg(j)=min(svd(Tg));
end
plot(gamma,minsvdTg)
[err,idx]=min(minsvdTg); gopt=gamma(idx);
%% Computation of Lopt(s)
g=gopt; bb=1i/g; Mnb=exp(-h*bb);   kg=(2/g)*sqrt(g^2-1);
Fga=-g*aa/(2*aa^2+kg*aa+1);        Fgb=-g*bb/(2*bb^2+kg*bb+1);
Dn=diag([Mna*Fga,Mnb*Fgb]);       Vp=[1,aa;1,bb]; Vm=[1,-aa;1,-bb];
Tg=[Vp,Dn*Vp;Dn*Vm,Vm];           [UU,SS,VV]=svd(Tg);
Psi=real(VV(:,4));    % coefficients of L are real, so throw away
                      % the imaginary parts; they are negligible
check_err=norm(Tg*Psi,2)  % this should be close to zero
Lopt=zpk(tf([Psi(4),Psi(3)],[Psi(2),Psi(1)]))
```

**Exercise 6.3.**

For the numerical values $k = 2$, $h = 0.2$, and $\alpha = 1$, construct the $2 \times 2$ matrices $(\Im_n \mathcal{W}_n \pm I)$ and $\mathcal{R}$ and plot their minimum singular values for $\gamma \in (1, 10)$, for two possible values of the sign. Verify the value of $\gamma_{\text{opt}} = 4.648$, and compute $L_{\gamma_{\text{opt}}}$ from the corresponding singular vector satisfying (6.19) with $\gamma = \gamma_{\text{opt}}$.

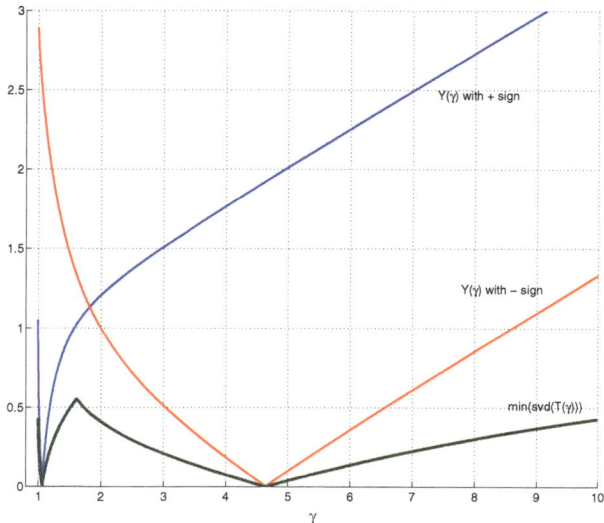

**Figure 6.2.** $\mathcal{Y}(\gamma)$ *and* $\texttt{min}(\texttt{svd}(\mathcal{T}(\gamma)))$ *versus* $\gamma$.

In [59], it has been illustrated that $\gamma_{\text{opt}}$ increases exponentially with increasing $h\alpha$. In fact, this is consistent with the conclusions of [230], where the level of difficulty of controlling an unstable plant is discussed in relation to Bode's integral constraints. Figure 6.3 quantifies these points: $\ln(\gamma_{\text{opt}})$ is approximately an affine function of $h$, and the slope of this function increases with $\alpha$.

As we have seen above, for the numerical values $k = 2$, $h = 0.2$, and $\alpha = 1$, the optimal performance level is computed as $\gamma_{\text{opt}} = 4.648$, and now the computation of $L(s)$ can be done by using (6.39) and (6.40) to obtain $\Phi_1 = -0.131$ and

$$L(s) = -\frac{\Phi_1 - s}{\Phi_1 + s} = \frac{s + 0.131}{s - 0.131}.$$

Then, the controller is given by

$$C_{\text{opt}}(s) = \left(\frac{1}{\gamma s}\right) \left(\frac{(1 + \gamma^2 s^2)(s - \alpha)(s - \Phi_1)}{(k\, s^2 + k_\gamma s + 1)(s + \Phi_1) - \gamma s\, (s - \Phi_1)\, e^{-hs}}\right), \quad (6.41)$$

with $\gamma = \gamma_{\text{opt}}$. Note that there are unstable pole-zero cancellations in the expression (6.41). Since the denominator of (6.41) is infinite dimensional, a straightforward model reduction is not possible. So, one should be able to find a stable way to implement this controller. For this purpose, by performing some algebraic manipulations, (6.41) is rewritten as

$$C_{\text{opt}}(s) = \frac{\gamma}{k} \left(1 - \frac{\Phi_1}{s}\right) \left(\frac{1}{1 + H_F(s)}\right), \quad (6.42)$$

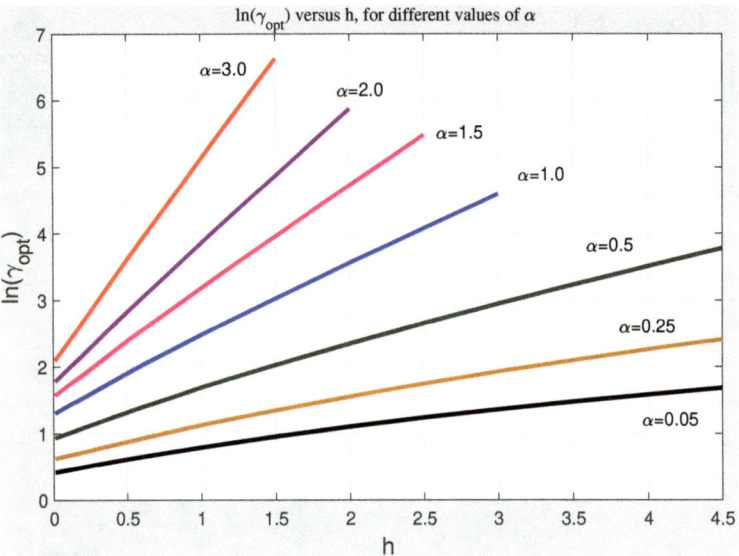

**Figure 6.3.** $\gamma_{\mathrm{opt}}$ *increases exponentially with* $h\alpha$.

where $\gamma = \gamma_{\mathrm{opt}}$, and

$$H_F(s) = \frac{as^2 + bs + c - s\,(s - \Phi_1)\,e^{-hs}\,\gamma/k}{(s^2 + 1/\gamma^2)\,(s - \alpha)},\tag{6.43}$$

with the constants $a$, $b$, $c$ determined from a partial fraction expansion, more precisely,

$$a = \Phi_1 + \alpha + k_\gamma/k,\tag{6.44}$$
$$b = (\Phi_1 k_\gamma + 1)/k - 1/\gamma^2,\tag{6.45}$$
$$c = \Phi_1/k + \alpha/\gamma^2,\tag{6.46}$$

and they are computed as $a = 1.846$, $b = 0.326$, $c = -0.02$ for the numerical example considered here. Note that the impulse response of $H_F(s)$, (6.43), is nonzero only on the time interval between 0 and $h$ (i.e., $H_F$ is a continuous-time FIR filter, which is infinite dimensional). In fact, on this time interval, the impulse response of $H_F$ coincides with the impulse response of the unstable third order system

$$H_o(s) := \frac{as^2 + bs + c}{(s^2 + 1/\gamma_{\mathrm{opt}}^2)(s - \alpha)}.\tag{6.47}$$

In Section 6.2 this observation will be extended to optimal $\mathcal{H}_\infty$ controllers designed for a more general class of plants with possibly higher order weights.

**Exercise 6.4.**
For the numerical values given in the above example, i.e., for $k = 2$, $h = 0.2$, $\alpha = 1$, we have computed $\gamma_{\mathrm{opt}} = 4.6848$ and $\Phi_1 = -0.131$, $k_\gamma = 1.9539$, and $a$, $b$, $c$ are determined as in (6.44)–(6.46). Now verify the FIR structure of the filter (6.43) from the following steps:
1. Let $H_o(s) = C(sI - A)^{-1}B$ be a minimal state-space realization of the transfer function defined in (6.47).

2. The impulse response of $H_o$ is $h_o(t) = Ce^{At}B\mathrm{u}(t)$, $t \geq 0$, where $\mathrm{u}(t)$ represents the unit step.

3. The restriction of $h_o(t)$ to the interval $t \in [0.2, \infty)$ is

$$h_r(t) := Ce^{At}B\mathrm{u}(t - 0.2) = Ce^{0.2A}e^{A(t-0.2)}B\mathrm{u}(t - 0.2), \quad t \geq 0.2.$$

4. If we subtract $h_r(t)$ from $h_o(s)$, we obtain $h_{FIR}(t)$; taking the Laplace transforms, we have

$$H_{FIR}(s) = C(sI - A)^{-1}B - Ce^{0.2A}(sI - A)^{-1}Be^{-hs}.$$

5. Show that this matches (6.43), i.e., we have

$$Ce^{0.2A}(sI - A)^{-1}B = \frac{(\gamma/k)\,s\,(s - \Phi_1)}{(s^2 + 1/\gamma^2)\,(s - \alpha)}.$$

Due to numerical round-off error in the values of $\gamma_{\mathrm{opt}}$, $\Phi_1$, $a, b, c$ we may get an imprecise result; with the above values MATLAB gives

$$Ce^{0.2A}(sI - A)^{-1}B = \frac{2.3247(s - 0.0032)(s + 0.1346)}{(s^2 + 0.04556)(s - 1)}.$$

The denominators match, but the exact values of the numerator should be $2.3424\,s\,(s + 0.131)$. Therefore, in the implementation of the controller such numerical errors should be kept in mind and necessary cancellations should be done.

**Remark 6.5.** In general, for a given finite dimensional system, if $H_o(s) = C(sI - A)^{-1}B$ is its transfer function, then

$$H_{FIR}(s) = C(sI - A)^{-1}B - Ce^{hA}(sI - A)^{-1}Be^{-hs}$$

is an FIR system, in the sense that its impulse response is the impulse response of $H_o$ restricted to the time interval $[0, h)$. These types of transfer functions appear frequently in the control of systems with I/O delays; see, e.g., [152, 159, 161, 279] and their references. As we shall see later in this chapter, similar FIR structures appear in the $\mathcal{H}_\infty$ control of a much larger class of SISO DPSs. See also Lemma 7.3 for a formal generalization to the MIMO case.

## 6.1.5 ▪ Connections with the Zhou–Khargonekar formula

By definition, (6.7), $L$ is all pass (not necessarily inner, because it may turn out to be unstable, as in the example of Sections 6.1.4), so it satisfies $L(-s) = 1/L(s)$. Since $M_n$ is an inner function, we also have $M_n(-s) = 1/M_n(s)$. Recall that $F_\gamma$ is defined as (6.4), where $G_\gamma$ is determined from the spectral factorization (6.5). These two equations imply that

$$F_\gamma(-s)F_\gamma(s) = \left(\left(\frac{W_1(-s)W_1(s)}{\gamma^2} - 1\right)\left(1 - \frac{W_2(-s)W_2(s)}{\gamma^2}\right) + 1\right)^{-1}.$$

Hence, for each $\beta_k$, a zero of $E_\gamma(s) = \left(\frac{W_1(-s)W_1(s)}{\gamma^2} - 1\right)$, we have

$$F_\gamma(-\beta_k) = 1/F_\gamma(\beta_k).$$

Thus, in addition to the interpolation conditions (6.19), the function $L(s)$ satisfies

$$1 + M_n(-\beta_k)F_\gamma(-\beta_k)L(-\beta_k) = 0 \qquad \forall\, k = 1, \ldots, n_1. \qquad (6.48)$$

This means that the function

$$\frac{1 + M_n(s)F_\gamma(s)L(s)}{M_d(s)E_\gamma(s)}$$

has no poles at the zeros of $M_d$ and $E_\gamma$.

Let $W_1(s) = C_1(sI - A_1)^{-1}B_1$ be a minimal realization (we consider a strictly proper weight here). Then, $E_\gamma^{-1}$ has a minimal realization in the form

$$E_\gamma^{-1}(s) = C_\gamma(sI - A_\gamma)^{-1}B_\gamma - 1,$$

where

$$A_\gamma = \begin{bmatrix} A_1 & B_1B_1^{\mathsf{T}}/\gamma \\ -C_1^{\mathsf{T}}C_1/\gamma & -A_1^{\mathsf{T}} \end{bmatrix}, \quad B_\gamma = \begin{bmatrix} -B_1/\sqrt{\gamma} \\ 0 \end{bmatrix}, \quad C_\gamma = \begin{bmatrix} 0 \\ B_1/\sqrt{\gamma} \end{bmatrix}^{\mathsf{T}}.$$

The zeros of $E_\gamma(s)$, namely $\beta_1, \ldots, \beta_{2n_1}$, are the eigenvalues of the Hamiltonian matrix $A_\gamma$. Since it is assumed that these eigenvalues are distinct and enumerated in such a way that $\beta_k = -\beta_{n_1+k} \in \overline{\mathbb{C}}_+$ for $k = 1, \ldots, n_1$, we can find a $2n_1 \times 2n_1$ invertible matrix $T_2$ such that

$$A_\gamma = T_2 \begin{bmatrix} \Lambda_\gamma^+ & 0 \\ 0 & -\Lambda_\gamma^+ \end{bmatrix} T_2^{-1},$$

where $\Lambda_\gamma^+$ is the diagonal matrix whose diagonal entries are $\beta_1, \ldots, \beta_{n_1}$.

Appending (6.48) to (6.19), after some matrix manipulations, we obtain

$$Q_1\Phi \pm \begin{bmatrix} I_{n_1+\ell} & 0_{n_1} \end{bmatrix} \begin{bmatrix} W_{MF}(A_d) & 0 \\ 0 & W_{MF}(A_\gamma) \end{bmatrix} \begin{bmatrix} Q_1 \\ Q_2 \end{bmatrix} \mathfrak{I}_{n_1+\ell}\Phi = 0, \quad (6.49)$$

where $W_{MF}(s) := M_n(s)F_\gamma(s)$, $I_m$ is the $m \times m$ identity matrix, $0_m$ is the $m \times m$ matrix whose entries are 0, and

$$\begin{bmatrix} Q_1 \\ Q_2 \end{bmatrix} := \begin{bmatrix} T_1 & 0 \\ 0 & T_2 \end{bmatrix} \begin{bmatrix} \mathcal{V}_\alpha^{n_1+\ell} \\ \mathcal{V}_\beta^{n_1+\ell} \\ \mathcal{V}_{-\beta}^{n_1+\ell} \end{bmatrix}, \qquad (6.50)$$

with $T_1$ being the invertible matrix which satisfies $A_d = T_1\Lambda_\alpha T_1^{-1}$, where $\Lambda_\alpha$ is the diagonal matrix whose entries are $\alpha_1, \ldots, \alpha_\ell$; the partitioning in (6.50) is such that $Q_1$ is an $(n_1 + \ell) \times (n_1 + \ell)$ square matrix and $Q_2$ is an $n_1 \times (n_1 + \ell)$ matrix.

Equation (6.49) shows the extension of [120], where mixed sensitivity minimization was considered for stable plants. We should note that [120] was an extension of the Zhou–Khargonekar formula [282], where the sensitivity minimization was considered for stable plants, i.e., $W_2 = 0$. In the stable case, $\ell = 0$, and $Q_1$ and $Q_2$ are square matrices of dimensions $n_1 \times n_1$. In that case, $M_n(A_\gamma)F_\gamma(A_\gamma)$, together with $Q_1$ and $Q_2$, determine $\gamma_{\mathrm{opt}}$ and the corresponding $\Phi$.

## 6.2 ▪ Stable implementation of the $\mathcal{H}_\infty$ optimal controller

The optimal $\mathcal{H}_\infty$ controller has unstable pole-zero cancellations due to the interpolation conditions. When the plant $P$ is finite dimensional, the cancellations can be performed and the resulting controller is obtained. On the other hand, when the plant $P$ is infinite dimensional, these cancellations have to be carefully incorporated into a stable implementation of the $\mathcal{H}_\infty$ optimal controller.

In this section, we show how to obtain a stable implementation of the optimal $\mathcal{H}_\infty$ controller for a general SISO plant. The unstable pole-zero cancellations in the optimal $\mathcal{H}_\infty$ controller are eliminated; it is shown that the cancellations correspond to a special structure in the optimal $\mathcal{H}_\infty$ controller.

### 6.2.1 ▪ On the structure of the $\mathcal{H}_\infty$ optimal controller

Let us recall the controller structure for infinite dimensional plants with finitely many poles in $\mathbb{C}_+$ and finite dimensional weights $W_1$ and $W_2$. First, write

$$W_1(s) = nW_1(s)/dW_1(s),$$

and for simplicity assume that the polynomials $nW_1$ and $dW_1$ are of the same degree, $n_1$. Then, define

$$E_\gamma(s) = \frac{W_1(s)W_1(-s)}{\gamma^2} - 1 =: \frac{nE_\gamma(s)}{\gamma^2\, dW_1(s)\, dW_1(-s)}.$$

Recall that $F_\gamma(s) = \gamma \frac{dW_1(-s)}{nW_1(s)} G_\gamma(s)$, where $G_\gamma \in \mathcal{H}_\infty$ is determined from the spectral factorization

$$G_\gamma(s)G_\gamma(-s) = \left(1 + \frac{W_2(s)W_2(-s)}{W_1(s)W_1(-s)} - \frac{W_2(s)W_2(-s)}{\gamma^2}\right)^{-1}.$$

The optimal controller given by (6.6) is

$$C_{opt}(s) = E_\gamma(s)M_d(s)\frac{F_\gamma(s)\, N_o^{-1}(s)}{L^{-1}(s) + M_n(s)F_\gamma(s)},$$

where $M_n, M_d, N_o$ are obtained from coprime and inner-outer factorizations of the plant; the dimension of $M_d$ is $\ell$ and its zeros in $\mathbb{C}_+$ are the unstable poles of the plant; and $L(s)$ is an all-pass transfer function of dimension $n_1 + \ell$, determined from the nontrivial solution of (6.19) for the largest $\gamma$.

Let us now define $M_1(s) := \frac{dW_1(-s)}{dW_1(s)}$, $K_{opt}(s) := L^{-1}(s)$, and the outer function $\widehat{G}(s)$ as the spectral factor of

$$\widehat{G}_\gamma(-s)\widehat{G}_\gamma(s) = \left(W_1(-s)W_1(s) - W_2(-s)W_2(s)E_\gamma(s)\right)^{-1}. \tag{6.51}$$

Then, we have

$$G_\gamma(s) = W_1(s)\widehat{G}_\gamma(s), \qquad F_\gamma(s) = \gamma\, M_1(s)\, \widehat{G}_\gamma(s).$$

With the above notation, the optimal controller can be rewritten as

$$C_{opt}(s) = \frac{W_1(s)}{\gamma^2}\left(\frac{\widehat{G}_\gamma(s)N_o^{-1}(s)}{\left(\frac{nW_1(s)\, dW_1(s)}{nE_\gamma(s)\, M_d(s)}\right)\left(\gamma^{-1}K_{opt}(s) + M_n(s)M_1(s)\widehat{G}_\gamma(s)\right)}\right). \tag{6.52}$$

Note that

$$R_o(s) := \frac{nW_1(s)\, dW_1(s)}{nE_\gamma(s)\, M_d(s)} \tag{6.53}$$

is a finite dimensional biproper transfer function. Moreover, as discussed in Section 6.1.5, in the product $R_o(s)(\gamma^{-1}K_{\text{opt}}(s) + M_n(s)M_1(s)\widehat{G}_\gamma(s))$ all the poles of $R_o(s)$ are canceled by the zeros of $(\gamma^{-1}K_{\text{opt}}(s) + M_n(s)M_1(s)\widehat{G}_\gamma(s))$ at the same locations. Since $\widehat{G}_\gamma$ is strictly proper, whenever $W_2$ is improper (which is the case in most practical problems), we have that $M_nM_1\widehat{G}_\gamma$ is strictly proper too. Let us define

$$d_\infty := \gamma^{-1} R_o(\infty) K_{\text{opt}}(\infty).$$

Since $R_o$ and $K_{\text{opt}}$ are biproper, $d_\infty \neq 0$. Then, the controller (6.52) can be put in the form

$$C_{\text{opt}}(s) = \frac{W_1(s)}{\gamma^2 d_\infty} \left( \frac{\widehat{G}_\gamma(s) N_o^{-1}(s)}{1 + \left( d_\infty^{-1} R_o(s)\left( \gamma^{-1} K_{\text{opt}}(s) + M_n(s)M_1(s)\widehat{G}_\gamma(s) \right) - 1 \right)} \right). \tag{6.54}$$

**Proposition 6.6.** *Suppose that $W_1$ is biproper and $W_2$ is an improper rational function. Then, the following decomposition holds:*

$$d_\infty^{-1} R_o(s)\left(\gamma^{-1}K_{\text{opt}}(s) + M_n(s)M_1(s)\widehat{G}_\gamma(s)\right) - 1 =: H_n(s) + H_d(s), \tag{6.55}$$

*where $H_n(s)$ and $H_d(s)$ are strictly proper transfer functions, $H_n$ is rational with all its poles in $\mathbb{C}_+$, and $H_d$ is in $\mathcal{H}_\infty$. The poles of $H_n$ are the poles of $K_{\text{opt}}$ in $\mathbb{C}_+$; if $K_{\text{opt}}$ does not have any poles in $\mathbb{C}_+$, then $H_n(s) \equiv 0$. When $M_n(s) = e^{-hs}$ for $h > 0$ (inner part of a time delay system), $H_d(s)$ can be decomposed into two parts, where the first part is a stable transfer function with finitely many poles in $\mathbb{C}_-$ and the second part is an FIR function, in the sense that the impulse response is restricted to the finite time interval $[0, h]$.*

***Proof.*** Since the function $K_{\text{opt}}$ is finite dimensional and $M_n$, $M_1$, $\widehat{G}$ are stable, the function $H(s)$, defined by

$$H(s) := d_\infty^{-1} R_o(s)\left(\gamma^{-1}K_{\text{opt}}(s) + M_n(s)M_1(s)\widehat{G}_\gamma(s)\right) - 1,$$

can have at most finitely many poles in $\mathbb{C}_+$. Moreover, by the interpolation conditions given in Theorem 6.1 all unstable poles of $R_o(s)$ are canceled. Therefore, the only unstable poles of $H(s)$ come from those of $K_{\text{opt}}$; once they are separated (e.g., using partial fraction expansion, i.e., the residue theorem) to define $H_n(s)$, the remaining term $H_d(s)$ is in $\mathcal{H}_\infty$. For time delay systems, using partial fractions, $H_d$ can be decomposed into two parts: (i) we have an FIR structure that cancels the unstable poles of $R_o$, and (ii) in some cases we may have an additional term which contains some of the stable poles of $\widehat{G}_\gamma$. For the details of this decomposition, see [91]. This is discussed in more detail in Section 6.2.3, and several examples are given in Section 6.3. $\quad\square$

By using the above result, the optimal controller can now be written as

$$C_{\text{opt}}(s) = \frac{W_1(s)}{\gamma^2 d_\infty} \left( \frac{C_0(s)}{1 + C_0(s)H_d(s)} \right) \widehat{G}_\gamma(s)\, N_o^{-1}(s), \tag{6.56}$$

where $C_0(s) := (1 + H_n(s))^{-1}$.

Recall that $W_1$, $\widehat{G}_\gamma$, and $C_0$ are finite dimensional; $N_o^{-1}$ and $H_d$ are (possibly) infinite dimensional. When it comes to implementation of the controller, the infinite dimensional terms need to be approximated.

We conclude this section with few remarks. For time delay systems, the FIR structure of $H_d(s)$ in (6.56) is the extension of well-known results for I/O time delays [154] to general SISO time delay systems. Note that the FIR system in time delay systems is generally not so easy to implement due to the sensitivity to the numerical errors. For the digital implementation of (6.56), see [277, 278, 160, 171]. If a rational $\mathcal{H}_\infty$ controller is desired in a practical implementation, the optimal $\mathcal{H}_\infty$ controller can be approximated by a rational controller; see [190, 240, 206]. The conditions on the approximation errors for guaranteed stability and $\mathcal{H}_\infty$ performance are derived in Section 6.4.

### 6.2.2 ▪ Example: stable implementation of the $\mathcal{H}_\infty$-optimal controller

For illustration of the above computations, let us consider the following mixed sensitivity minimization problem involving an unstable plant with a pseudorational inner part:

$$W_1(s) = \frac{1 + \varepsilon s}{s + \varepsilon}, \quad W_2(s) = k\,s,$$

$$M_n(s) = \frac{0.2 - se^{-hs}}{s + 0.2e^{-hs}}, \quad N_o(s) = \frac{1}{s + 1}, \quad M_d(s) = \frac{s - a}{s + a},$$

with the parameters $\varepsilon = 0.025$, $k = 0.5$, and $a = 1$. Note that the plant has a pole at $s = 1$ and its zeros are the roots of $0.2 = se^{-hs}$. For $h = \ln(5) \approx 1.6$, there is an unstable pole-zero cancellation; in this case the plant is not stabilizable. Since $n_1 = 1$ and $\ell = 1$, the optimal $K_{\text{opt}}(s)$ is first order in the form

$$K_{\text{opt}}(s) = \pm \frac{\phi + s}{\phi - s}.$$

The value of $\phi$ and the sign is determined from the largest $\gamma$ which gives a solution of the equations

$$\gamma^{-1} K_{\text{opt}}(\beta) + M_n(\beta) M_1(\beta) \widehat{G}_\gamma(\beta) = 0, \tag{6.57}$$

$$\gamma^{-1} K_{\text{opt}}(a) + M_n(a) M_1(a) \widehat{G}_\gamma(a) = 0, \tag{6.58}$$

where $a = 1$, the unstable pole of the plant, and

$$\beta = j \sqrt{\frac{1 - \gamma^2 \varepsilon^2}{\gamma^2 - \varepsilon^2}}$$

is one of the complex conjugate pair of roots of $nE_\gamma(s)$. The function $\widehat{G}_\gamma(s)$ is determined from the spectral factorization, (6.51), and computed as

$$\widehat{G}_\gamma(s) = \frac{(s + \varepsilon)}{k \sqrt{1 - (\varepsilon/\gamma)^2}\, s^2 + k_\gamma\, s + 1},$$

with $k_\gamma = \sqrt{2k\sqrt{1 - (\varepsilon/\gamma)^2} + \varepsilon^2 + k^2\varepsilon^2 - (k/\gamma)^2}$. For all values of $h \in [0, \ln(5))$, we find that the optimal $\gamma$ is computed with the $+$ sign, and as $h \to \ln(5)$ we have $\gamma_{\text{opt}} \to \infty$; see Figure 6.4.

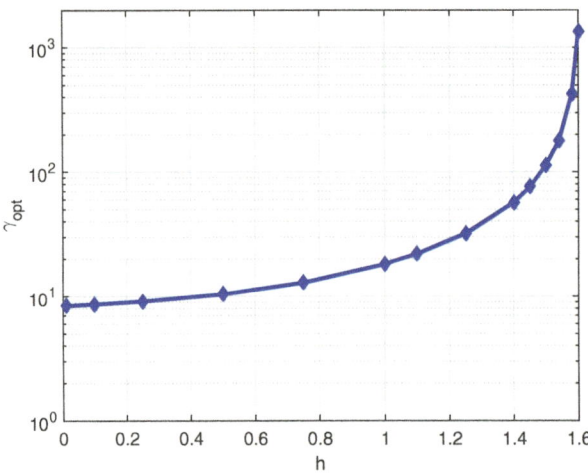

**Figure 6.4.** *As $h \nearrow \ln(5) \approx 1.6$, the minimal mixed sensitivity level $\gamma_{\mathrm{opt}} \to \infty$.*

Let us now take $h = 0.1$ and compute $\gamma_{\mathrm{opt}}$ and the corresponding $\phi$ from (6.57) and (6.58); $\gamma_{\mathrm{opt}} = 8.6279$ and $\phi = 0.3354$. The decomposition in Proposition 6.6 leads to

$$H_n(s) = -\frac{-156.37}{s - \phi}, \quad C_0(s) = \frac{1}{1 + H_n(s)} = \frac{s - \phi}{s - (\phi + 156.37)}.$$

A third order approximation of the stable infinite dimensional term $H_d(s)$ is

$$H_{da}(s) = -372.3 \frac{(1 - s/0.78)(1 + s/0.71)}{(1 + s/0.203)(1 + 1.4(s/1.34) + (s/1.34)^2)}.$$

The magnitude and phase graphs of $H_d$ and $H_{da}$ are shown in Figure 6.5. Such an approximation can be obtained by the MATLAB command `fitfrd` or other stable transfer function approximation methods; see, e.g., [88].

The optimal controller, (6.56), leads to the minimal mixed sensitivity level

$$\Upsilon_\infty(\omega) := \frac{\sqrt{|W_1(j\omega)|^2 + |W_2(j\omega)P(j\omega)C_{\mathrm{opt}}(j\omega)|^2}}{|1 + P(j\omega)C_{\mathrm{opt}}(j\omega)|} \equiv \gamma_{\mathrm{opt}} = 8.6279.$$

In the controller structure (6.56), if $H_d$ is replaced by the third order $H_{da}$ given above, then a sixth order controller $C_a(s)$ is obtained:

$$C_a(s) = \frac{-3.6357(1 + \frac{s}{40})(1 + s)(1 + \frac{s}{0.203})(1 - \frac{s}{0.3354})(1 + \frac{1.4\,s}{1.34} + (\frac{s}{1.34})^2)}{(1 + \frac{s}{94.82})(1 - \frac{s}{5.341})(1 + \frac{s}{0.908})(1 + \frac{s}{0.02523})(1 + \frac{1.412\,s}{1.414} + (\frac{s}{1.414})^2)};$$

it leads to a suboptimal mixed sensitivity level

$$\Upsilon_6(\omega) := \frac{\sqrt{|W_1(j\omega)|^2 + |W_2(j\omega)P(j\omega)C_a(j\omega)|^2}}{|1 + P(j\omega)C_a(j\omega)|},$$

$$\max_\omega \Upsilon_6(\omega) = 10.82 \approx 1.2541\gamma_{\mathrm{opt}},$$

as shown in Figure 6.6. In other words, the performance degradation is about 25%; a better performance can be obtained by using higher order approximations of $H_d(s)$ in the controller expression; see Section 6.4.1.

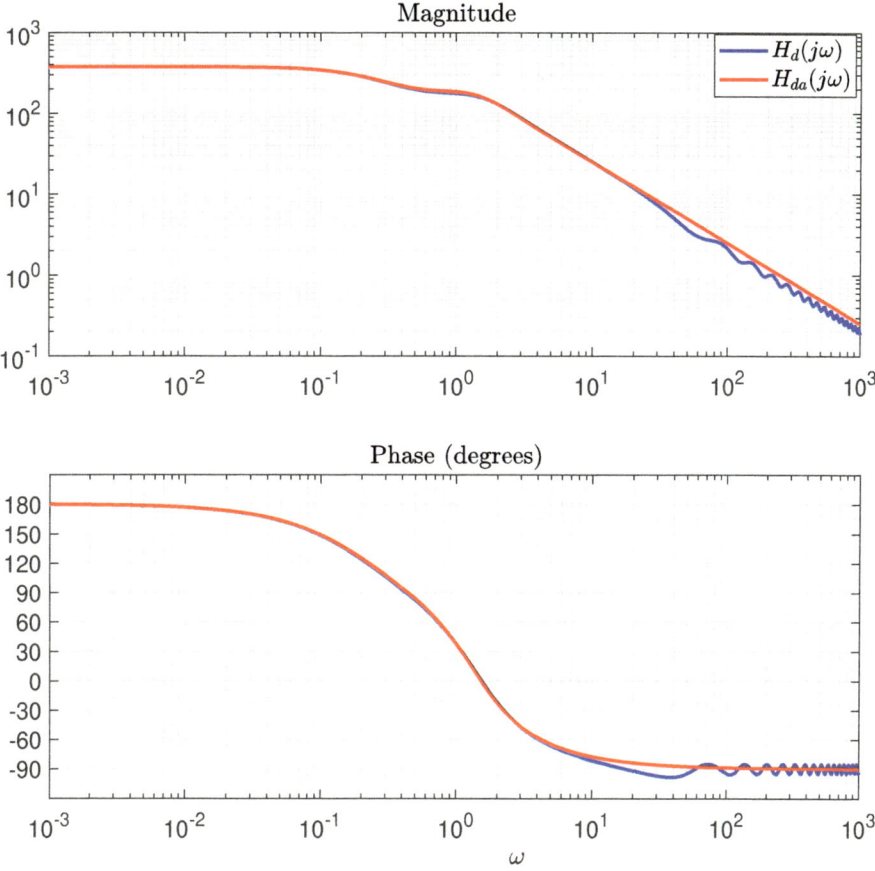

**Figure 6.5.** *Magnitude and phase plots of $H_d(j\omega)$ and $H_{da}(j\omega)$.*

### 6.2.3 ▪ FIR feedback in the optimal $\mathcal{H}_\infty$ controller for time delay systems

As described in Proposition 6.6, the optimal $\mathcal{H}_\infty$ controller has unstable pole-zero cancellations, and these result in a stable decomposition within the controller structure. In this section, we show that for time delay plants this stable decomposition in (6.56) yields an $H_d(s)$ which has an FIR structure.

Recall the controller expression given by (6.54), where the denominator term is

$$R_o(s)\left(\gamma^{-1}K_{\mathrm{opt}}(s) + M_n(s)M_1(s)\widehat{G}_\gamma(s)\right),$$

where $R_o(s)$, given by (6.53), is finite dimensional and has poles in $\mathbb{C}_+$, but all the poles of $R_o$ are canceled by the zeros of

$$D_{co}(s) := \gamma_{\mathrm{opt}}^{-1}K_{\mathrm{opt}}(s) + M_n(s)M_1(s)\widehat{G}_{\gamma_{\mathrm{opt}}}(s)$$

due to the interpolation conditions of Theorem 6.1.

When the plant is such that $M_n$ is a finite product of Moebius transformations containing time delays, we have a special structure for $M_n(s) = N_{Mn}(s)/D_{Mn}(s)$, where

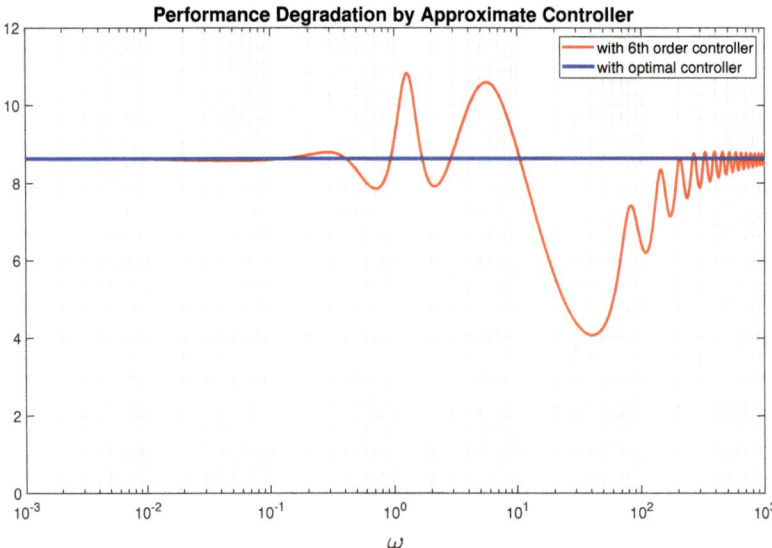

**Figure 6.6.** *Mixed sensitivity functions $\Upsilon_\infty$ and $\Upsilon_6$.*

both $N_{Mn}$ and $D_{Mn}$ are in the form

$$\sum_{k=1}^{v} \Theta_k(s)e^{-h_k s}, \tag{6.59}$$

where each $\Theta_k(s)$ is a stable finite dimensional transfer function and the delays satisfy $0 < h_1 < h_2 < \cdots < h_v$. Then, we can write $D_{co}(s) = D_{con}(s)/D_{Mn}(s)$, with the numerator term given by

$$D_{con}(s) := \gamma_{\mathrm{opt}}^{-1}K_{\mathrm{opt}}(s)D_{Mn}(s) + N_{Mn}(s)M_1(s)\widehat{G}_{\gamma_{\mathrm{opt}}}(s) =: \sum_{k=1}^{v} D_k(s)e^{-h_k s}, \tag{6.60}$$

where $D_k(s)$ for $k = 1,\dots,v$ are proper rational transfer functions. Moreover, the roots of $D_{con}(s)$ are disjoint from those of $D_{Mn}(s)$.

We show that unstable pole-zero cancellations between the time delay system $D_{con}(s)$, in the form (6.60), and the rational transfer function $R_o(s)$ yield a transfer function whose impulse response has a finite support, known as an FIR filter. In the proposition below we use the notation $G_0(s) = 1/R_o(s)$; recall that when the weight $W_1$ is biproper, so is $R_o$ and hence in this case $G_0$ is biproper. The zeros of $G_0(s)$ are the poles of $R_o(s)$, and they coincide with the zeros of $E_{\gamma_{\mathrm{opt}}}(s)$ and the zeros of $M_d(s)$. By the interpolation conditions of Theorem 6.1, these zeros are also the zeros of $D_{con}(s)$.

**Proposition 6.7.** *Let $D_{con}(s)$ be a time delay system of the form* (6.60), *$G_0(s)$ be a biproper, rational system, and $S_z^+$ be the nonempty set of common $\mathbb{C}_+$ zeros of $D_{con}(s)$ and $G_0(s)$. If the transfer function $\frac{D_{con}(s)}{G_0(s)}$ is decomposed as $\widehat{V}(s) + \widehat{F}(s)$ such that $\widehat{V}(s)$ has no poles in $S_z^+$ and the poles of $\widehat{F}(s)$ are the elements of $S_z^+$, then $\widehat{V}(s)$ has no unstable pole-zero cancellations from $S_z^+$ and $\widehat{F}(s)$ is an FIR filter with a support $t \in [0, h_v]$.*

**Proof.** For simplicity assume that $z_1, z_2, \ldots, z_{n_z} \in \mathcal{S}_z^+$ are distinct. We decompose $D_{con}(s)/G_0(s)$ into two terms as

$$\frac{D_{con}(s)}{G_0(s)} = \sum_{i=1}^{v} \frac{D_i(s)}{G_0(s)} e^{-h_i s} = \widehat{V}(s) + \widehat{F}(s)$$

$$= \left( \sum_{i=1}^{v} \widehat{V}_i(s) e^{-h_i s} \right) + \left( \sum_{i=1}^{v} \widehat{F}_i(s) e^{-h_i s} \right), \qquad (6.61)$$

where $\widehat{F}_i(s)$ is strictly proper and its poles are all elements of $\mathcal{S}_z^+$ obtained by the partial fraction expansion $D_i(s)/G_0(s) = \widehat{V}_i(s) + \widehat{F}_i(s)$ for $i = 1, \ldots, v$. By construction, $\widehat{V}(s)$ has no poles in $\mathcal{S}_z^+$ and $\widehat{F}(z_k)$ is finite for $k = 1, \ldots, n_z$. We prove the assertion if we can show that the inverse Laplace transform of $\widehat{F}(s)$ has a compact support and its support is equal to $[0, h_v]$. Since the function $\widehat{F}(s)$ is entire and it can be bounded by an exponential function as $|\widehat{F}(s)| \le Ce^{\delta|s|}$ for $\delta > 0$, the support of its inverse Laplace transform is compact and inside $[-\delta, \delta]$ by the Paley–Wiener theorem [219]. In order to show that its impulse response $\widehat{f}(t) \equiv 0$ for $t > h_v$, the inverse Laplace transform of $\widehat{F}(s)$ is written as

$$\widehat{f}(t) = \sum_{k=1}^{n_z} \left[ \sum_{i=1}^{v} \text{Res}(\widehat{F}_i(s)) \big|_{s=z_k} e^{z_k(t-h_i)} \text{u}(t - h_i) \right],$$

where $\text{u}(t)$ and $\text{Res}(\cdot)$ are the unit step function and the residue of the function, respectively. For $t > h_v$, we have

$$\widehat{f}(t) = \sum_{k=1}^{n_z} e^{z_k t} \left[ \sum_{i=1}^{v} \text{Res}(\widehat{F}_i(s)) \big|_{s=z_k} e^{-h_i z_k} \right].$$

Using $\text{Res}(\widehat{F}_i(s)) \big|_{s=z_k} = \dfrac{D_i(z_k)}{\text{Res}(G_0(s)) \big|_{s=z_k}}$ for $i = 1, \ldots, v$ and $k = 1, \ldots, n_z$, the impulse response of $\widehat{F}(s)$, for $t > h_v$, is equal to

$$\widehat{f}(t) - \sum_{k=1}^{n_z} \frac{e^{z_k t}}{\text{Res}(G_0(s)) \big|_{s=z_k}} \left[ \sum_{i=1}^{v} D_i(z_k) e^{-h_i z_k} \right]$$

$$= \sum_{k=1}^{n_z} \frac{e^{z_k t}}{\text{Res}(G_0(s)) \big|_{s=z_k}} D_{con}(z_k) \equiv 0.$$

The last equivalence follows from the fact that $z_k$, $k = 1, \ldots, n_z$, are the zeros of $D_{con}(s)$, i.e., $D_{con}(z_k) = 0$ and $\text{Res}(G_0(s)) \big|_{s=z_k} \neq 0$. By similar arguments, the results can be proven for common roots with multiplicities in $\mathcal{S}_z^+$.   □

In conclusion, for plants whose numerator inner part has the structure $M_n(s) = N_{Mn}(s)/D_{Mn}(s)$, where both $N_{Mn}$ and $D_{Mn}$ are in the form (6.59), the $\mathcal{H}_\infty$ optimal controller contains a factor $\widehat{V}^{-1}(1 + \widehat{V}^{-1}\widehat{F})^{-1}$, which means that the controller has a feed-forward term $\widehat{V}^{-1}$ with an internal FIR feedback $\widehat{F}$.

**Example 6.8.** Consider the time delay system

$$D_{con}(s) = 1 - \frac{(5s + 1)e^{-hs}}{2(s + 0.5)(s + 1)(1 - s/2)}, \qquad h = \ln(2),$$

and the biproper transfer function

$$G_0(s) = \frac{s(s-1)(s+2)}{2(s+0.5)(s+1)(1-s/2)}.$$

The common $\mathbb{C}_+$ zeros of $D_{con}(s)$ and $G_0(s)$ are $\mathcal{S}_z^+ = \{0, 1\}$. Let us decompose the transfer function $\frac{D_{con}(s)}{G_0(s)}$ as $\widehat{V}(s) + \widehat{F}(s)$ such that $\widehat{V}(s)$ has no poles in $\mathcal{S}_z^+$ and the poles of $\widehat{F}(s)$ are the elements of $\mathcal{S}_z^+$, i.e.,

$$
\begin{aligned}
\frac{D_{con}(s)}{G_0(s)} &= \frac{-2(s+0.5)(s+1)(1-s/2) - (5s+1)e^{-hs}}{s(s-1)(s+2)} \\
&= \left(\frac{-2(s+0.5)(s+1)(1-s/2)}{s(s-1)(s+2)}\right) - \left(\frac{(5s+1)e^{-hs}}{s(s-1)(s+2)}\right)e^{-hs} \\
&= \left(-1 + \frac{1}{s-1} - \frac{0.5}{s} + \frac{1}{s+2}\right) - \left(\frac{2}{s-1} - \frac{0.5}{s} - \frac{1.5}{s+2}\right)e^{-hs} \\
&= \left(-1 + \frac{1}{s+2} + \frac{1.5}{s+2}e^{-hs}\right) + \left(\frac{1}{s-1} - \frac{0.5}{s} - \left(\frac{2}{s-1} - \frac{0.5}{s}\right)e^{-hs}\right) \\
&= \underbrace{\frac{-s-1+1.5e^{-hs}}{s+2}}_{\widehat{V}(s)} + \underbrace{\frac{s+1-(3s+1)e^{-hs}}{2s(s-1)}}_{\widehat{F}(s)}.
\end{aligned}
$$

We can see that all terms whose poles are in $\mathcal{S}_z^+$ are collected in $\widehat{F}(s)$, and this term is an FIR filter with a support $t \in [0, \ln(2)]$ by Proposition 6.7. The impulse response of $\widehat{F}(s)$ is given in Figure 6.7, where its finite support is illustrated. Note that due to this decomposition, $\widehat{V}(s)$ is a transfer function with time delays and has no unstable pole-zero cancellations (see Figure 6.7 for its impulse response). ∎

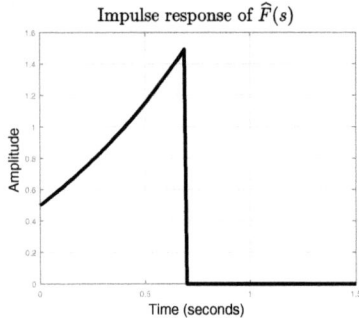
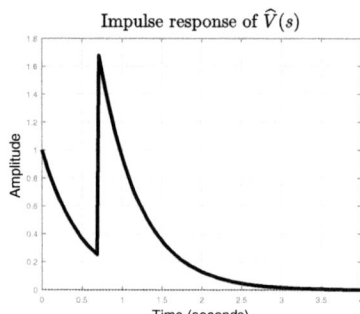

**Figure 6.7.** *The impulse responses of $\widehat{F}$ and $\widehat{V}$.*

## 6.3 ▪ Examples

### 6.3.1 ▪ Retarded delay systems

In this section, two $\mathcal{H}_\infty$ controller design examples will be given for retarded time delay systems: the first one is an academic example involving an unstable plant; the second example is taken from [213], where an $\mathcal{H}_\infty$ control–based active queue management (AQM) scheme is developed; see also Section 3.1.2 for a brief description of the plant model. In both of the examples, the control problem dealt with is the mixed sensitivity minimization problem (5.28).

**An unstable retarded time delay system**

In order to illustrate the computational method described in Section 6.1, here we consider the following plant:

$$P(s) = \frac{(s-1)(s+4)\,e^{-hs}}{(s^2+8s+17)(s+1-3e^{-2hs})}, \quad h = \ln(2) \approx 0.693. \tag{6.62}$$

First, compute the location of the poles in $\mathbb{C}_+$ using available numerical tools for finding the roots of quasi-polynomials; see, e.g., [225] for references. It is a simple exercise to show that $P(s)$ has only one pole in $\mathbb{C}_+$, at $s = 0.5$ (for larger values of $h$ the number of unstable poles of $P$ may be higher). Now, the plant can be factored as follows:

$$P(s) = \frac{M_n(s)N_o(s)}{M_d(s)}, \tag{6.63}$$

where

$$M_n(s) = \frac{s-1}{s+1}\,e^{-hs} \quad \text{and} \quad M_d(s) = \frac{s-0.5}{s+0.5}$$

are all-pass (inner) transfer functions and

$$N_o(s) = \frac{(s+4)(s+1)}{(s^2+8s+17)(s+0.5)} \left( \frac{s-0.5}{s+1-3e^{-2hs}} \right)$$

is a minimum-phase (outer) transfer function. Note that

$$\frac{s-0.5}{s+1-3e^{-2hs}} = \frac{1}{1+H_F(s)}, \quad H_F(s) = 1.5\,\frac{1-e^{-2h(s-0.5)}}{s-0.5}. \tag{6.64}$$

The impulse response of $H_F$ is $h_F(t) = 1.5e^{t/2}$ when $t \in [0, 2h]$ and $h_F(t) = 0$ otherwise. The stability of $N_o$ can also be verified from the Nyquist graph of $H_F$. Also, note that $N_o(s)$ can be factored as $N_o(s) = N_1(s)N_2(s)$, where

$$N_1(s) = \frac{(s+4)(s+1)}{(s^2+8s+17)} \left( \frac{1}{1+H_F(s)} \right), \quad N_2(s) = \frac{1}{s+0.5}, \tag{6.65}$$

with $N_1, N_1^{-1} \in \mathcal{H}_\infty$ and $N_2$ first order. The above steps illustrate *coprime factorizations* and *inner-outer factorizations* for systems with retarded time delays.

Let us now consider the mixed sensitivity minimization problem with the weights (6.22), i.e., $W_1(s) = \frac{1}{s}$ and $W_2(s) = k\,s$, with $k = 2$. Note that since $W_1$ is strictly proper, $R_o(s)$ defined by (6.53) is strictly proper, so $d_\infty = 0$, and hence Proposition 6.6 is not directly applicable. Nevertheless, it is possible to separate the controller into several cascade and parallel connections of stable blocks as illustrated below.

In this example, $\ell = 1$ and $n_1 = 1$, with $\alpha_1 = 0.5$, $\beta_1 = j/\gamma$. The largest $\gamma$ which makes $\mathcal{T}(\gamma)$ singular is $\gamma_{\text{opt}} = 17.846$, and the coefficients of the corresponding $L(s)$ are computed from the SVD of $\mathcal{T}(\gamma_{\text{opt}})$:

$$L(s) = \frac{\psi s + 1}{\psi s - 1}, \quad \text{with} \quad \psi = 23.284.$$

Note that zeros of $E_\gamma(s)M_d(s)$ in $\overline{\mathbb{C}}_+$ appear as roots of the equation

$$1 + M_n(s)F_\gamma(s)L(s) = 0.$$

Hence, there are *internal* unstable pole-zero cancellations in the controller expression; this leads to the FIR structure in the implementation. The optimal control is written as in (6.56), resulting in a numerically stable implementation,

$$C_{\text{opt}}(s) = -0.5\,\gamma_{\text{opt}} \left(\frac{\psi s + 1}{\psi s}\right) \left(\frac{1}{1 + H_d(s)}\right) N_1^{-1}(s),$$

where $H_d$ is an FIR structure,

$$H_d(s) = \frac{\gamma_{\text{opt}}^2}{2\psi} \left( \frac{(\psi s - 1)(2s^2 + 1.997s + 1) + \gamma_{\text{opt}}\, s\,(\psi s + 1)M_n(s)}{(\gamma_{\text{opt}}^2 s^2 + 1)(s - 0.5)} \right) - 1.$$

The frequency responses of $H_d$ and $C_{\text{opt}}$ are shown in Figures 6.8 and 6.9, respectively. Note that around the frequency range 3 rad/s to 10 rad/s the frequency response of $C_{\text{opt}}$ is dominated by the frequency response of $N_1^{-1}$. Check that the first five rightmost roots of the quasi-polynomial $(s + 1 - 3e^{-2hs})$ are

$$0.5000, \quad -0.1454 \pm j\,3.5688, \quad -0.7043 \pm j\,7.9584.$$

The first one is the unstable pole of the plant which is factored out; the other pairs lead to relatively large magnitude and phase variations near the frequencies 3.5 rad/s and 8 rad/s.

**Exercise 6.9.**
1. Verify that the feedback system is stable; for this purpose note that the sensitivity function is in the form

$$S_{\text{opt}}(s) = (1 + PC_{\text{opt}})^{-1} = \frac{1 + H_d(s)}{1 + Z_{\text{opt}}(s)},$$

where

$$Z_{\text{opt}}(s) = H_d(s) - 0.5\,\gamma_{\text{opt}} \left(\frac{\psi s + 1}{\psi s}\right) \left(\frac{1}{s - 0.5}\right) M_n(s).$$

Note that $Z_{\text{opt}}$ has only one pole in the open right half plane; now one can verify the feedback system stability using the Nyquist stability criterion by drawing the graph of $Z_{\text{opt}}(j\omega)$.
2. Plot the Nyquist graph of $H_d$ to find out that $C_{\text{opt}}$ has two poles in the open right half plane. Then, plot the Nyquist graph of the loop gain $PC_{\text{opt}}$ and observe that $-1$ is encircled three times in the counterclockwise direction; this is an alternative proof for feedback system stability.
3. Plot the graph of

$$\Upsilon(j\omega) = \sqrt{|W_1(j\omega)S_{\text{opt}}(j\omega)|^2 + |W_2(j\omega)T_{\text{opt}}(j\omega)|^2},$$

and verify that it is equal to $\gamma_{\text{opt}}^2$ for all $\omega$.

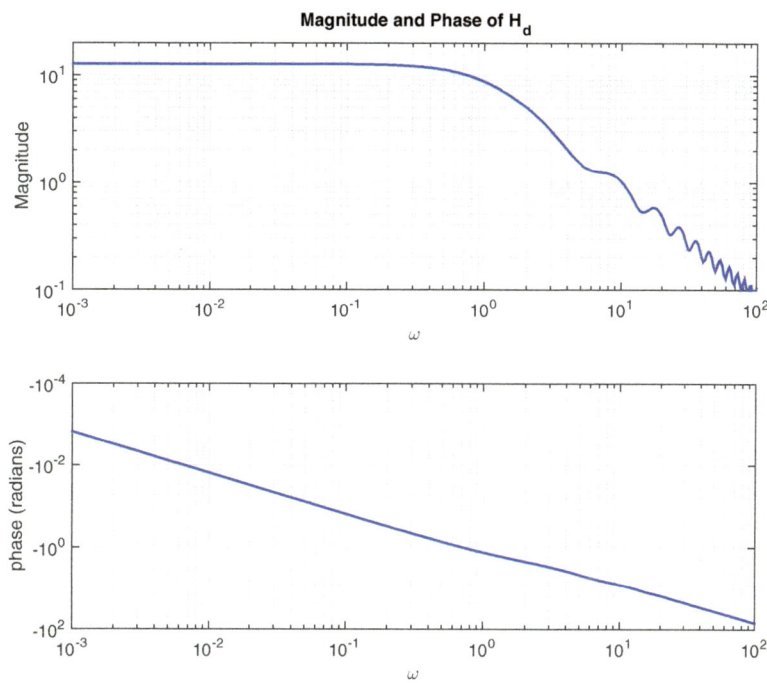

**Figure 6.8.** *Magnitude and phase plots of $H_d(j\omega)$.*

## Design of AQM scheme supporting TCP flows

In Section 3.1.2, a transfer function (3.5) is derived for the linearized model of AQM supporting TCP flows. The nominal values of the linearized model are the number of homogeneous TCP sources $N = 50$, the router's transmission capacity $C = 300$ packets/sec, and the round trip delay $R_o = 0.5$ sec; then we have $W_o - R_oC/N - 3$, and

$$P(s) = \frac{0.5\, N\, W_o^3\, e^{-R_o s}}{W_o\, R_o^2\, s^2 + (W_o + 1)\, R_o\, s + 2 + R_o\, s\, e^{-R_o s}} = \frac{337.5\, Z(s)\, e^{-0.5\, s}}{1 + 0.25\, Z(s)\, s\, e^{-0.5\, s}},$$

where

$$Z(s) = \frac{1}{0.375\, s^2 + s + 1}.$$

Since $\|sZ(s)\|_\infty = 1$, by the small gain theorem the function

$$N_o(s) = \frac{337.5\, Z(s)}{1 + 0.25\, Z(s)\, s\, e^{-0.5\, s}}$$

is outer. Hence the plant is stable, and with $M_n(s) = e^{-0.5\, s}$, the inner outer factorization of the plant is $P = M_n N_o$.

### Exercise 6.10.
Find the rightmost pole of $N_o(s)$ defined above. That is, we need to find the root $r$ of the quasi-polynomial

$$q_o(s) = 0.375s^2 + s + 1 + 0.25\, s\, e^{-0.5s},$$

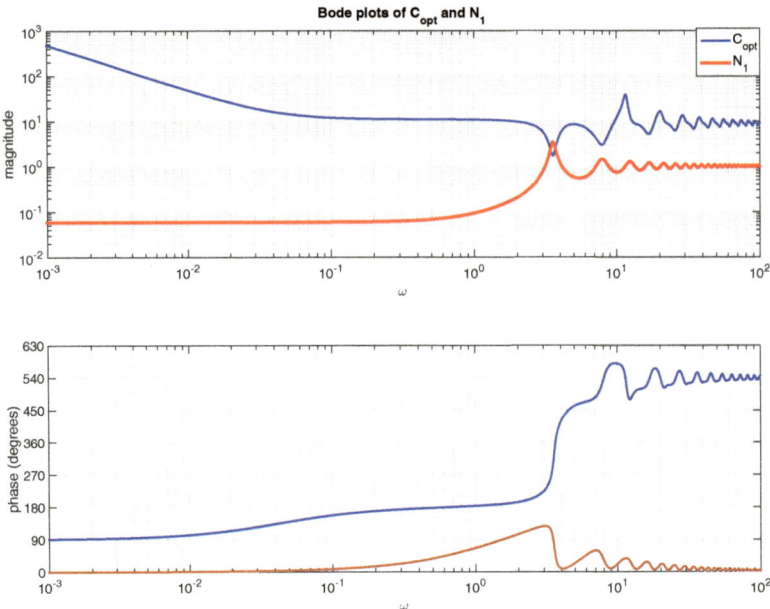

**Figure 6.9.** *Magnitude and phase plots of* $C_{\mathrm{opt}}(j\omega)$ *and* $N_1(j\omega)$.

with the property that $\mathrm{Re}(r) \geq \mathrm{Re}(r_k)$, where $r_k$ is any other root.
*Hint*: You may use YALTA [11, 10] or QPmR [248, 249].
The answer is $r = -0.95743$.

We design a robust controller for this system minimizing the $\mathcal{H}_\infty$ cost function in (6.1). The first weight $W_1(s) = \frac{100}{s}$ is chosen to track step-like reference inputs such as constant desired queue size. The second weight $W_2(s) = 20(s^2 + 6.25s + 5.75)$ is calculated to bound the multiplicative plant uncertainty due to variations in the number of homogeneous TCP sources, $\Delta_N = 10$ TCP sessions, round trip delay, $\Delta_{R_o} = 0.1$ sec, and the router's transmission capacity, $\Delta_C = 50$ packets/sec; see [213] for details.

Given weights and the nominal plant, the optimal $\mathcal{H}_\infty$ cost is computed as $\gamma_{\mathrm{opt}} = 171.33$ (note that if we scale both of the weights by a factor 0.01, then $\gamma_{\mathrm{opt}}$ is also scaled by the same factor). The optimal $\mathcal{H}_\infty$ controller's terms in (6.21) for this example are $L_{\gamma_{\mathrm{opt}}} = -1$, $E_{\gamma_{\mathrm{opt}}} = -(s^2 + 0.34067)/s^2$, and

$$F_{\gamma_{\mathrm{opt}}} = \frac{8.5665\,s}{(s + 5.133)(s^2 + 1.524\,s + 0.7221)},$$

and the optimal $\mathcal{H}_\infty$ controller has unstable pole-zero cancellations at interpolation points $s = \pm j0.58367$. These cancellations are eliminated by rewriting the controller as in (6.56),

$$C_{\mathrm{opt}}(s) = \left( \frac{8.5665\,(0.375\,s^2 + s + 1 + 0.25\,s\,e^{-0.5s})}{337.5\,s\,(s + 5.133)} \right) \left( \frac{1}{1 + H_d(s)} \right),$$

where

$$H_d(s) = \frac{(s + 5.133)(1.524\,s + 0.38143) - 8.5665\,s\,e^{-0.5s}}{(s + 5.133)(s^2 + 0.34067)}.$$

The Bode plots of $H_d$ and the optimal controller $C_{\text{opt}}$ are given in Figure 6.10. Note that $H_d$ consist of two parts: an FIR filter plus a first order stable term. In order to illustrate this, first we write

$$\frac{8.5665\,s}{(s+5.133)(s^2+0.34067)} = \frac{a}{s+5.133} + \frac{b\,s+c}{s^2+0.34067},$$

where $a = -1.6476$, $b = 1.6476$, and $c = 0.10935$. Then,

$$H_d(s) = \frac{(1.524\,s+0.38143)-(b\,s+c)\,e^{-0.5s}}{s^2+0.34067} - \frac{a\,e^{-0.5s}}{s+5.133}.$$

We define

$$H_{FIR}(s) = \frac{(1.524\,s+0.38143)-(b\,s+c)\,e^{-0.5s}}{s^2+0.34067}$$

as the FIR part of $H_d$.

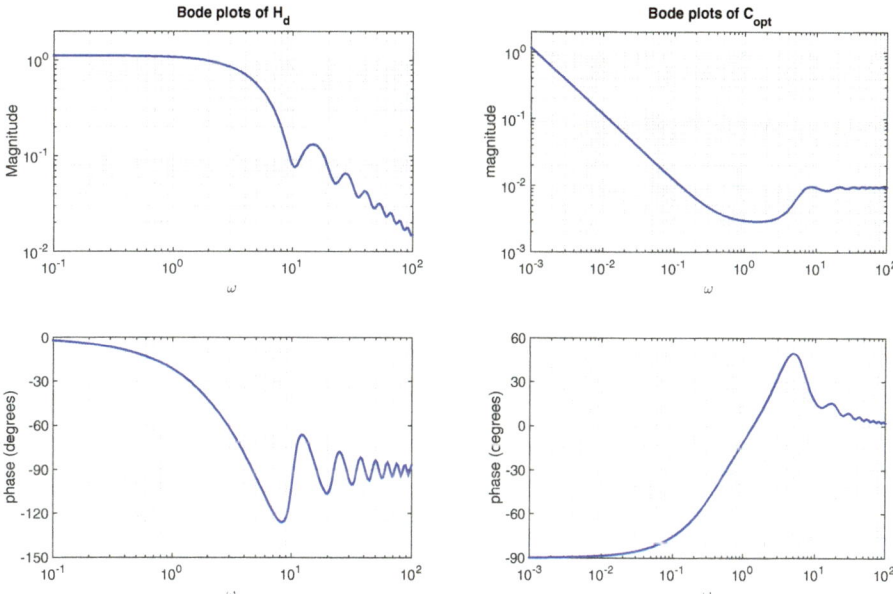

**Figure 6.10.** *Magnitude and phase plots of $H_d(j\omega)$ and $C_{\text{opt}}(j\omega)$.*

**Exercise 6.11.**
1. Verify that $(1.524\,s+0.38143)-(b\,s+c)\,e^{-0.5s}$ vanishes (up to a certain numerical precision error, which is on the order of $10^{-5}$) at $s = \pm j\sqrt{0.34067}$; hence, $H_{FIR}$ is a stable transfer function. The inverse Laplace transform of $H_{FIR}$ is the impulse response of the filter $\left(\frac{1.524\,s+0.38143}{s^2+0.34067}\right)$ restricted to the time interval $[0, 0.5]$. Based on this information, determine the "exact" values of $b$ and $c$ and compare them with the numerical values computed above.
2. From the Bode plots of $H_d$, conclude that $(1+H_d)^{-1}$, and hence the optimal controller, do not have any poles in the open right half plane. Then, using the Nyquist plot of $P(j\omega)C_{\text{opt}}(j\omega)$ prove that the feedback system $(C_{\text{opt}}, P)$ is stable.

3. Draw $\Upsilon(j\omega) = \sqrt{|W_1(j\omega)S_{\text{opt}}(j\omega)|^2 + |W_2(j\omega)T_{\text{opt}}(j\omega)|^2}$, and observe that it is equal to $\gamma_{\text{opt}}$ for all $\omega$.

## 6.3.2 • Neutral delay systems

### A system with infinitely many zeros in $\mathbb{C}_+$

Consider the plant described by a state-space representation in the form

$$
\begin{aligned}
\dot{x}_1(t) &= x_2(t), \\
\dot{x}_2(t) &= x_1(t - h_2) + u(t), \\
\dot{x}_3(t) &= 3x_1(t) + x_2(t) - x_3(t) - 2x_1(t - h_1) + 2x_2(t - h_1), \\
y(t) &= x_3(t),
\end{aligned}
$$

where $h_1 = 0.4$ and $h_2 = 0.19$. The plant is a special case of the general form

$$
\begin{aligned}
\dot{x}(t) &= A_0 x(t) + A_1 x(t - h_1) + A_2 x(t - h_2) + Bu(t), \\
y(t) &= Cx(t),
\end{aligned}
$$

with particular values of the matrices $B = [0 \ \ 1 \ \ 0]^{\mathrm{T}}$, $C = [0 \ \ 0 \ \ 1]$, and

$$
A_0 = \begin{bmatrix} 0 & 1 & 0 \\ 0 & 0 & 0 \\ 3 & 1 & -1 \end{bmatrix} \quad A_1 = \begin{bmatrix} 0 & 0 & 0 \\ 0 & 0 & 0 \\ -2 & 2 & 0 \end{bmatrix} \quad A_2 = \begin{bmatrix} 0 & 0 & 0 \\ 1 & 0 & 0 \\ 0 & 0 & 0 \end{bmatrix}.
$$

The transfer function $Y(s)/U(s)$ is

$$
P(s) = C(sI - A_0 - A_1 e^{-h_1 s} - A_2 e^{-h_2 s})^{-1} B = \frac{s + 3 + 2(s - 1)e^{-h_1 s}}{(s + 1)(s^2 - e^{-h_2 s})}, \quad (6.66)
$$

where the numerator and denominator are neutral and retarded quasi-polynomials, respectively. The plant has infinitely many unstable zeros and one unstable pole at $\rho = 0.9166$. The numerator quasi-polynomial is the same as

$$
q_2(s) = s + 3 + 2(s - 1)e^{-h_1 s}, \quad (6.67)
$$

considered earlier in Example 4.26. Recall that the conjugate function of $q_2(s)$ is

$$
\bar{q}_2(s) = -q_2(-s)e^{-h_1 s} = 2(s + 1) + (s - 3)e^{-h_1 s} \quad (6.68)
$$

and this quasi-polynomial has one unstable root at $\varrho = 0.247$. Then, the plant factorization can be done as in (4.38), $P = M_n N_o / M_d$, where

$$
M_n(s) = \frac{q_2(s)\,(s - \varrho)}{\bar{q}_2(s)\,(s + \varrho)}, \quad M_d(s) = \frac{(s - \rho)}{(s + \rho)},
$$

$$
N_o(s) = \frac{\bar{q}_2(s)\,(s + \varrho)\,(s - \rho)}{(s - \varrho)\,(s + \rho)\,(s + 1)\,(s^2 - e^{-h_2 s})}.
$$

Let us design mixed sensitivity weights such that the closed-loop system tracks step-like references and tolerates 10% uncertainty in time delays. We achieve the first goal by setting $W_1(s) = 1/s$. For the second goal, since the number of unstable poles of the plant does not change for $h_2$ values between 0.4 and 0.44, it is sufficient to calculate a weight function $W_2$ covering the multiplicative plant uncertainties, i.e.,

$$\left| \frac{P(j\omega) - P_\Delta(j\omega)}{P(j\omega)} \right| < |W_m(j\omega)| \quad \forall \, \omega, \tag{6.69}$$

where

$$P_\Delta(s) = \frac{s + 3 + 2(s-1)e^{-\tilde{h}_1 s}}{(s+1)(s^2 - e^{-\tilde{h}_2 s})} + \frac{\delta \, s}{0.1 \, s + 1}, \tag{6.70}$$

where $\tilde{h}_1 \in [0.4, 0.44]$, $\tilde{h}_2 \in [0.19, 0.21]$, and the high-pass filter multiplying $\delta \in [0, 0.01]$ represents a bound on possible high frequency dynamics that are not taken into account in the model. We compute the weight function $W_m(s) = 0.12 \, s \, (0.1 \, s + 1)$, which satisfies the inequality (6.69) as shown in Figure 6.11. Since the relative degree of $P$ is 2, in order to obtain a strictly proper controller, which is easy to approximate, we will take $W_2(s) = (0.1s + 1) \, W_m(s)$, which makes the relative degree of $W_2^{-1}$ equal to 3, and hence the optimal controller should have a relative degree $3 - 2 = 1$.

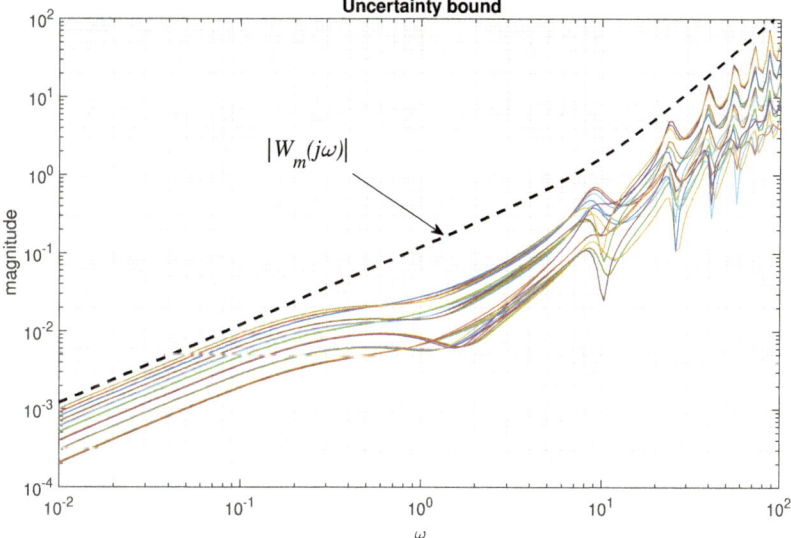

**Figure 6.11.** *Weight function $W_m$ satisfying (6.69) for $h_1 \in [0.4, 0.44]$, $h_2 \in [0.19, 0.21]$, and $\delta \in [0, 0.01]$, with $P$ as in (6.66) and $P_\Delta$ as in (6.70).*

With the above plant and weights, the $\mathcal{H}_\infty$ cost is $\gamma_{\mathrm{opt}} = 0.845663$ and the optimal controller is written as in (6.71), where unstable pole-zero cancellations are eliminated,

$$C_{\mathrm{opt}}(s) = \left( \frac{s + \phi}{s} \right) \frac{Z_o(s)}{1 + H_d(s)}, \tag{6.71}$$

where $\phi = 0.39885$, $Z_o$ is a strictly proper outer function

$$Z_o(s) = \frac{704.72}{(s + \rho) \, d_2(s)} \, N_o^{-1}(s),$$

and

$$H_d(s) = \cfrac{(s - \phi)\, d_1(s) - \cfrac{704.72\, s\, (s + \phi)}{d_2(s)} M_n(s)}{(s^2 + \gamma_{\text{opt}}^{-2})\,(s - \rho)} - 1, \qquad (6.72)$$

with the second order stable polynomials

$$d_1(s) = (s^2 + 20.0822\, s + 100.9893) \quad \text{and} \quad d_2(s) = (s^2 + 3.71326\, s + 8.2517)$$

constructed from the poles of $F_{\gamma_{\text{opt}}}(s)$, defined by (6.4). The frequency responses of $H_d$ and the optimal controller, $C_{\text{opt}}$ are given in Figures 6.12 and 6.13.

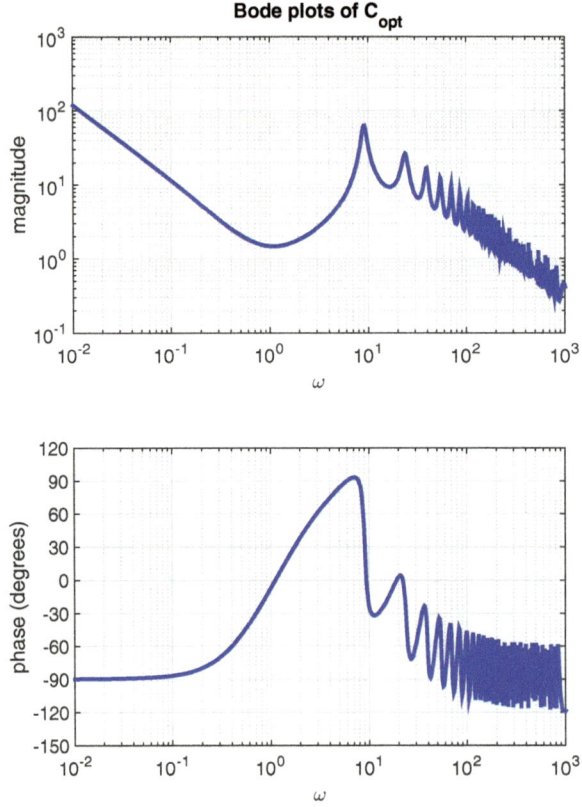

**Figure 6.12.** *Magnitude and phase plots of $C_{opt}(j\omega)$ given by (6.71).*

**Exercise 6.12.**

1. Prove that $Z_o/(1 + H_d)$ is stable from the following arguments: it is clear that $Z_o$ is outer; now use the magnitude and phase plots of $H_d$ to prove the stability of $(1 + H_d)^{-1}$.

2. Draw the Nyquist graph of $P(j\omega)C_{\text{opt}}(j\omega)$, and observe that there is one counter-clockwise encirclement of $-1$. Since the plant has one unstable pole and the controller does not have a pole in the open right half plane (its only unstable pole is at $s = 0$, which comes from the pole of $W_1$), the Nyquist stability criterion implies that the feedback system is stable.

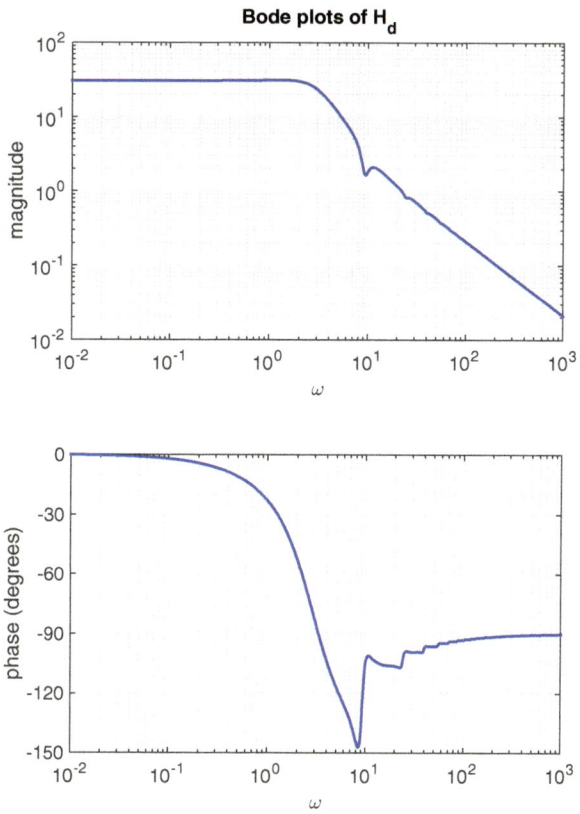

**Figure 6.13.** *Magnitude and phase plots of $H_d(j\omega)$ given by (6.72).*

## A system with infinitely many poles in $\mathbb{C}_+$

Now consider the plant

$$P(s) = \frac{10\left(s^2 - e^{-h_2 s}\right)}{(10\,s + 1)(s + 3 + 2(s - 1)e^{-h_1 s})}, \tag{6.73}$$

with $h_1 = 0.4$ and $h_2 = 0.19$ as in the previous example. From the roots of the numerator and denominator quasi-polynomials, we know that $P$ has one zero in $\mathbb{C}_+$ and infinitely many poles in $\mathbb{C}_+$ (see the example given above). The plant defined in (6.73) is proper but not strictly proper (recall that it is not possible to stabilize a strictly proper plant with infinitely many poles in $\mathbb{C}_+$ by a proper controller). It has a factorization in the form $P = M_n N_o / M_d$, where

$$M_d(s) = \frac{q_2(s)\,(s - \varrho)}{\bar{q}_2(s)\,(s + \varrho)}, \quad M_n(s) = \frac{(s - \rho)}{(s + \rho)}, \tag{6.74}$$

$$N_o(s) = \frac{10\,(s + \rho)\left(s^2 - e^{-h_2 s}\right)(s - \varrho)}{(10\,s + 1)\,(s - \rho)\,(s + \varrho)\,\bar{q}_2(s)}, \tag{6.75}$$

where $\bar{q}_2(s)$ is defined by (6.68), and as before, $\rho = 0.9166$ and $\varrho = 0.247$ are the $\mathbb{C}_+$ roots of the quasi-polynomials $\left(s^2 - e^{-h_2 s}\right)$ and $\bar{q}_2(s)$, respectively.

We will consider robust stabilization of all plants in the form

$$P_\Delta(s) = \frac{10\,(s^2 - e^{-h_2 s})}{(10\,s + \delta)(s + 3 + 2(s - 1)e^{-h_1 s})} + \frac{\delta\,s\,(s - \rho)}{(s + 1)(s + \rho)}, \qquad (6.76)$$

with $\delta \in [0.01, 0.03]$. Note that $\delta$ appears as an uncertainty of a pole location and as the magnitude of a high-pass term which is neglected in the nominal plant model. Recall from the discussion in Section 5.1.3 that in order to apply Theorem 5.2, $P_\Delta$ must have the same unstable poles as $P$; furthermore, because the multiplicative perturbation is assumed to be stable, the unstable zeros of $P$ must also appear in the additive uncertainty. The multiplicative uncertainty weight

$$W_m(s) = \frac{0.12\,(s + 1)}{(s + 0.001)} \qquad (6.77)$$

satisfies the assumption which defines $\mathscr{P}$ in (5.7), i.e., $W_m$ is an outer function and $|W_m(j\omega)| \geq |P_\Delta(j\omega) - P(j\omega)|/|P(j\omega)|$; see Figure 6.14. Moreover, $\Delta_o$ resulting from this definition of $P_\Delta$ is stable and scaled by $W_m$ to have norm less than or equal to 1. By using the sufficiency part of Theorem 5.2, to guarantee robust stability we will design a controller $C$ which satisfies $\|W_m T\|_\infty < 1$, where $T = PC(1 + PC)^{-1}$.

As far as reference tracking is concerned, we will require the plant to track a sinusoidal signal whose period is $\pi$ seconds. So, we choose a sensitivity weight which has a large amplitude at 2 rad/sec,

$$W_s(s) = \frac{0.2\,k_s\,(2.5\,s\,+\,1)(0.25\,s\,+\,1)}{(0.25\,s^2\,+\,0.0025\,s\,+\,1)}; \qquad (6.78)$$

see Figure 6.14. Our objective will be to find a controller $C$ stabilizing the feedback system and achieving $\|W_s S\|_\infty < 1$ for the largest possible $k_s > 0$, where $S = (1 + PC)^{-1}$. For a given fixed $k_s$, if a stabilizing controller $C$ satisfies $\|[W_m T \ \ W_s S]^T\|_\infty < 1$, then both performance and robust stability conditions are met, i.e., $\|W_s S\|_\infty < 1$ and $\|W_m T\|_\infty < 1$.

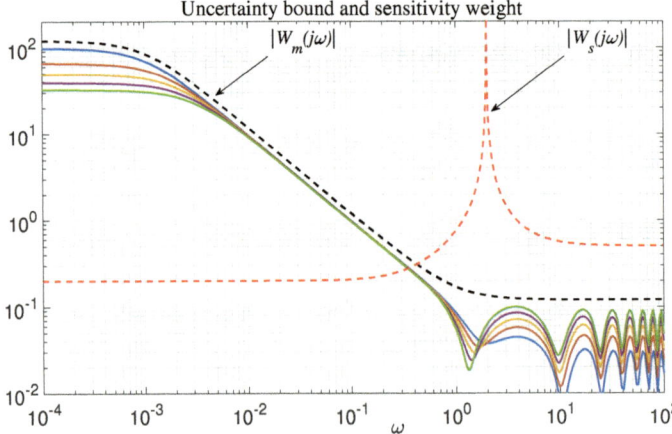

**Figure 6.14.** *Weight function $W_m$ satisfying $|W_m(j\omega)| \geq |\frac{P_\Delta(j\omega)}{P(j\omega)} - 1|$ for values of $\delta \in [0.01, 0.03]$, with $P$ as in (6.73) and $P_\Delta$ as in (6.76); $W_s$ is given in (6.78) with $k_s = 1$.*

Let us now take $k_s = 1$ and compute

$$\gamma_{\text{opt}} = \inf_{(C,P)\text{ stable}} \left\| \begin{bmatrix} W_m\,PC(1 + PC)^{-1} \\ W_s\,(1 + PC)^{-1} \end{bmatrix} \right\|_\infty.$$

Since $N_o, N_o^{-1} \in \mathcal{H}_\infty$, we can define $C = C_1^{-1} N_o^{-1}$, where $C_1$ is a stabilizing controller for $P_1 = M_d/M_n$. Note that $P_1$ has one pole in $\mathbb{C}_+$, as in the previous example. The key element for the solution of this problem comes from controller design using duality (Proposition 4.31), which implies that

$$\gamma_{\text{opt}} = \inf_{(C_1, P_1) \text{ stable}} \left\| \begin{bmatrix} W_m \, (1 + P_1 C_1)^{-1} \\ W_s \, P_1 C_1 (1 + P_1 C_1)^{-1} \end{bmatrix} \right\|_\infty.$$

This is the mixed sensitivity minimization problem (6.1) with weights $W_1 = W_m$ and $W_2 = W_s$ and the plant $P_1 = M_d/M_n$.

Applying the computations leading to (6.21), we find that $\gamma_{\text{opt}} = 0.9107$, and the corresponding optimal controller is in the form

$$C_{\text{opt}} = \left( \frac{F_{\text{opt}}^{-1} K_{\text{opt}} + M_d}{E_{\text{opt}} M_n} \right) N_o^{-1}, \tag{6.79}$$

where $K_{\text{opt}} = \frac{1 - \eta \, s}{1 + \eta \, s}$, with $\eta = 12.5406$,

$$F_{\text{opt}}^{-1}(s) = \frac{0.14 \, (s + 3.83254) \, (s^2 + 0.8073 \, s + 0.2396)}{(s - 0.001) \, (0.25 \, s^2 + 0.0025 \, s + 1)},$$

$$E_{\text{opt}}(s) = -0.98264 \, \frac{(s^2 + 0.01767)}{(s - 0.001)(s + 0.001)}.$$

The Bode plots of $C_{\text{opt}}$ are shown in Figure 6.15.

**Exercise 6.13.**

1. Verify that the $\mathcal{H}_\infty$ optimal controller (6.79) can be written as

$$C_{\text{opt}}(s) = \left( \frac{1}{0.98264} \left( \frac{0.14}{0.25} - M_d(s) \right) + R_0(s) \right) N_o^{-1}(s),$$

with

$$R_0(s) = \frac{a_1 s + a_0}{s^2 + 0.01s + 4} + \frac{b_0}{s + \eta^{-1}} + H_0(s),$$

where $H_0(s)$ is a strictly proper stable transfer function, and $a_1, a_0, b_0$ are constants determined from partial fraction expansion.

2. Check that the characteristic function $U = (M_d + M_n N_o C_{\text{opt}})$ of this system is in the form

$$U(s) = M_d(s) + \left( 1 - \frac{2\rho}{s + \rho} \right) \left( \frac{1}{0.98264} \left( \frac{0.14}{0.25} - M_d(s) \right) + R_0(s) \right)$$

$$= 0.57 \left( 1 - R_1(s) \right),$$

where

$$R_1(s) = 0.031 M_d(s) + \frac{3.51\rho}{s + \rho} \left( \frac{1}{0.98264} \left( \frac{0.14}{0.25} - M_d(s) \right) + R_0(s) \right).$$

Now, draw the Nyquist graph of $R_1(j\omega)$ and conclude that $(1 - R_1)^{-1} \in \mathcal{H}_\infty$. Note that as $\omega \to \infty$ the function $R_1(j\omega)$ converges to a circle whose center is the origin and

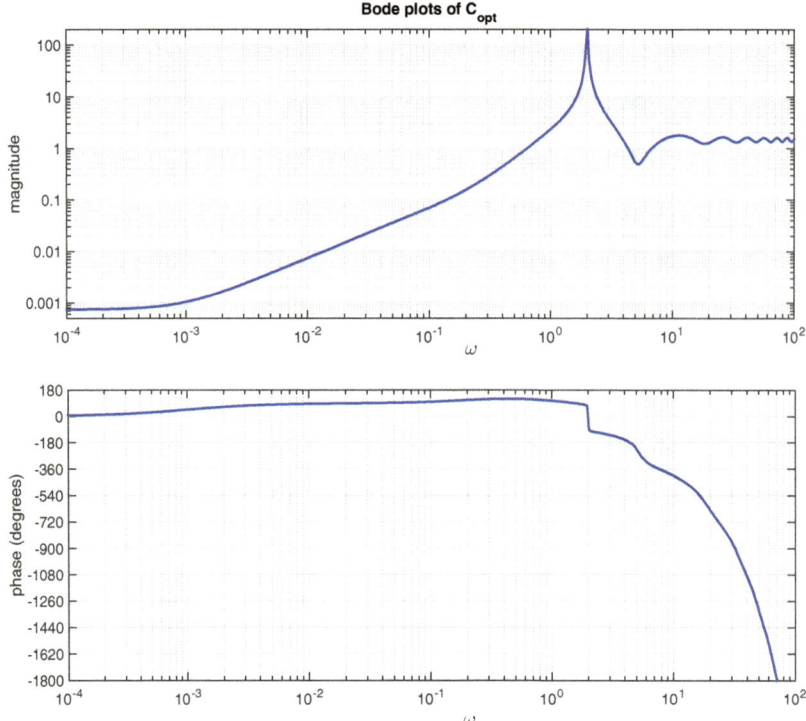

**Figure 6.15.** *Bode plots of $C_{\mathrm{opt}}$ given by* (6.79).

the radius is 0.031, so it is possible to count the number of encirclements of $-1$. On the other hand, it is impossible to conclude feedback system stability from the Nyquist graph of $PC_{\mathrm{opt}}$, because it encircles $-1$ in the counterclockwise direction infinitely many times (as expected, since the plant has infinitely many poles in $\mathbb{C}_+$).

3. Since for $k_s = 1$ we obtain $\gamma_{\mathrm{opt}} = 0.9107$, there is still some room to increase $k_s$. Find the largest $k_s$ in (6.78) for which we have $\gamma_{\mathrm{opt}} < 1$.

### 6.3.3 ▪ Fractional order systems

Let us now recall the unstable fractional order model of a nonlaminated magnetic suspension system, defined by (3.25):

$$P(s) = \frac{e^{-hs}}{s^{2.5} + s^2 - c}, \qquad (6.80)$$

where $h > 0$ is the I/O delay, and $c > 0$. The fractional order of this system is $\alpha = \frac{1}{2}$. By the observation noted in (3.25), the inner-outer factorization of this plant can be done as follows:

$$M_n(s) = e^{-hs},$$

$$M_d(s) = \frac{(s - p^2)}{(s + p^2)},$$

$$N_o(s) = \frac{(s^\alpha + p)}{(s + p^2)(s^\alpha - p_1)(s^\alpha - p_2)(s^\alpha - p_3)(s^\alpha - p_4)},$$

where $p > 0$ and $p_1, \dots, p_4$ are the roots of the fifth order equation

$$\zeta^5 + \zeta^4 - c = 0, \tag{6.81}$$

with $p_i = r_i e^{j\theta_i}$ satisfying $|\theta_i| > \frac{\pi}{4}$ for all $i = 1, \dots, 4$. For example, when $c = 2$ the roots of (6.81) are $p_{1,2} \approx -1.1898 \pm j0.6028$, $p_{3,4} \approx 0.1898 \pm j1.0432$, and $p = 1$. Since $|\theta_{1,2}| = 2.6726 > \pi/4$ and $|\theta_{3,4}| = 1.3908 > \pi/4$, by Matignon's theorem [149], we have that $N_o$ is indeed stable (and outer).

Let us now consider the mixed sensitivity minimization problem for this plant with the weights from Section 6.2.2,

$$W_1(s) = \frac{1 + \varepsilon s}{s + \varepsilon}, \quad W_2(s) = k\, s,$$

where $\varepsilon = 0.025$ and $k = 0.5$. As far as the computation of the optimal performance level and subsystems of the optimal controller are concerned, the only difference between the example of Section 6.2.2 and the problem at hand are the definitions of $M_n$ and $N_o$. Therefore, $\gamma_{\mathrm{opt}}$ is computed from (6.57)–(6.58), where $M_n(s)$ is replaced by $e^{-hs}$; all other terms have the same structure.

For $h = 0.1$ we compute that

$$\gamma_{\mathrm{opt}} = 1.47023 \quad \text{and} \quad K_{\mathrm{opt}} = \frac{(s - 0.3060)}{(s + 0.3060)}.$$

Since $K_{\mathrm{opt}}$ is stable, we have that $H_n(s) = 0$ and hence $C_0(s) = 1$. In this case, $H_d$, obtained from (6.55), is a stable infinite dimensional transfer function whose frequency response is shown in Figure 6.16 with solid lines. The same figure also shows a fourth order approximation,

$$H_{da}(s) = \frac{36.134(s + 63.49)(s + 0.9835)(s + 0.01352)}{(s + 45.99)(s + 0.3059)(s^2 + 1.882s + 2)}, \tag{6.82}$$

which is obtained from the `fitfrd` command of MATLAB.

**Exercise.** Let `om=logspace(-4,3,1000);` be the frequency points $\omega$ selected to generate the plots shown in Figure 6.16. Evaluate $H_d(j\omega)$ using (6.55) with the numerical values given above. Then, define the frequency response data `Hdfreq` and a fourth order approximation with relative degree 1. Using the commands

```
>> Hdfreq=frd(Hd,om);
>> Hdasys=fitfrd(Hdfreq,4,1);
>> zpk(Hdasys)
```

observe that the answer is the transfer function given by (6.82) and its magnitude and phase plots are as shown in Figure 6.16 with dashed lines. Note that in the controller expression, if one replaces $H_d$ with $H_{da}$, the resulting performance level becomes $\gamma = 1.62$, which is about 10% worse than the optimal level $\gamma_{\mathrm{opt}} = 1.47$. Another point to note is that the optimal controller is improper, because $N_o$ has a relative degree 2.5 and $\widehat{G}_{\gamma_{\mathrm{opt}}}$ has a relative degree 1. In order to obtain a strictly proper controller, $W_2^{-1}$ should be chosen to have a relative degree 3 or higher. Clearly, for such fractional order systems, where the outer part of the plant has a noninteger relative degree, it is not possible to obtain a biproper controller using finite dimensional weights. It is important to keep in mind that replacing $H_d$ with $H_{da}$ does not lead to a finite dimensional controller, because the other infinite dimensional term, $N_o^{-1}$, remains to be approximated. This issue is discussed in Section 6.4. □

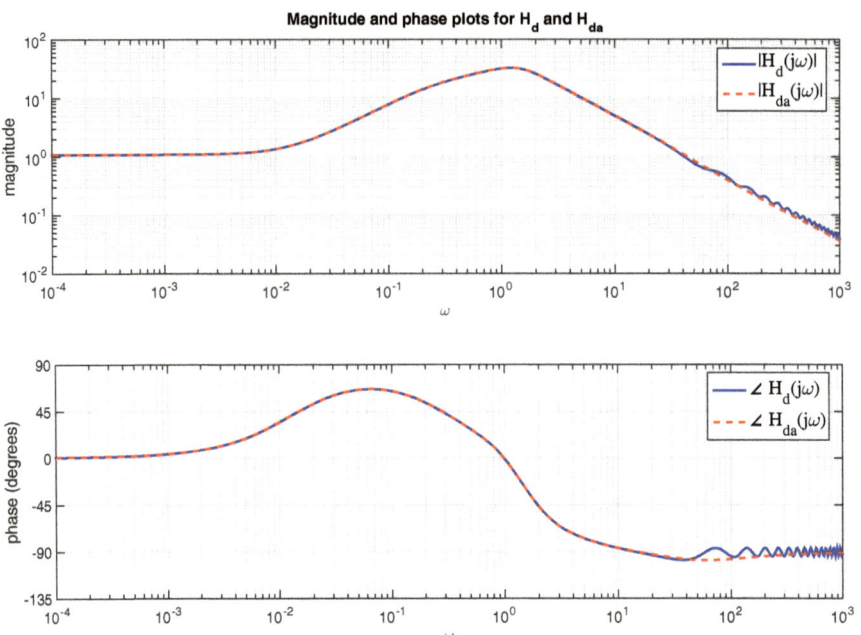

**Figure 6.16.** *Magnitude and phase plots of $H_d(j\omega)$ and $H_{da}(j\omega)$.*

In view of the above remarks, to obtain a controller which is strictly proper and closer to having an integral action, let us modify the weights as follows:

$$W_1(s) = \frac{1 + \varepsilon s}{s + \varepsilon}, \quad W_2(s) = k\, s\, (0.2\, s + 1)^2,$$

where $\varepsilon = 0.001$ and $k = 0.5$. Note that having a third order $W_2$ will lead to a strictly proper controller whose relative degree is $3 - 2.5 = 0.5$, because the plant has relative degree 2.5. Indeed, performing the computations of Section 6.2.1 we obtain $\gamma_{\mathrm{opt}} = 2.5203$, and

$$K_{\mathrm{opt}}(s) = \frac{\phi - s}{\phi + s}, \quad \text{with} \quad \phi = 0.2141,$$

which is stable, so once again $H_n(s) = 0$ and $C_0(s) = 1$. The Bode plots of the resulting $H_d(s)$ are shown in Figure 6.17. Using the same procedure described above, $H_d$ can be approximated with a very high accuracy by the following fifth order transfer function

$$H_{da}(s) = \frac{1014.7(s+1)(s+0.0008428)(s^2 + 11.86s + 65.32)}{(s+0.2141)(s^2 + 9.877s + 24.68)(s^2 + 1.87s + 1.981)}. \tag{6.83}$$

Another interesting point about $H_d$ is that from its Bode plot we see that the internal feedback loop $1/(1+H_d)$ is stable. Since all other subsystems of the optimal controller, (6.56), are also stable, i.e., $W_1$ and the product $\widehat{G}N_o^{-1}$ are stable, we have that $C_{\mathrm{opt}}$ is stable as well.

The Bode plot of the optimal controller is shown in Figure 6.18 with solid lines. We can approximate this infinite dimensional stable controller directly by using the `fitfrd`

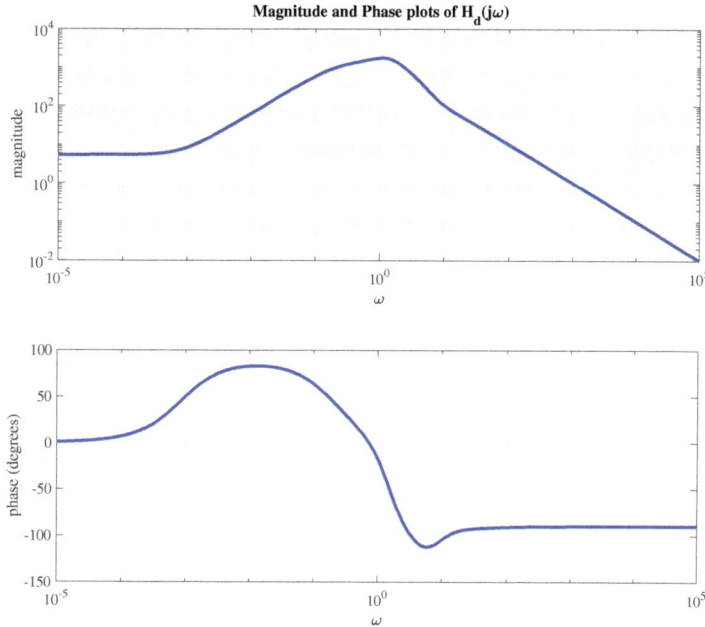

**Figure 6.17.** *Magnitude and phase plots of $H_d(j\omega)$.*

command of MATLAB. A fourth order approximation with relative degree 1 is

$$C_{app}(s) = \frac{1871.7(s + 7.438)(s + 0.8928)(s + 0.2366)}{(s + 66.4)(s + 0.001)(s^2 + 7.853s + 55.64)}, \qquad (6.84)$$

and its magnitude and phase plots are shown in Figure 6.18 with the dashed lines. The resulting performance of the fourth order approximate controller $C_a$ is obtained by measuring the peak value of

$$\Upsilon_4(j\omega) = \frac{\sqrt{|W_1(j\omega)|^2 + |W_2(j\omega)P(j\omega)C_{app}(j\omega)|^2}}{|1 + P(j\omega)C_{app}(j\omega)|},$$

which is 3.032; see Figure 6.19. This means that the performance of $C_{app}$ is about 20% worse than that of the optimal controller ($\gamma_{opt} = 2.52$). For a performance closer to the optimum, higher order approximations of the controller are needed; see further discussion in Section 6.4 on alternative ways to derive finite dimensional controllers.

### 6.3.4 ▪ Strong stabilization of systems with time delays

Let us now recall the problem defined in Section 5.4, with the plant given as (5.40). We are interested in finding a relationship between the plant parameters $\zeta$, $\rho$, $\sigma$, and $h$ such that $\gamma_{opt} < 1$ is satisfied in (5.46). We will take $\omega_o = 1$ for frequency scaling. Clearly, as $\sigma$ increases, the constraint on the control magnitude gets relaxed, and it should be easier to solve the problem at hand. Moreover, the relation between $\zeta$ and $\rho$ should be such that the PIP should not be "close to violation" by small perturbations of these parameters. Furthermore, we expect that as $h$ increases, $\gamma_{opt}$ increases; hence, there

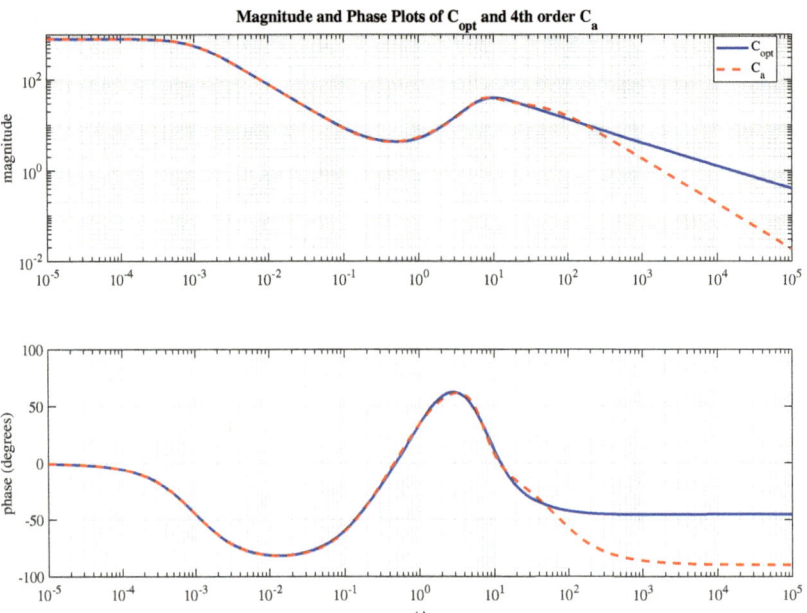

**Figure 6.18.** *Bode plots of $C_{\mathrm{opt}}(j\omega)$ and $C_a(j\omega)$ defined in (6.84).*

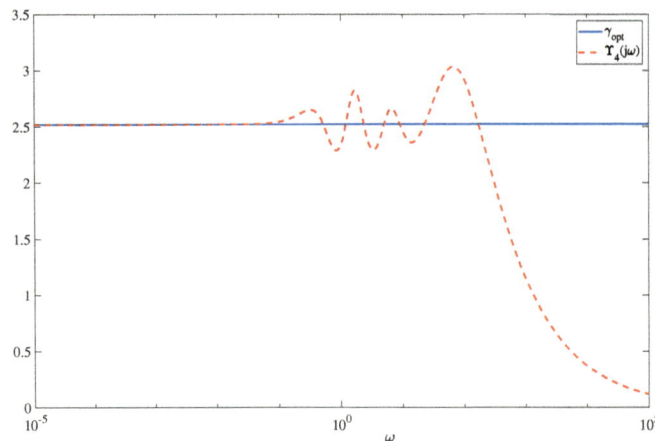

**Figure 6.19.** *Performance comparisons between the optimal controller and the fourth order controller* (6.84).

should be an upper limit on $h$ to have $\gamma_{\mathrm{opt}} < 1$, once all other parameters are fixed. These are the guiding principles in parameter selections.

The mixed sensitivity minimization problem to be solved has plant $P = M_n N_o / M_d$, with

$$M_n(s) = e^{-hs}\frac{(s^2 - 2\zeta s + 1)}{(s^2 + 2\zeta s + 1)}, \quad M_d(s) = \frac{s - \rho}{s + \rho}$$

and the weights

$$W_1(s) = \left(\frac{1}{1+\epsilon}\right)\left(\frac{\epsilon s + (2+\epsilon)\rho}{s+\rho}\right), \quad W_2(s) = \left(\frac{\epsilon s + (2+\epsilon)\rho}{\sigma}\right)\frac{(s+\delta)(s+1)^2}{(s^2+2\zeta s+1)}$$

(clearly, $N_o$ does not play a role in the computation of $\gamma_{\text{opt}}$, but it appears in the definition of $W_2$ and the optimal solution $Q_{\text{opt}}$).

By taking $\sigma = 10$, $\epsilon = 10^{-6}$, and $\delta = 10^{-6}$, we obtain the numerical values of $\gamma_{\text{opt}}$ listed in Table 6.1.

**Table 6.1.** $\gamma_{\text{opt}}$ *values as $\rho$, $\zeta$, and $h$ vary.*

| $\rho$ | 0.15 | 0.1 | 0.25 | 0.25 | 0.1 | 0.1 | 0.41 | 0.25 | 0.25 |
|---|---|---|---|---|---|---|---|---|---|
| $\zeta$ | 0.15 | 0.8 | 0.1 | 0.25 | 0.1 | 0.8 | 0.1 | 0.25 | 0.39 |
| $h$ | 0.1 | 0.1 | 0.1 | 0.1 | 5.07 | 2.375 | 0.1 | 0.70 | 0.1 |
| $\gamma_{\text{opt}}$ | 0.399 | 0.568 | 0.614 | 0.797 | 0.998 | 0.999 | 0.992 | 0.996 | 0.996 |

**Exercise.**
1. Let $h = 0$; then, on the $\rho - \zeta$ plane, within the rectangle $\rho < 1$ and $\zeta < 1$, find the feasible region in which we have $\gamma_{\text{opt}} \leq 1$.
2. For the values of $(\rho, \zeta)$ in the feasible region found in part 1, compute $h_{\max}(\rho, \zeta)$ and obtain a 3D graph for this function. Verify the values given in Table 6.1.

## 6.3.5 ▪ $\mathcal{H}_\infty$-Optimal estimation under delayed measurements

Let us now consider the $\mathcal{H}_\infty$ optimal estimation problem posed in Section 5.3. It was shown that minimizing $\gamma(Q)$, defined by (5.39), over $Q \in \mathcal{H}_\infty$ is equivalent to solving a mixed sensitivity minimization problem defined by the plant $P(s) = e^{-hs}$ and the weights $W_1(s) = P_w(s)$, $W_2(s) = W_v(s)$. Once we find the optimal controller $C_{\text{opt}}$ solving this problem, the optimal $Q$ minimizing $\gamma(Q)$ is then computed as

$$Q_{opt}(s) = \frac{C_{\text{opt}}(s)}{1 + e^{-hs}C_{\text{opt}}(s)}. \tag{6.85}$$

Typically, the process $P_w$ is a low-pass filter, and the measurement noise is generated by a high-pass filter $W_v$. Accordingly, let us take

$$W_1(s) = P_w(s) = \frac{0.01s+1}{s+1}, \quad W_2(s) = W_v(s) = \frac{s+0.1}{s+100}$$

and analyze the effect of time delay $h > 0$ on the achievable performance level

$$\gamma_{\text{opt}} = \inf_{Q \in \mathcal{H}_\infty} \gamma(Q).$$

The results are listed in Table 6.2, where $Q_k(s)$ refers to the $k$th order approximation of $Q_{\text{opt}}$, such that $\|Q_{\text{opt}} - Q_k\|_\infty \leq 0.1$.

$$Q_1(s) = \frac{0.1364(s+110.8)}{(s+17.54)},$$

$$Q_2(s) = \frac{0.215(s+81.32)(s+16.62)}{(s^2+22.18s+368.8)},$$

$$Q_4(s) = \frac{0.319(s+91.2)(s+8.876)(s^2+13.75s+163.3)}{(s^2+5.343s+132.8)(s^2+27.29s+465.3)}.$$

**Table 6.2.** $\gamma_{\text{opt}}$ *values as $h$ varies.*

| $h =$ | 0.1 sec | 0.25 sec | 0.5 sec | 1.0 sec |
|---|---|---|---|---|
| $\gamma_{\text{opt}} \approx$ | 0.13854 | 0.21106 | 0.31726 | 0.47984 |
| $Q_{\text{opt}} \approx$ | $Q_1(s)$ | $Q_2(s)$ | $Q_4(s)$ | $Q_7(s)$ |

**Exercise.** For each of the above cases, obtain $Q_{\text{opt}}(j\omega)$ for $\omega \in (10^{-3}, 10^5)$ and draw the magnitude and phase as functions of $\omega$. Show that $\|Q_{\text{opt}} - Q_1\|_\infty < 0.009$, $\|Q_{\text{opt}} - Q_2\|_\infty < 0.044$, and $\|Q_{\text{opt}} - Q_4\|_\infty < 0.05$. By using the MATLAB commands below, where om refers to the vector of $\omega$ points and Qopt is the vector of values of $Q(j\omega)$ at those points,

```
>> Qoptfreq=frd(Qopt,om);
>> Qappsys=fitfrd(Qoptfreq,7,0);
>> zpk(Qappsys);
```

obtain an expression for the seventh order approximation of $Q_{\text{opt}}$,

$$Q_7(s) = \frac{0.47872(s+103.8)(s^2+9.237s+27.73)}{(s+22.83)(s^2+2.125s+47.6)}$$
$$\times \frac{(s^2+7.494s+76.76)(s^2+4.77s+196.3)}{(s^2+4.279s+164.2)(s^2+9.596s+223.6)},$$

and show that the approximation error satisfies $\|Q_{\text{opt}} - Q_7\|_\infty < 0.1$. Observe that as $h$ increases, the filter $Q_{\text{opt}}$ moves from a low-pass filter to a band-pass filter (see the Bode plots of $Q_{\text{opt}}$, $Q_2$ for $h = 0.25$, and $Q_{\text{opt}}$, $Q_7$ for $h = 1.0$, shown in Figures 6.20 and 6.21, respectively). We have assumed the process $W_1 = P_w$ to be a low-pass filter with cut-off at 1rad/sec, and the noise becomes significant for frequencies above 10rad/sec (draw the Bode plots of $P_w$ and $W_v$); hence, the delay introduces somewhere between $h$rad and 10$h$rad phase shift within this frequency band. When $h = 0.1$sec this phase shift is on the order of $5.73°$ to $57.3°$, which is relatively small. The phase shift becomes more and more significant as $h$ increases. □

## 6.4 • Approximations

This section is devoted to the discussion of the computation of suboptimal controllers from two alternative methods:

- approximate the infinite dimensional parts of the $\mathcal{H}_\infty$ optimal controller;

- approximate the infinite dimensional plant, and design $\mathcal{H}_\infty$ optimal controller for the approximate plant.

In each of these methods, performance degradation due to approximations needs to be determined. There have been numerous paper published on this topic. One of the earliest results was in [201], where the structure of the optimal controller was identified, and it was proposed to use a finite dimensional $M_{na}(s)$ to replace $M_n(s)$ in (6.21); it was noted that such $M_{na}$ must take the same values as $M_n$ at the interpolation points $\beta_1, \ldots, \beta_{n_1}$ and $\alpha_1, \ldots, \alpha_\ell$ (the paper [201] considered stable plants and sensitivity minimization problem only, but the basic idea extends to unstable plants and mixed sensitivity minimization). Since then, many improvements have been reported in the

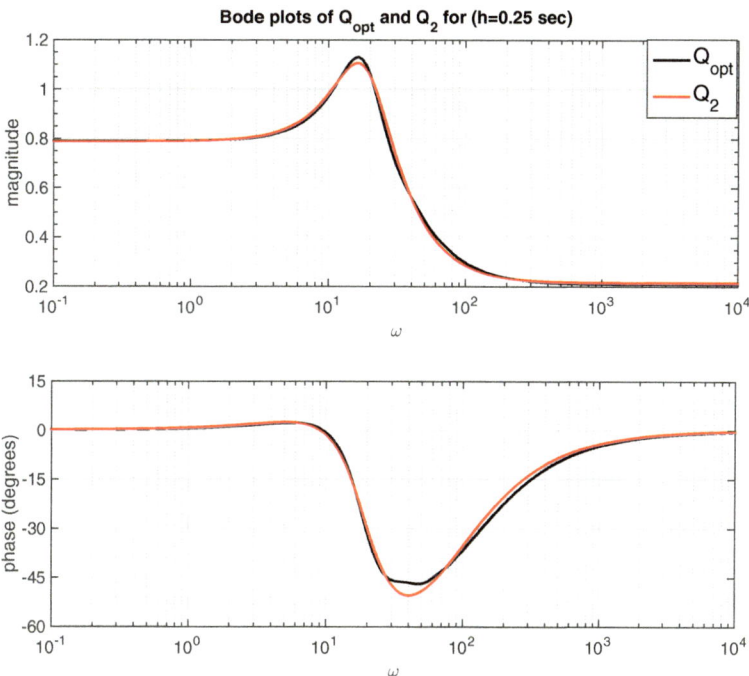

**Figure 6.20.** *Magnitude and phase plots of $Q_{\mathrm{opt}}(j\omega)$ and $Q_2(j\omega)$ for $h = 0.25$.*

literature. Most of the results given in this section are based on recent publications [195, 265, 266].

There is also a direct way of obtaining fixed-structure $\mathcal{H}_\infty$ controllers for systems with time delays [93]. This is useful in order to compute low order $\mathcal{H}_\infty$ controllers, such as PID controllers or lead-lag controllers. However, this approach does local optimization, and therefore the resulting controller is in general suboptimal. Finding the optimal fixed-structure $\mathcal{H}_\infty$ controller for DPSs is an open problem. Once a given infinite dimensional plant is approximated by a rational transfer function, a fixed order $\mathcal{H}_\infty$ controller for this reduced order plant can also be computed using optimization-based tuning methods, e.g., HIFOO [92]. However, even for finite dimensional systems, optimal fixed order $\mathcal{H}_\infty$ controller computation is an unsolved problem.

## 6.4.1 ▪ Approximations of the $\mathcal{H}_\infty$ optimal controller

The optimal $\mathcal{H}_\infty$ controller is computed in Section 6.1, and its stable implementation is given in Section 6.2. Recall that when the plant has finitely many unstable poles and the weights are finite dimensional, the optimal controller can be written as

$$C_{\mathrm{opt}}(s) = \frac{C_0(s)}{1 + C_0(s)H_d(s)}\, C_1(s), \qquad (6.86)$$

where

$$C_0(s) = (1 + H_n(s))^{-1}$$

is finite dimensional, $H_d(s)$ is an infinite dimensional transfer function in $\mathcal{H}_\infty$, and

$$C_1(s) = (d_\infty \gamma_{\mathrm{opt}})^{-1} G_{\mathrm{opt}}(s) N_o^{-1}(s).$$

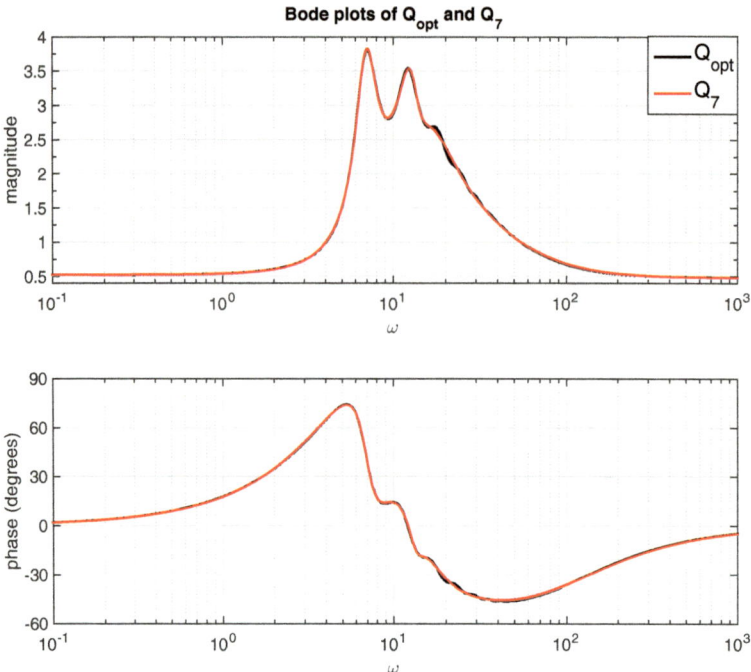

**Figure 6.21.** *Magnitude and phase plots of $Q_{\mathrm{opt}}(j\omega)$ and $Q_7(j\omega)$ for $h = 1.0$.*

The transfer function $G_{\mathrm{opt}} \in \mathcal{H}_\infty$ is determined from the weights using a spectral factorization

$$G_{\mathrm{opt}}(s)G_{\mathrm{opt}}(-s) = \left(1 + \frac{W_2(s)W_2(-s)}{W_1(s)W_1(-s)} - \frac{W_2(s)W_2(-s)}{\gamma_{\mathrm{opt}}^2}\right)^{-1},$$

and $N_o$ is the outer part of the plant. If $N_o$ is infinite dimensional, then so is $C_1$. When the plant is a retarded or neutral time delay system, the impulse response of $H_d$ is restricted to the time interval $[0, \tau_{\max}]$, where $\tau_{\max}$ is the maximum time delay in the inner part of the plant. In this case, various interesting approximation techniques can be used in sampled-data implementation; see, e.g., [171].

Let us now consider an "approximately optimal" controller in the form

$$C_a(s) = \frac{C_0(s)}{1 + C_0(s)H_{da}(s)}\, C_{1a}(s), \tag{6.87}$$

where $H_{da}$ is an approximation of $H_d$, and

$$C_{1a}(s) = (d_\infty \gamma_{\mathrm{opt}})^{-1} G_{\mathrm{opt}}(s) N_{oa}^{-1}(s),$$

with $N_{oa}^{-1}$ being an outer approximation of $N_o^{-1}$.

The effect of these approximations on the feedback system stability and the resulting performance level in terms of deviations in the optimal sensitivity and complementary sensitivity functions must be analyzed. It will be clear from the analysis below that we want $N_o C_{1a} \in \mathcal{H}_\infty$; this imposes a constraint on $N_{oa}$ that its relative degree should be less than or equal to the that of $N_o$. Moreover, having an outer $N_{oa}$ does not introduce artificial unstable poles that are not present in the optimal controller.

Under the controller $C_a$ defined in (6.87) the resulting sensitivity is

$$S_a = \frac{M_d(1 + C_0 H_{da})}{M_d(1 + C_0 H_{da}) + M_n N_o C_0 C_{1a}}, \tag{6.88}$$

where $P = M_n N_o/M_d$ is the plant factorization. By adding and subtracting the term $M_d(1 + C_0 H_d)$ to the numerator and denominator of (6.88) and adding and subtracting $M_n N_o C_1 C_0$ to the denominator, it is possible to extract the optimal sensitivity $S_{\text{opt}}$ in (6.88). That leads to

$$S_a = \frac{S_{\text{opt}} + \Delta_1}{1 + \Delta_1 + \Delta_2} \quad \text{and} \quad T_a = \frac{T_{\text{opt}} + \Delta_2}{1 + \Delta_1 + \Delta_2}, \tag{6.89}$$

where $T_a = 1 - S_a$,

$$\Delta_1 = S_{\text{opt}} C_{\text{opt}} (H_{da} - H_d)/C_1 = S_{\text{opt}} \frac{C_0(H_{da} - H_d)}{1 + C_0 H_d},$$

$$\Delta_2 = T_{\text{opt}}(N_{oa}^{-1} N_o - 1),$$

and $S_{\text{opt}} = (1 + P C_{\text{opt}})^{-1}, T_{\text{opt}} = 1 - S_{\text{opt}}$. Thus, $S_a \in \mathcal{H}_\infty$ if

$$\delta_a := (\delta_1 + \delta_2) < 1, \tag{6.90}$$

where

$$\delta_1 := \|\Delta_1\|_\infty = \|S_{\text{opt}} \frac{C_0(H_{da} - H_d)}{1 + C_0 H_d}\|_\infty, \tag{6.91}$$

$$\delta_2 := \|\Delta_2\|_\infty = \|T_{\text{opt}}(N_{oa}^{-1} N_o - 1)\|_\infty. \tag{6.92}$$

**Exercise**: Prove that when $S_a \in \mathcal{H}_\infty$ we have $C_a S_a \in \mathcal{H}_\infty$ and $P S_a \in \mathcal{H}_\infty$. With this fact we conclude that the feedback system $(C_a, P)$ is stable when $S_a \in \mathcal{H}_\infty$.
*Hint*: First write a coprime factorization $C_0 = N_{C0}/D_{C0}$, where $N_{C0}$ and $D_{C0}$ are in $\mathcal{H}_\infty$. Then, $S_a$ can be written as

$$S_a = \frac{N_{Sa}}{D_{Sa}}, \qquad \begin{aligned} N_{Sa} &:= M_d(D_{C0} + N_{C0} H_{da}), \\ D_{Sa} &:= M_d(D_{C0} + N_{C0} H_{da}) + N_{C0} M_n N_o C_{1a}. \end{aligned}$$

Clearly, $N_{Sa}$ and $D_{Sa}$ are strongly coprime in $\mathcal{H}_\infty$. Now it is easy to see that

$$C_a S_a = \frac{M_d N_{C0} C_{1a}}{D_{Sa}}, \qquad P S_a = \frac{M_n N_o(D_{C0} + N_{C0} H_{da})}{D_{Sa}}$$

are strongly coprime factorizations.                                                    $\square$

From (6.89) it is clear that for small deviation from the optimal performance level, we need to design $H_{da}$ and $N_{oa}^{-1}$ so that $\delta_a$ is much smaller than 1. In order to estimate the resulting performance level under the controller $C_a$, we note that

$$\begin{bmatrix} W_1 S_a \\ W_2 T_a \end{bmatrix} = \begin{bmatrix} W_H & 0 \\ 0 & W_N \end{bmatrix} \begin{bmatrix} W_1 S_{\text{opt}} \\ W_2 T_{\text{opt}} \end{bmatrix} \frac{1}{1 + \Delta_1 + \Delta_2}, \tag{6.93}$$

where

$$W_H := 1 + \frac{\Delta_1}{S_{\text{opt}}} = \frac{1 + C_0 H_{da}}{1 + C_0 H_d}, \tag{6.94}$$

$$W_N := 1 + \frac{\Delta_2}{T_{\text{opt}}} = N_{oa}^{-1} N_o. \tag{6.95}$$

The above derivations lead to the following result.

**Theorem 6.14.** *The controller $C_a$ given by (6.87) stabilizes the feedback system, with the plant $P = M_n N_o / M_d$, if $\delta_a$ defined by (6.90) and (6.91)–(6.92) satisfies $\delta_a < 1$. Moreover, in this case the resulting performance level is estimated by the following inequality:*

$$\gamma_a := \left\| \begin{bmatrix} W_1 S_a \\ W_2 T_a \end{bmatrix} \right\|_\infty \leq \gamma_{\text{opt}} \frac{1 + \varepsilon_a}{1 - \delta_a}, \tag{6.96}$$

*where*

$$1 + \varepsilon_a := \max\{\|W_H\|_\infty, \|W_N\|_\infty\},$$

*with $W_H$ and $W_N$ defined by (6.94) and (6.95), respectively.*

Typically, $N_o$ and $H_d$ are low-pass or band-pass filters; therefore, it is sufficient to approximate them on a finite frequency band. The width of this frequency band is mainly determined by the bandwidths of $T_{\text{opt}}$ and $S_{\text{opt}} C_0 H_d (1 + C_0 H_d)^{-1}$. More precisely, $H_{da}$ and $N_{oa}^{-1}$ are designed in such a way that the peaks of $|W_H(j\omega)|$ and $|W_N(j\omega)|$ are close to unity for $\omega \in \mathbb{R}$. Also, since we take $W_N$ to be in $\mathcal{H}_\infty$, the relative order of $N_{oa}$ should be less than or equal to that of $N_o$.

An upper bound of $\delta$ can be found by noting that $|W_1(j\omega) S_{\text{opt}}(j\omega)| \leq \gamma_{\text{opt}}$ and $|W_2(j\omega) T_{\text{opt}}(j\omega)| \leq \gamma_{\text{opt}}$ for all $\omega$:

$$\delta_1 \leq \gamma_{\text{opt}} \|W_1^{-1} \frac{C_0}{1 + C_0 H_d} (H_{da} - H_d)\|_\infty,$$

$$\delta_2 \leq \gamma_{\text{opt}} \|W_2^{-1} (N_{oa}^{-1} N_o - 1)\|_\infty.$$

These inequalities give guidelines on how $H_d$ and $N_o$ should be approximated for a small deviation in the $\mathcal{H}_\infty$ performance level.

## 6.4.2 • Approximations of the plant

In this section we obtain a finite dimensional $\mathcal{H}_\infty$ controller by approximating the infinite dimensional parts of the plant and designing a controller minimizing the mixed sensitivity for this approximate plant.

The notation to be used is as follows: recall that the original plant has a factorization in the form $P = M_n N_o / M_d$, where $M_d$ is inner and finite dimensional; accordingly, assume that the numerator of the plant is approximated in such a way that its inner outer factorization yields

$$P_\nu(s) = \frac{M_{n\nu}(s) N_{o\nu}(s)}{M_d(s)}, \tag{6.97}$$

where $M_{n\nu}$ is inner and $N_{o\nu}$ is outer and they are both finite dimensional. Let us define

$$\gamma_\nu^o := \inf_{C \in \mathscr{C}(P_\nu)} \left\| \begin{bmatrix} W_1(1 + P_\nu C)^{-1} \\ W_2 P_\nu C (1 + P_\nu C)^{-1} \end{bmatrix} \right\|_\infty, \tag{6.98}$$

where $\mathscr{C}(P_\nu)$ denotes the set of all controllers stabilizing the feedback system $(C, P_\nu)$. Recall that, by Corollary 6.2, $\gamma_\nu^o$ is the largest $\gamma$ which makes $\mathcal{T}_\nu(\gamma)$ singular, where $\mathcal{T}_\nu(\gamma)$ is the same as $\mathcal{T}(\gamma)$, defined by (6.13), except that the terms $M_n(\alpha_i) F_\gamma(\alpha_i)$ and $M_n(\beta_j) F_\gamma(\beta_j)$ are replaced by $M_{n\nu}(\alpha_i) F_\gamma(\alpha_i)$ and $M_{n\nu}(\beta_j) F_\gamma(\beta_j)$, respectively. This observation leads to the following result.

**Proposition 6.15.** *Let $W_1$, $W_2$, and $M_n$ be such that function $W_{MF}(s) = M_n(s) F_\gamma(s)$ can be approximated by a sequence of functions $\{W_{MF1}, W_{MF2}, \ldots\}$, $W_{MF\nu} \in \mathcal{H}_\infty$*

*for all $\nu = 1, 2, \ldots$, in such a way that*

$$\|W_{MF} - W_{MF\nu}\|_\infty \to 0 \quad as \quad \nu \to \infty. \tag{6.99}$$

*Then, $\gamma_\nu^o \to \gamma_{\mathrm{opt}}$ as $\nu \to \infty$.*

**Proof.** Under the condition (6.99) we have that

$$\mathcal{T}_\nu(\gamma) \to \mathcal{T}(\gamma) \quad as \quad \nu \to \infty \tag{6.100}$$

for each fixed $\gamma$. In particular, this means that all singular values and vectors of $\mathcal{T}_\nu(\gamma)$ converge to those of $\mathcal{T}(\gamma)$. $\square$

Since $F_\gamma$ is finite dimensional and directly determined from $W_1$ and $W_2$, for the approximation $W_{MF\nu}$ we can take

$$W_{MF\nu}(s) = M_{n\nu}(s)F_\gamma(s).$$

In most practical applications, $W_1$ is an outer low-pass transfer function, and so is $W_2^{-1}$. In this case, $F_\gamma(s)$ is a low-pass transfer function. This means that in order to guarantee a performance level $\gamma_\nu^o$ close to $\gamma_{\mathrm{opt}}$, it is sufficient to approximate the $M_n(s)$ within a finite frequency band by an inner function $M_{n\nu}(s)$.

Let the optimal controller resulting from (6.98) be denoted by $C_\nu$. Note that it is in the form

$$C_\nu(s) = M_d E_\nu \frac{F_\nu L_\nu N_{ov}^{-1}}{1 + M_{n\nu}F_\nu L_\nu}, \tag{6.101}$$

where $E_\nu$ and $F_\nu$ denote $E_\gamma$ and $F_\gamma$, respectively, for the value of $\gamma$ equal to $\gamma_\nu^o$; $L_\nu$ is determined from the singular vectors of $\mathcal{T}_\nu(\gamma_\nu^o)$, similar to the computation of $L$ from the singular vectors of $\mathcal{T}(\gamma_{\mathrm{opt}})$. Assuming that as $\nu \to \infty$ the properties (6.99) and (6.100) hold, then

$$E_\nu \to E_{\gamma_{\mathrm{opt}}}, \quad F_\nu \to F_{\gamma_{\mathrm{opt}}}, \quad L_\nu \to L$$

as $\nu \to \infty$. Each convergence above is in the sense of the convergence of the coefficients of the finite dimensional systems in question.

Now we need to investigate the stability and performance of the closed-loop system formed by the finite dimensional controller $C_\nu$ and the original infinite dimensional plant $P$. For this purpose, let us first examine the resulting sensitivity function

$$S_\nu := (1 + PC_\nu)^{-1} = \left(1 + P_\nu C_\nu + C_\nu(P - P_\nu)\right)^{-1},$$

which can be rewritten as

$$S_\nu = S_\nu^o \left(1 + T_\nu^o \left(\frac{P}{P_\nu} - 1\right)\right)^{-1}, \tag{6.102}$$

where

$$S_\nu^o := (1 + P_\nu C_\nu)^{-1} \quad \text{and} \quad T_\nu^o := 1 - S_\nu^o. \tag{6.103}$$

Since $P = M_n N_o/M_d$ and $P_\nu = M_{n\nu}N_{ov}/M_d$ have the same number of unstable poles, and the feedback system $(C_\nu, P_\nu)$ is stable, we have that $(C_\nu, P)$ is stable if

$$\left\| T_\nu^o \left(\frac{M_n N_o}{M_{n\nu}N_{ov}} - 1\right) \right\|_\infty < 1.$$

By design, we have that $|W_2(j\omega)T_\nu^o(j\omega)| \leq \gamma_\nu^o$ for all $\omega \in \mathbb{R}$. So, a sufficient condition for the stability of the feedback system $(C_\nu, P)$ is

$$\delta_\nu := \left\| W_2^{-1}\left( \frac{M_n N_o}{M_{n\nu} N_{o\nu}} - 1 \right) \right\|_\infty < \frac{1}{\gamma_\nu^o}. \tag{6.104}$$

Thus, the approximating functions $M_{n\nu}$ and $N_{o\nu}$ should be designed to satisfy (6.104).

Let us now assume that the stability condition (6.104) is satisfied and examine the performance of the feedback system $(C_\nu, P)$. For this purpose we note that (6.102) implies

$$S_\nu = \frac{S_\nu^o}{1 + T_\nu^o \Delta_\nu}, \quad T_\nu = \frac{T_\nu^o + T_\nu^o \Delta_\nu}{1 + T_\nu^o \Delta_\nu}, \tag{6.105}$$

where

$$\Delta_\nu = \frac{M_n N_o}{M_{n\nu} N_{o\nu}} - 1. \tag{6.106}$$

The resulting performance for the feedback system $(C_\nu, P)$ is determined from

$$\begin{bmatrix} W_1 S_\nu \\ W_2 T_\nu \end{bmatrix} = \begin{bmatrix} 1 & 0 \\ 0 & 1 + \Delta_\nu \end{bmatrix} \begin{bmatrix} W_1 S_\nu^o \\ W_2 T_\nu^o \end{bmatrix} \frac{1}{1 + T_\nu^o \Delta_\nu}. \tag{6.107}$$

This leads us to the following result.

**Theorem 6.16.** *The controller $C_\nu$ given by (6.101) stabilizes the feedback system, with the plant $P = M_n N_o / M_d$, if $\delta_\nu \gamma_\nu^o < 1$, i.e., (6.104) holds. Moreover, in this case the resulting performance level is estimated by the following inequality:*

$$\gamma_\nu := \left\| \begin{bmatrix} W_1 S_\nu \\ W_2 T_\nu \end{bmatrix} \right\|_\infty \leq \gamma_\nu^o \frac{1 + \varepsilon_\nu}{1 - \delta_\nu}, \tag{6.108}$$

*where*

$$1 + \varepsilon_\nu := \| N_{o\nu}^{-1} N_o \|_\infty.$$

*Furthermore, if a sequence of approximations $M_{n\nu} N_{o\nu}$, $\nu = 1, 2, \ldots$, is designed in such a way that*

$$\delta_\nu = \left\| W_2^{-1}\left( \frac{M_n N_o}{M_{n\nu} N_{o\nu}} - 1 \right) \right\|_\infty \to 0 \quad as \quad \nu \to \infty,$$

*and $M_{n\nu} F_\gamma \to M_n F_\gamma$ for all $\gamma \in [\gamma_{\min}, \gamma_{\max}]$, then $\gamma_\nu \to \gamma_{\text{opt}}$ as $\nu \to \infty$.*

**Proof.** The result follows from the identity (6.107) and Proposition 6.15. $\quad\square$

### 6.4.3 ▪ Examples of approximations

In this section, we give examples for the two approaches analyzed above:

(i) approximate the infinite dimensional parts of the optimal controller, and implement this on the original plant;

(ii) approximate the plant, compute the finite dimensional optimal controller for the new system, and then implement it on the original plant.

Both of these methods require approximations of infinite dimensional transfer functions in $\mathcal{H}_\infty$: in the first approach, infinite dimensional parts of the controller to be approximated are $H_d$ and $N_o^{-1}$; in the second approach, assuming the original plant is factored as $P = N/D$, we need approximations of $N$ and $D$ (in most cases studied, $D$ is already finite dimensional). We do not discuss or compare several existing methods for finding an $\mathcal{H}_\infty$ approximation of a given infinite dimensional system; there is a large literature on this topic. See, for example, [22, 88, 90, 150, 186, 185, 206, 254, 257] for alternative approximation methods and related results. In particular, in our case, the infinite dimensional term $H_d$ is an FIR filter. Rational approximations of these types of systems have attracted special attention in recent years; see, e.g., [279, 157] and their references. Naturally, sampled-data implementation of such an FIR block uses the values of the impulse response at sampling instants. From a practical implementation point of view, high order digital FIR filters can be desirable, compared to other rational approximations that assume no structure. See [170, 171] and [279] for detailed discussion on sampled-data implementations.

In what follows we use the MATLAB built-in function `fitfrd` to approximate stable infinite dimensional transfer functions whose frequency responses are obtained by evaluating the functions at $j\omega_k$, $k = 1, \ldots, N$, with sufficiently large $N$ capturing essential characteristics of the function. Another MATLAB command to estimate the transfer function from a set of finite number of frequency response data is `tfest`, which was introduced recently in R2016a and shown to give successful results on some benchmark examples [203].

**A retarded time-delay system**

In this section we return to the unstable retarded time delay system example, where the plant was given by (6.62). For the mixed sensitivity weights taken as $W_1(s) = 1/s$ and $W_2(s) = 2\,s$, the optimal controller was computed as

$$C_{\text{opt}}(s) = -0.5\,\gamma_{\text{opt}}\,\left(\frac{\psi\,s + 1}{\psi\,s}\right)\,\left(\frac{1}{1 + H_d(s)}\right)\,N_1^{-1}(s),$$

where $\gamma_{\text{opt}} = 17.846$, $\psi = 23.28$, the Bode plots of $H_d$ are as shown in Figure 6.8, and $N_1$, given by (6.65), is infinite dimensional,

$$N_1(s) = \frac{(s+4)(s+1)}{(s^2 + 8s + 17)}\,\left(\frac{1}{1 + H_F(s)}\right),$$

with the FIR filter $H_F(s)$ determined in (6.64). Therefore, the indirect controller approximation is in the form

$$C_a(s) = -0.5\,\gamma_{\text{opt}}\,\left(\frac{(\psi\,s + 1)\,(s^2 + 8s + 17)}{\psi\,s\,(s+4)\,(s+1)}\right)\,\left(\frac{1 + H_{Fa}(s)}{1 + H_{da}(s)}\right),$$

where $H_{Fa}$ and $H_{da}$ are finite dimensional approximations of $H_F$ and $H_d$, respectively. By using the `fitfrd` command of MATLAB, we determine a seventh order $H_{da}$ from the frequency response of $H_d$; similarly, we determine a fifth order $H_{Fa}$ from the Bode plots of $H_F$:

$$H_{da}(s) = \frac{11(s^2 - 5s + 31)(s^2 - 5s + 206)(s^2 - 18s + 580)}{(s+1)(s^2 + 18s + 84)(s^2 + 14s + 132)(s^2 + 9s + 290)},$$

$$H_{Fa}(s) = \frac{4.26(s^2 - s + 21)(s^2 - 3s + 125)}{(s + 3.6)(s^2 + 6s + 20)(s^2 + 5s + 52)}.$$

The Bode plots of $H_d$ and $H_{da}$ are shown in Figure 6.22. In the above expressions, the coefficients are rounded off to integers from the initial outputs of `fitfrd` for $H_{da}$ and $H_{Fa}$.

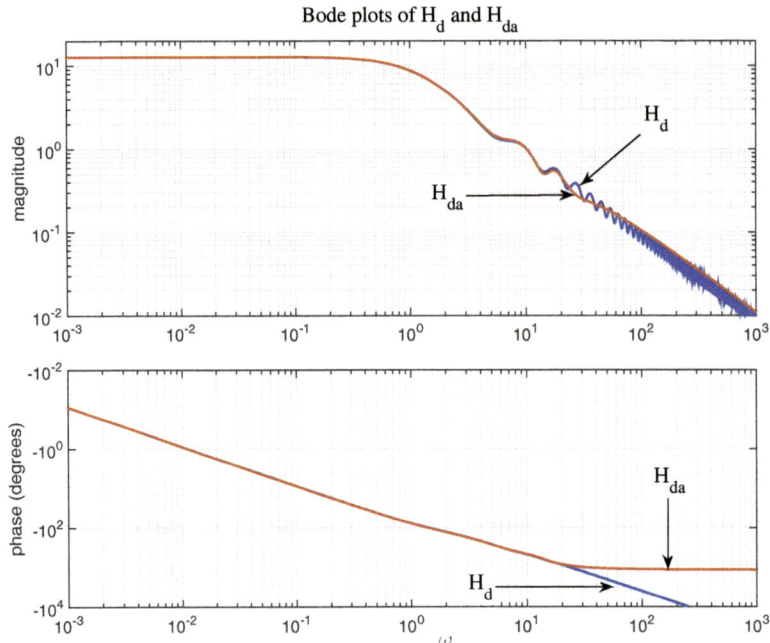

**Figure 6.22.** *Magnitude and phase plots of $H_d(j\omega)$ and $H_{da}(j\omega)$.*

Inserting the above transfer function into the formula for $C_a$ and then performing a model reduction (cancellation of poles and zeros that are in $\mathbb{C}_-$ and significantly close to each other), we obtain the following 13th order controller:

$$C_a(s) = \frac{-9(s + 0.043)(s^2 + 8s + 17)(s^2 + 13s + 50)(s^2 + 0.3s + 12.5)}{s(s + 34.8)(s + 4)(s^2 + 8s + 19)(s^2 - 1.8s + 13.8)}$$
$$\times \frac{(s^2 + 1.5s + 65)(s^2 + 3.45s + 100)(s^2 + 15s + 250)}{(s^2 + 6.5s + 34)(s^2 + 2s + 81)(s^2 + 2.5s + 140))}.$$

The Bode plots of $C_{\text{opt}}$ and $C_a$ are shown in Figure 6.23. Once this controller is implemented, the suboptimal performance level $\gamma_a$, defined as the peak value of $\Upsilon_a(j\omega)$, can be computed:

$$\Upsilon_a(j\omega) = \frac{\sqrt{|W_1(j\omega)|^2 + |W_2(j\omega)P(j\omega)C_a(j\omega)|^2}}{|1 + P(j\omega)C_a(j\omega)|}. \qquad (6.109)$$

Figure 6.24 shows that $\gamma_a = 19.68$, which is about 10% higher than $\gamma_{\text{opt}} = 17.846$.

At this point we should also verify feedback system stability under the controller $C_a$. For this purpose we draw the Nyquist graph of $PC_a$; see Figure 6.25. Observe that $PC_a$ encircles $-1$ in the counterclockwise direction three times; this matches the number of right half plane poles of $PC_a$; the plant has one and the controller has two poles in $\mathbb{C}_+$. Thus, the feedback system is stable. Also note that the graph of $PC_a$ is close to the graph of $PC_{\text{opt}}$. The optimal controller also has two poles in the open right half plane.

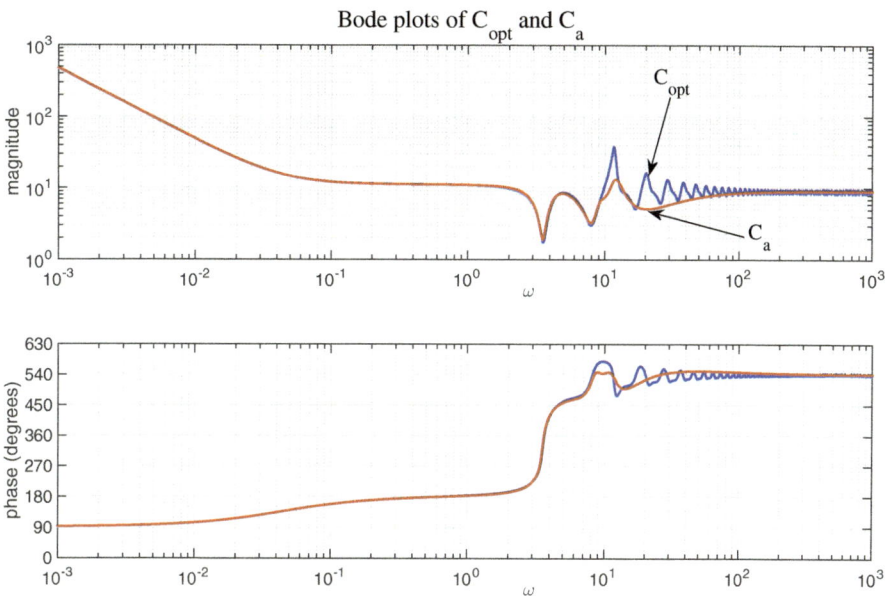

**Figure 6.23.** *Magnitude and phase plots of $C_{\mathrm{opt}}(j\omega)$ and $C_a(j\omega)$.*

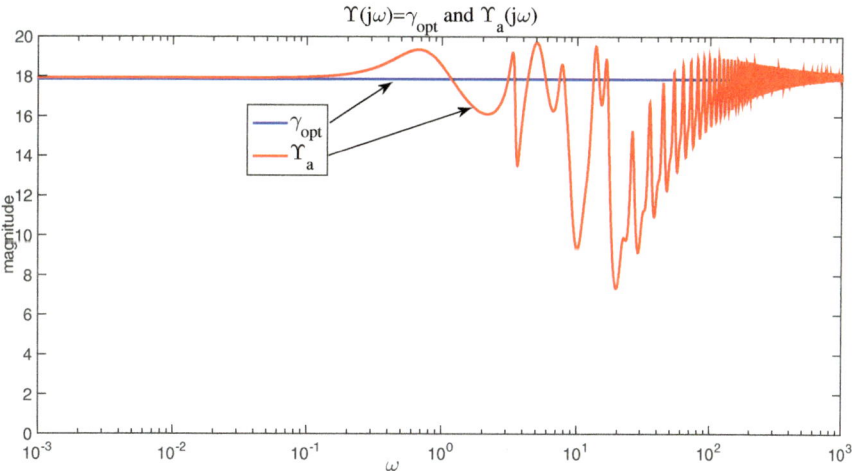

**Figure 6.24.** *The optimal performance level $\gamma_{\mathrm{opt}}$ and $\Upsilon_a(j\omega)$.*

**Exercise 6.17.**

1. In this case, the approximations of $H_d$ and $H_F$ are not sufficiently high order,[2] which leads to an inconclusive performance estimate given in Theorem 6.14. Verify that the above choices lead to $\delta_1 = 0.67$ and $\delta_2 = 0.77$, yielding $\delta_a = (\delta_1 + \delta_2) > 1$; hence the sufficient condition of Theorem 6.14 is not satisfied. Also, check that $\varepsilon_a = 0.554$.

---

[2]For example, we can see from Figure 6.22 that $H_{da}$ starts to deviate from $H_d$ around the frequency interval 20 rad/sec to 30 rad/sec where the magnitude of $H_d$ is in the order of 0.2 to 0.5, so the relative error is significant after this frequency range. Obviously, one has to go to much higher order approximations for a sufficiently small relative approximation error.

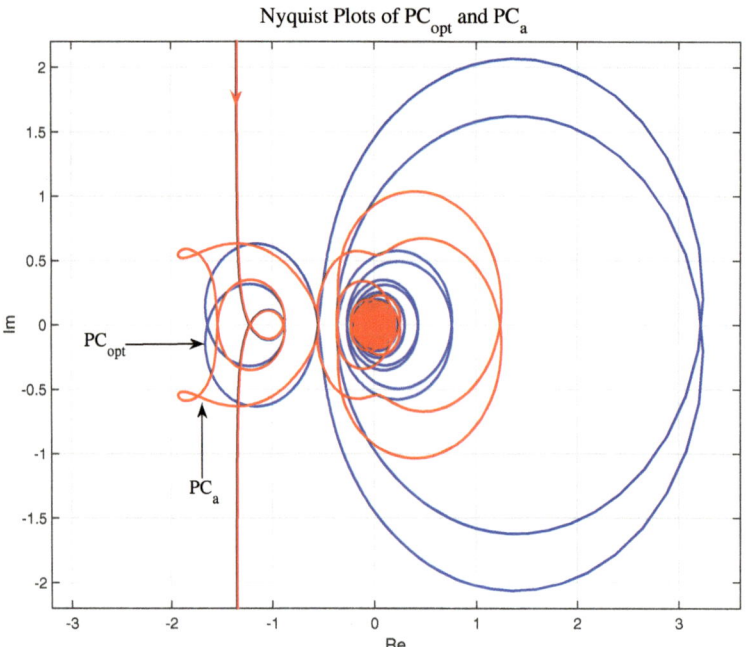

**Figure 6.25.** *Nyquist plots of* $PC_{\mathrm{opt}}$ *and* $PC_a$.

Despite large values of $\delta_a$ and $\varepsilon_a$ the actual performance level $\gamma_a$ is still within 10% of $\gamma_{\mathrm{opt}}$.

2. An alternative approach for this particular example would be to approximate the infinite dimensional term $\frac{1+H_F}{1+H_d}$ in a single step, rather than using separate approximations of $H_F$ and $H_d$. Check that the same difficulty mentioned in part 1 appears in this approach as well.

3. Now consider the approximate plant in the form

$$P_a(s) = \left( \frac{D_h(s)}{1 + H_{Fa}(s)} \right) \frac{(s-1)(s+4)}{(s-0.5)(s^2 + 8s + 17)},$$

where $D_h(s)$ is a second order Padé approximation of $e^{-hs}$ and $H_{Fa}(s)$ is given above. Design a mixed sensitivity minimizing controller $C_{aa}(s)$ using the weights $W_1(s) = 1/s$ and $W_2(s) = 2\ s$. Check whether the feedback system $(C_{aa}, P)$ is stable; then compute $\gamma_{aa}$ as the peak value of $\Upsilon_{aa}(\omega)$, which is obtained by replacing $C_a$ with $C_{aa}$ in (6.109). Increase the Padé approximation order one by one, and repeat the above computations.

**Exercise 6.18.**
Consider the plant

$$P(s) = M_n(s)\, \frac{1}{(s-\alpha)}, \quad \text{where } \alpha = 0.1 \text{ and } M_n(s) = \frac{1 - se^{-hs}}{s + e^{-hs}}, \quad h = \frac{\pi}{8} \text{ sec.}$$

Note that $M_n$ is the sum of two stable pseudorational transfer functions in the form (4.33), as discussed in Example 4.13. The mixed sensitivity weights are $W_1(s) = \frac{1}{s}$ and $W_2(s) = 2\ s$.

1. Find
$$\gamma_{\text{opt}} = \inf_{C \in \mathcal{C}(P)} \left\| \begin{bmatrix} W_1(1 + PC)^{-1} \\ W_2 PC(1 + PC)^{-1} \end{bmatrix} \right\|_\infty$$

and the corresponding optimal controller, $C_{\text{opt}}$ (draw its Bode plots). Express the optimal controller in the form $C_{\text{opt}}(s) = C_{pi}(s)(1 + H(s))^{-1}$, where $C_{pi}$ is a PI controller and $H(s)$ is an FIR filter.

2. By using the `fitfrd` command of MATLAB on $H(j\omega)$, find an approximation of the optimal controller, $C_{\text{app}}(s) = C_{pi}(s)(1 + H_a(s))^{-1}$, where $H_a$ is a third order approximation of $H$, with relative degree 1. Show that $(C_{\text{app}}, P)$ is a stable feedback system and $\gamma_{\text{app}} \leq 1.05 \, \gamma_{\text{opt}}$, where

$$\gamma_{\text{app}} = \left\| \begin{bmatrix} W_1(1 + PC_{\text{app}})^{-1} \\ W_2 PC_{\text{app}}(1 + PC_{\text{app}})^{-1} \end{bmatrix} \right\|_\infty.$$

Note: express $C_{\text{app}}(s)$ in `zpk` form.

3. Let $D_{h,2}(s)$ be a second order Padé approximation of the delay term $e^{-hs}$, and define

$$M_{na}(s) = \frac{1 - sD_{h,2}(s)}{s + D_{h,2}(s)}.$$

Now define the approximate plant as $P_a(s) = M_{na}(s)/(s - \alpha)$. Consider an approximate sensitivity weight $W_{1a}(s) = 1/(s + 10^{-8})$, and define an additive uncertainty weight $W_{3a}(s) = 2\,s/(s + \alpha)$. By using the command

```
>>[Ka,CLa,ga]=mixsyn(Pa,W1a,W3a,[])
```

determine ga and compare it with $\gamma_{\text{opt}}$ found above. Also, compare the controller Ka (express it in `zpk` form) with $C_{\text{app}}$. Check that the feedback loop (Ka, $P$) is stable, and compute the resulting suboptimal performance level

$$\widehat{\gamma}_{\text{app}} = \left\| \begin{bmatrix} W_1(1 + PKa)^{-1} \\ W_2 PKa(1 + PKa)^{-1} \end{bmatrix} \right\|_\infty.$$

Compare this result with $\gamma_{\text{app}}$.

## A fractional order system

In this section we revisit the mixed sensitivity minimization problem for the plant

$$P(s) = \frac{e^{-0.1\,s}}{s^2(\sqrt{s} + 1) - 2}$$

with the weights

$$W_1(s) = \frac{1 + \varepsilon s}{s + \varepsilon}, \quad W_2(s) = 0.5\,s\,(0.2s + 1)^2, \quad \varepsilon = 10^{-3},$$

that was solved in Section 6.3.3. We first apply the method described in Section 6.4.1 to compute a finite dimensional controller. Recall that the optimal controller is in the form

$$C_{\text{opt}}(s) = \left( \frac{C_0(s)}{1 + C_0(s)H_d(s)} \right) \frac{G_{\text{opt}}(s)}{N_o(s)},$$

where $C_0(s) = 1$, $H_d(s)$ is an infinite dimensional system whose Bode plots are as given in Figure 6.17,

$$G_{\mathrm{opt}}(s) = \frac{W_1(s)}{\gamma_{\mathrm{opt}}^2 d_\infty} \widehat{G}_{\gamma_{\mathrm{opt}}}(s) = \frac{\gamma_{\mathrm{opt}}(s + \varepsilon^{-1})}{0.02s^4 + 0.2382s^3 + 0.9198s^2 + 1.3417s + 1},$$

and $N_o(s)$ is an infinite dimensional outer function whose Bode plots are shown in Figure 6.26 with solid lines. A third order minimum phase approximation $N_{oa}$ is obtained using `fitfrd` and enforcing the relative degree to be 2; see the dashed lines in Figure 6.26. Note that the performance deviation bound given in Theorem 6.14 depends on the error term $(\|N_{oa}^{-1}N_o\|_\infty - 1)$, so the relative degree of $N_{oa}$ should be smaller than that of $N_o$.

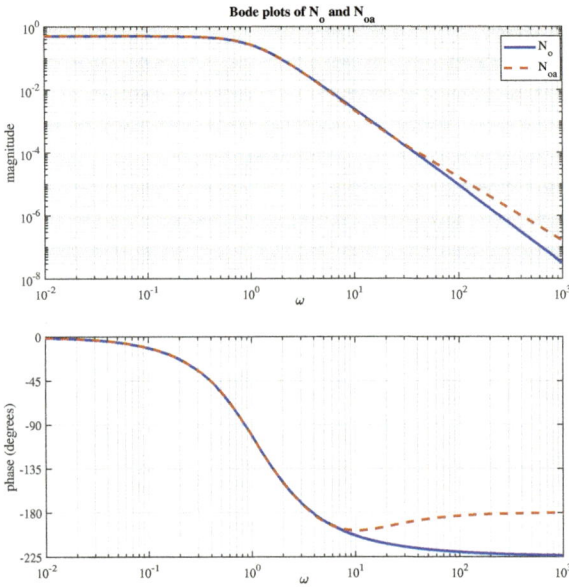

**Figure 6.26.** *Magnitude and phase plots of $N_o(j\omega)$ and third order $N_{oa}(j\omega)$.*

An indirect way to obtain a finite dimensional controller is

$$C_a(s) = \left( \frac{C_0(s)}{1 + C_0(s)H_{da}(s)} \right) \frac{G_{\mathrm{opt}}(s)}{N_{oa}(s)}, \tag{6.110}$$

where $H_{da} \in \mathcal{H}_\infty$ is the approximation given in (6.83) and $N_{oa}$ is determined as above. Since $C_0(s) = 1$, the orders of $G_{\mathrm{opt}}$, $H_{da}$, and $N_{oa}$ are 4, 5, and 3, respectively, and we expect that $C_a(s)$ is a 12th order controller. On the other hand, this controller has poles and zeros that are close to each other, and, after going through a model reduction, a fifth order controller can be obtained. The resulting controller expression (6.110) after model reduction becomes

$$C_a(s) = \frac{2}{2.52} \frac{(0.24s + 1)(4.75s + 1)(1.04s^2 + 1.8s + 1)}{(s + 0.001)(s + 1)(0.082s + 1)(0.016s^2 + 0.18s + 1)}.$$

The low-frequency behavior of this controller is approximately similar to the low-frequency behavior of $(N_{oa}(0)\ \gamma_{\mathrm{opt}}\ (s + \varepsilon))^{-1}$. This means that the low-frequency behavior is dominated by the pole of $W_1$, and the DC gain of the controller is inversely proportional to $\gamma_{\mathrm{opt}}\ N_{oa}(0)$.

Since the optimal controller is stable, it can be approximated directly; a fifth order approximation is determined using `fitfrd`:

$$C_{app}(s) = \frac{2}{2.52}\ \frac{(0.04s + 1)(4.6s + 1)(0.48s^2 + 1.3s + 1)}{(s + 0.001)(0.01s + 1)(0.28s + 1)(0.012s^2 + 0.16s + 1)}.$$

The Bode plots of the optimal controller, and those of $C_a$ and $C_{app}$, are shown in Figure 6.27.

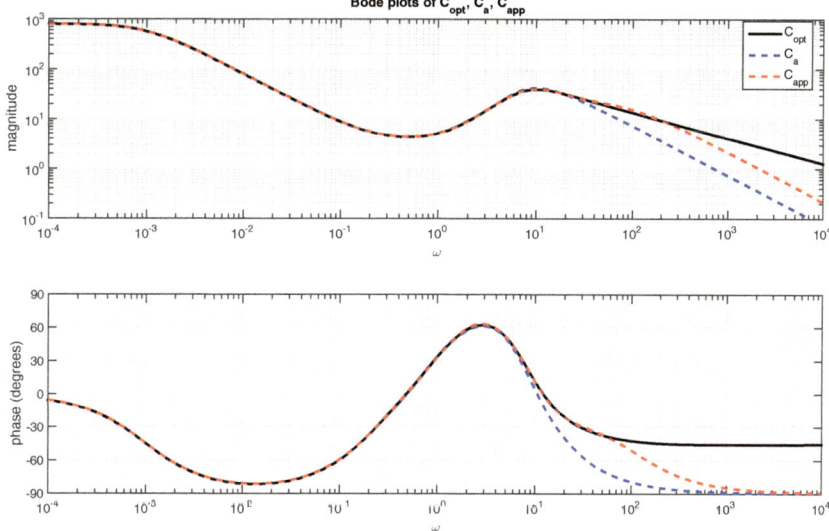

**Figure 6.27.** *Magnitude and phase plots of $C_{\mathrm{opt}}$ and alternative fifth order approximations.*

The performance deviation bound, (6.96), given in Theorem 6.14 can now be computed:

$$\gamma_a \leq 2.935,$$

whereas the optimal performance level is $\gamma_{\mathrm{opt}} = 2.52$. On the other hand, (6.96) is simply a conservative bound which sheds light on how individual approximations should be done. The actual performance levels corresponding to $C_a$ and $C_{app}$ are 2.63 and 2.84, respectively; i.e., the performance of $C_a$ is within 5% of the optimal performance level; for $C_{app}$ this deviation is about 13%; see Figure 6.28 for the plots of

$$\Upsilon_a(j\omega) = \frac{\sqrt{|W_1(j\omega)|^2 + |W_2(j\omega)P(j\omega)C_a(j\omega)|^2}}{|1 + P(j\omega)C_a(j\omega)|} \tag{6.111}$$

and $\Upsilon_{app}(j\omega)$, which is obtained by replacing $C_a$ with $C_{app}$ in the above expression.

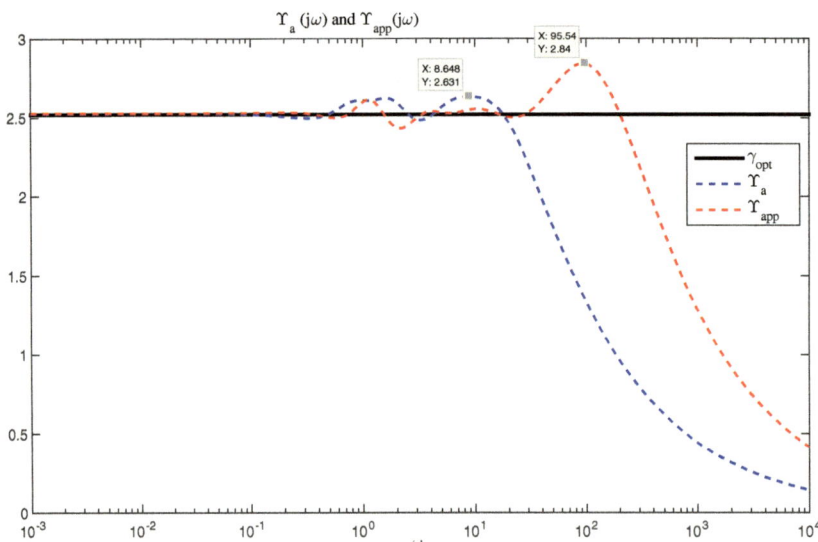

**Figure 6.28.** *Performance comparisons for fifth order controllers: plots of* $\Upsilon_a(j\omega)$ *and* $\Upsilon_{app}(j\omega)$.

Let us now apply the approach outlined in Section 6.4.2 to derive an alternative finite dimensional controller for the above problem involving an unstable fractional order plant. Recall that $P$ can be factored as

$$P(s) = \frac{e^{-0.1\,s}}{s^2(\sqrt{2}+1)-2} = \frac{M_n N_o}{M_d},$$

where $M_n(s) = e^{-0.1s}$, $M_d(s) = \frac{s-1}{s+1}$, and $N_o(s)$ is the infinite dimensional outer function whose Bode plots are shown in Figure 6.26. We define

$$P_{m_h,m_o}(s) = \frac{M_{m_h}(s)N_{m_o}(s)}{M_d(s)},$$

where $M_{m_h}$ is an $m_h$th order Padé approximation of $e^{-0.1s}$, and $N_{m_o}$ is an $m_o$th order approximation of $N_o$. The optimal controller for this plant is

$$C_{\mathrm{opt},m_h,m_o}(s) = \left(\frac{C_0(s)}{1+C_0(s)H_{d,m_h}(s)}\right)\frac{G_{\mathrm{opt},m_h}(s)}{N_{m_o}(s)},$$

where $C_0(s)$, $H_{d,m_h}(s)$, $G_{\mathrm{opt},m_h}(s)$, and the resulting performance level $\gamma_{\mathrm{opt},m_h,m_o}$ are computed from $W_1$, $W_2$, $M_{m_h}$, and $M_d$, i.e., once the weights are defined, these terms do not depend on the outer part of the plant. For $m_h = 1$ and $m_o = 3$ the above controller becomes

$$C_{\mathrm{opt},1,3} = \frac{0.795(0.05s+1)(0.24s+1)(4.67s+1)(1.04s^2+1.8s+1)}{(0.04755s+1)(s+0.001)(s+1)(0.082s+1)(0.0155s^2+0.18s+1)},$$

which is very close to $C_a$ and $C_{app}$ obtained above. The extra term appearing here, $\frac{(0.05s+1)}{(0.04755s+1)}$, is close to 1, and it is due to Padé approximation. The low-frequency gain $0.795$ is close to the original value $2/\gamma_{\mathrm{opt}} = 0.79365$. On the actual plant, this controller yields a performance level of $2.70$ which is about $7\%$ worse than the optimal performance level. If we further perform a pole-zero cancellation and use

$$C_{\mathrm{opt},1,3} \approx \frac{0.795(0.24s+1)(4.67s+1)(1.04s^2+1.8s+1)}{(s+0.001)(s+1)(0.082s+1)(0.0155s^2+0.18s+1)}, \tag{6.112}$$

the resulting performance level is $2.71$, which means that this method also gives a fifth order controller whose performance is within $10\%$ of the optimum performance level.

**Exercise 6.19.**
Verify that the mixed sensitivity level for the controller defined in (6.112) is indeed $2.71$ by drawing the function $\Upsilon(j\omega)$, (6.111), with this controller.

# Chapter 7

# Controllers for MIMO Plants with Scalar Inner Parts

In 1989, Doyle et al. [52] established the standard two algebraic Riccati equation (ARE) approach to $\mathcal{H}_\infty$ control problems for MIMO LTI, finite dimensional, and continuous-time systems. As a parallel to [52], two *operator-valued* Riccati equations are derived for the infinite dimensional counterpart by Curtain, Zwart, van Keulen, and others [43, Chapters 7 and 8], [244, Chapter 5]. However, from a computational point of view, for certain types of infinite dimensional systems, the approach described in Chapter 6 is more attractive in that it yields a finite rank condition for optimality, though the class of systems may be limited. In this chapter, we investigate the relationship between these two alternative approaches. The key building block is the Zhou–Khargonekar formula [282] (briefly discussed in Section 6.1.5), which is a state-space formula for the one-block problem associated with a delay system. The class of plants considered in this chapter are in the form $P(s) = m(s)P_f(s)$, where $P_f(s)$ is a MIMO finite dimensional system and $m(s)$ is inner and infinite dimensional. This is the simplest class of plants that satisfy a special factorization assumed in [66, Chapter 8], where the MIMO $\mathcal{H}_\infty$ control problem has been considered for DPSs. Also, for the special case where $m(s) = e^{-hs}$, one-, two-, and four-block versions of the $\mathcal{H}_\infty$ control problems have been considered in many publications, such as [152, 153, 154, 162, 234, 279, 282]. The results presented in this chapter are based on [119, 122, 124].

In order to give the main results of this chapter, we need to first define the notation and provide some building blocks in the next section. It is also helpful for the reader to review some of the special operators defined earlier in Chapter 2.

## 7.1 ▪ Mathematical preliminaries for matrix transfer functions

### 7.1.1 ▪ Notation for feedback systems

Let $\Theta(s)$ be a $(2 \times 2)$-block matrix with square diagonal blocks. Define

$$J = \left[ \begin{array}{c|c} I & 0 \\ \hline 0 & -I \end{array} \right],$$

where $J$ is partitioned according to $\Theta$. Then, $\Theta$ is said to be *J-unitary* if $\Theta^{\sim} J \Theta = J$, where $\Theta^{\sim}$ denotes the matrix transfer function $\Theta^{\mathsf{T}}(-s)$.

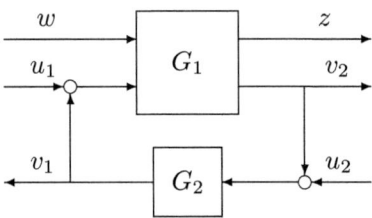

**Figure 7.1.** *Block diagram for the definition of the internal stability.*

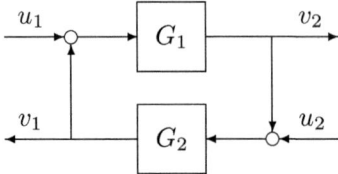

**Figure 7.2.** *Block diagram for the definition of the internal stability when $m_1 = p_1 = 0$.*

The transfer matrix whose realization is $(A, B, C, D)$ is denoted by

$$\left[ \begin{array}{c|c} A & B \\ \hline C & D \end{array} \right] := C(sI - A)^{-1}B + D.$$

For transfer matrices

$$P = \left[ \begin{array}{c:c} P_{11} & P_{12} \\ \hdashline P_{21} & P_{22} \end{array} \right]$$

and $K$ of appropriate dimensions, $\mathcal{F}_l(P, K)$ denotes the lower linear fractional transform of $P$ and $K$, that is,

$$\mathcal{F}_l(P, K) := P_{11} + P_{12}K(I - P_{22}K)^{-1}P_{21}.$$

Let $G_1$ and $G_2$ be transfer matrices whose sizes are $(m_1 + m_2) \times (p_1 + p_2)$ and $p_2 \times m_2$, respectively. We say that $G_2$ internally stabilizes $G_1$ if the nine transfer matrices from $w$, $u_1$, and $u_2$ to $z$, $v_1$, and $v_2$ in Figure 7.1 are all in $\mathcal{H}_\infty$. In particular, when $m_1 = p_1 = 0$, we say that $G_2$ internally stabilizes $G_1$ when the four transfer matrices from $u_1$ and $u_2$ to $v_1$ and $v_2$ in Figure 7.2 all belong to $\mathcal{H}_\infty$.

In the finite dimensional standard $\mathcal{H}_\infty$ control problem, the following rational transfer matrix $P$ is called the *generalized plant*:

$$P(s) = \left[ \begin{array}{c:c} P_{11} & P_{12} \\ \hdashline P_{21} & P_{22} \end{array} \right] = \left[ \begin{array}{c|c:c} A & B_1 & B_2 \\ \hline C_1 & D_{11} & D_{12} \\ \hdashline C_2 & D_{21} & D_{22} \end{array} \right]. \qquad (7.1)$$

The following assumption (which is generically satisfied) is usually imposed for the generalized plant.

**Assumption 7.1.**

(A1) $(C_2, A, B_2)$ is stabilizable and detectable.

(A2) $\begin{bmatrix} A - j\omega I & B_2 \\ C_1 & D_{12} \end{bmatrix}$ and $\begin{bmatrix} A - j\omega I & B_1 \\ C_2 & D_{21} \end{bmatrix}$ are of row- and column-full rank for any $\omega \in \mathbb{R}$, respectively.

(A3) $D_{12}^{\mathsf{T}} \begin{bmatrix} C_1 & D_{12} \end{bmatrix} = \begin{bmatrix} 0 & I \end{bmatrix}$ and $\begin{bmatrix} B_1 \\ D_{21} \end{bmatrix} D_{21}^{\mathsf{T}} = \begin{bmatrix} 0 \\ I \end{bmatrix}$.

(A4) $D_{11} = 0$ and $D_{22} = 0$.

These assumptions, except for (A1), can be easily removed by standard techniques [281, Chapter 16].

## 7.1.2 ▪ Matrix functions and the Dunford integral

When a scalar function $f(s)$ is analytic in a neighborhood of any eigenvalue of a matrix $X \in \mathbb{R}^{n \times n}$, we can define a matrix $f(X) \in \mathbb{R}^{n \times n}$. There are several equivalent ways of defining $f(X)$. Among these, the most elegant approach may be the one via the Dunford integral [82, Section 11.1.1]. Take a closed contour $\Delta$ such that $\Delta$ encircles all eigenvalues of $X$ and that $f(s)$ is analytic inside $\Delta$. Since the spectrum of $X$ is a finite set, this is possible. Define $f(X)$ by

$$f(X) := \frac{1}{2\pi j} \int_\Delta f(s)(sI - X)^{-1} ds. \qquad (7.2)$$

This definition can be interpreted as a matrix version of the Cauchy integral formula. For computational issues of such matrix functions, see [82, Chapter 11], [75, Chapter V]. In particular, when $f(s) = 1$, $f(X) = I$ for any $X$. Another important property of this matrix function is that $X$ and $f(X)$ mutually commute. Then, we have

$$(sI - X)^{-1} f(X) = f(X)(sI - X)^{-1}. \qquad (7.3)$$

If $X$ and $Y$ are similar to each other, then so are $f(X)$ and $f(Y)$; that is, $Y = U^{-1} X U$ with a nonsingular $U$ yields

$$f(Y) = U^{-1} f(X) U. \qquad (7.4)$$

Let $f(s)$ be an analytic function in $\mathcal{H}_\infty$ with real coefficients, so that $\overline{f(\bar{s})} = f(s)$. The *parahermitian conjugate* $f^\sim(s)$ is the function defined by $f(-s)$. For $s = j\omega$, with $\omega \in \mathbb{R}$, we have $f^\sim(j\omega) = \overline{f(j\omega)} = f(-j\omega)$. In particular, if $m(s)$ is an inner function in $\mathcal{H}_\infty$, then $m^\sim(s)$ becomes equal to $1/m(s)$. The set of matrices $X$ in $\mathbb{R}^{n \times n}$, such that $f^\sim(s)$ is analytic in a neighborhood of any eigenvalue of $X$, is denoted by $\mathcal{M}_f^{n \times n}$. The superscript $n \times n$ is often dropped when it is clear from the context. The matrix function $f^\sim(X)$ is well-defined for $X \in \mathcal{M}_f^{n \times n}$.

The following lemma plays a crucial role in the next section.

**Lemma 7.2.** *Let $f(s)$ be a scalar function in $\mathcal{H}_\infty$. Consider the matrices $A \in \mathcal{M}_f^{n \times n}$, $M_1, M_2 \in \mathbb{R}^{n \times p}$, $C \in \mathbb{R}^{q \times n}$, and define*

$$\Phi(s) := (sI - A)^{-1}(f^\sim(s) M_1 - M_2).$$

*Suppose that the pair $(C, A)$ is observable. Then, $C\Phi(s)$ is analytic in a neighborhood of every eigenvalue of $A$ if and only if*

$$f^\sim(A) M_1 = M_2.$$

This lemma allows us to introduce the following (possibly infinite dimensional) stable system, which plays a central role in this chapter.

**Lemma 7.3.** *Let* $W(s) = C(sI - A)^{-1}B$, *and let* $m(s)$ *be an inner function satisfying* $A \in \mathcal{M}_m$. *Define*

$$W^{(m)} := C(sI - A)^{-1}\tilde{m}(A)B \tag{7.5}$$

*and*

$$\pi^m [W] := W - mW^{(m)}, \tag{7.6}$$

*neither of which depends on the realization of* $W$. *Then,* $\pi^m [W]$ *belongs to both* $\mathcal{H}(m)$ *and* $\mathcal{H}_\infty$.

This lemma states that even if $W$ is not stable, the definition of $W^{(m)}$ is such that $(W - mW^{(m)})$ is stable. Hereafter, we refer to $\pi^m [\cdot]$ as the *generalized m-truncation*.

**Example 7.4.** When $m(s) = e^{-hs}$ for $h > 0$, $\pi^m [\cdot]$ is the same as the *h-truncation* defined in [153, 159], which is the operator truncating the impulse response to its restriction on $[0, h]$; see also Remark 6.5. ∎

**Example 7.5.** Consider

$$m(s) = \frac{0.2 - se^{-hs}}{s + 0.2e^{-hs}}, \quad W(s) = \frac{1}{s + \varepsilon}, \quad \varepsilon = 0.01, \tag{7.7}$$

for $h \in [0, \frac{5\pi}{2})$. When $h = 0$, for which $m(s)$ is a Blashke factor, we have

$$
\begin{aligned}
\pi^m [W] (s) &= \frac{1}{s + \varepsilon} - \frac{0.2 - s}{s + 0.2} \cdot \frac{0.2 + \varepsilon}{0.2 - \varepsilon} \cdot \frac{1}{s + \varepsilon} \\
&= \frac{(s + 0.2)(0.2 - \varepsilon) - (0.2 - s)(0.2 + \varepsilon)}{(s + 0.2)(s + \varepsilon)(0.2 - \varepsilon)} \\
&= \frac{0.4(s + \varepsilon)}{(s + 0.2)(s + \varepsilon)(0.2 - \varepsilon)} = \left( \frac{0.4}{0.2 - \varepsilon} \right) \frac{1}{s + 0.2}.
\end{aligned}
$$

We can observe that the pole of $W$ at $s = -\varepsilon$ disappears and that the only pole arises from $m(s)$. In general, for $h > 0$, the above computations lead to

$$\pi^m [W] (s) = \frac{1}{(s + \varepsilon)} \left( \frac{m(-\varepsilon) - m(s)}{m(-\varepsilon)} \right).$$

Again, note that $\pi^m [W]$ does not have a pole at $s = -\varepsilon$, because the term inside the parentheses vanishes at $s = -\varepsilon$. The magnitude plots of $\pi^m [W] (j\omega)$, with $\varepsilon = 0.01$ and for five different values of $h$, are shown in Figure 7.3, which implies that $\| \pi^m [W] \|_\infty \leq 10$ for these values of the parameters. ∎

**Exercise 7.6.**
Consider the same inner function $m(s)$ defined in (7.7), with $h = 2.0$, and a $2 \times 2$ transfer function for $W(s)$, given as

$$
W(s) = \begin{bmatrix} \frac{1}{s-1} & \frac{s+2}{s^2-2s+2} \\ 0 & \frac{1}{s^2-2s+2} \end{bmatrix}.
$$

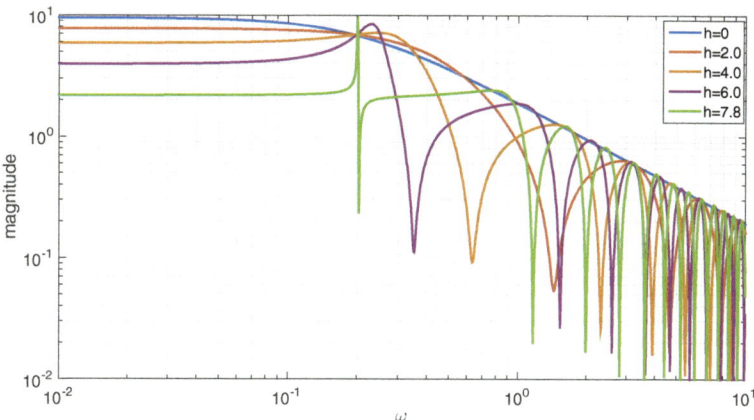

**Figure 7.3.** *Magnitude plots of* $\pi^m [W] (j\omega)$ *for* $h = 0$, $h = 2$, $h = 4$, $h = 6$, *and* $h = 7.8 < \frac{5\pi}{2}$.

Find an expression for $\pi^m [W] (s)$, and numerically estimate the norm $\|\pi^m [W] \|_\infty$, as was done in the above example. Note that although $W$ is unstable, the transfer function $\pi^m [W]$ is in $\mathcal{H}_\infty$; illustrate this fact for the special case $h = 0$.

## 7.2 ▪ The Nehari problem revisited

We have already seen that many interesting $\mathcal{H}_\infty$ control problems can be transformed to the so-called one-block problem; see Chapters 2 and 5. The one-block problem can be seen as a model matching problem where a stable approximation(s), in the sense of $\mathcal{L}_\infty$, of a given unstable system is sought. This is precisely the *suboptimal Nehari problem*, which can be stated as follows: given $F \in \mathcal{L}_\infty$, find all $\phi \in \mathcal{H}_\infty$ such that

$$\|F + \phi\|_{\mathcal{L}_\infty} < 1. \tag{7.8}$$

Nehari's theorem says that a solution $\phi \in \mathcal{H}_\infty$ satisfying (7.8) exists if and only if $\|\mathbf{\Gamma}_F\| < 1$, where $\mathbf{\Gamma}_F$ is the Hankel operator whose symbol is $F$. In this section, we assume that $F \in \mathcal{L}_\infty$, with $\|\mathbf{\Gamma}_F\| < 1$, is given, and we derive a parameterization of solutions $\phi \in \mathcal{H}_\infty$ of (7.8).

The Nehari problem has been studied in the control community for various classes of $F$, and many different solution techniques have been developed, depending on the assumptions of $F$; see, e.g., operator equation–based results [43], Adamjan, Arov, and Krein (AAK) theory [2], and continuous-time FIR systems [153]. In this section, a unified formula is given for a class of $F$ that captures the properties employed in the preceding chapters.

Let us assume that $F$ is square integrable on the imaginary axis, i.e., $F \in \mathcal{L}^2(j\mathbb{R})$. Under this assumption, $\mathbf{\Pi}_- [F]$ is well-defined; accordingly, we can define the Hankel operator $\mathbf{\Gamma}_F$. In what follows, we drop the subscript $F$ and write $\mathbf{\Gamma}$ whenever the symbol is clear from the context. Note that if $\|\mathbf{\Gamma}\| < 1$, then both $I - \mathbf{\Gamma}^*\mathbf{\Gamma}$ and $I - \mathbf{\Gamma}\mathbf{\Gamma}^*$ are invertible in $\mathcal{H}_2$ and $\mathcal{H}_2^\perp$, respectively, and consequently we can define

$$\eta := (I - \mathbf{\Gamma}\mathbf{\Gamma}^*)^{-1}(\mathbf{\Pi}_- [F]).$$

In other words, there exist unique $\xi \in \mathcal{H}_2$ and $\eta \in \mathcal{H}_2^{\perp}$ such that

$$\begin{cases} \boldsymbol{\Gamma}\xi + \boldsymbol{\Pi}_- [F] &= \eta, \\ \boldsymbol{\Gamma}^*\eta &= \xi. \end{cases} \qquad (7.9)$$

Dually, there exist unique $\acute{\xi} \in \mathcal{H}_2$ and $\acute{\eta} \in \mathcal{H}_2^{\perp}$ such that

$$\begin{cases} \boldsymbol{\Gamma}^*\acute{\eta} + \boldsymbol{\Pi}_+ [F\check{} ] &= \acute{\xi}, \\ \boldsymbol{\Gamma}\acute{\xi} &= \acute{\eta}. \end{cases} \qquad (7.10)$$

Under these definitions, a solution to the Nehari problem can be given as follows, where $B\mathcal{L}_{\infty}$ and $B\mathcal{H}_{\infty}$ denote the unit ball in $\mathcal{L}_{\infty}$ and $\mathcal{H}_{\infty}$, respectively, i.e.,

$$B\mathcal{L}_{\infty} = \{G \in \mathcal{L}_{\infty} \,:\, \|G\|_{\infty} < 1\}, \quad B\mathcal{H}_{\infty} = \{G \in \mathcal{H}_{\infty} \,:\, \|G\|_{\infty} < 1\}.$$

**Theorem 7.7 (see [124]).** *Let* $F \in \mathcal{L}_{\infty} \cap \mathcal{L}^2(j\mathbb{R})$ *such that* $\|\boldsymbol{\Gamma}\| < 1$*. Suppose that the unique solutions to operator equations* (7.9) *and* (7.10)*,* $\xi, \acute{\xi} \in \mathcal{H}_2$ *and* $\eta, \acute{\eta} \in \mathcal{H}_2^{\perp}$*, belong to* $\mathcal{L}_{\infty}$*. Then all* $\phi \in \mathcal{H}_{\infty}$ *such that* $F + \phi \in B\mathcal{L}_{\infty}$ *are given by*

$$\{\phi = (\Theta_{11}U + \Theta_{12})(\Theta_{21}U + \Theta_{22})^{-1} - F \,:\, U \in B\mathcal{H}_{\infty}\},$$

*where* $\Theta \in \mathcal{L}_{\infty}$ *is defined by*

$$\Theta := \begin{bmatrix} \Theta_{11} & \Theta_{12} \\ \Theta_{21} & \Theta_{22} \end{bmatrix} := \begin{bmatrix} \acute{\eta} + I & \eta \\ \acute{\xi} & \xi + I \end{bmatrix}. \qquad (7.11)$$

This parameterization requires us to solve the operator equations (7.9) and (7.10), which cannot be easily solved for general $F$. Next, we revisit the Zhou–Khargonekar formula to reveal the structure of $\Theta$ in Theorem 7.7.

## 7.3 ▪ Relations with the Zhou–Khargonekar formula

The following Zhou–Khargonekar formula gives an elegant state-space formula for the one-block problem for a stable rational weight and a pure delay.

Let $W := C(sI - A)^{-1}B$ be a minimal realization, where $A$ is Hurwitz. Then,

$$\gamma_{\text{opt}} := \inf_{\phi \in \mathcal{H}_{\infty}} \|W - e^{-hs}\phi\|_{\infty} \qquad (7.12)$$

is given by the maximal $\gamma > 0$ such that the $(2,2)$-block of $e^{hH_{\gamma}}$ is singular, where

$$H_{\gamma} := \begin{bmatrix} A & \gamma^{-1}BB^{\mathsf{T}} \\ -\gamma^{-1}C^{\mathsf{T}}C & -A^{\mathsf{T}} \end{bmatrix}. \qquad (7.13)$$

It should be emphasized that this criterion does not require us to solve any operator-valued equation despite the infinite dimensionality of the delay.

In order to make connections of this formula with the generalized $m$-truncation, we make the following observation.

**Lemma 7.8.** *Let* $W(s) = C(sI - A)^{-1}B$ *be a stable transfer matrix, and let* $m(s)$ *be a scalar inner function satisfying* $A \in \mathcal{M}_m$*. Then, we have*

$$\inf_{\phi \in \mathcal{H}_{\infty}} \|W - m\phi\|_{\infty} = \inf_{\phi \in \mathcal{H}_{\infty}} \|\pi^m [W] - m\phi\|_{\infty}.$$

***Proof.*** In this case, $W^{(m)}$ defined by (7.5) is a stable transfer matrix. Therefore,

$$
\inf_{\phi \in \mathcal{H}_\infty} \|W - m\phi\|_\infty = \inf_{\phi_1 \in \mathcal{H}_\infty} \|W - m(\phi_1 - W^{(m)})\|_\infty
$$
$$
= \inf_{\phi_1 \in \mathcal{H}_\infty} \|\pi^m [W] - m\phi_1\|_\infty,
$$

where $\phi_1 = \phi + W^{(m)} \in \mathcal{H}_\infty$ if and only if $\phi \in \mathcal{H}_\infty$.  $\square$

In view of Lemma 7.8, let us consider the following specific form of one-block problem where the stable system to be approximated is a generalized $m$-truncation: let $m(s)$ be an inner function and $(A, B, C, 0)$ with $A \in \mathcal{M}_m$ be a minimal realization of $W(s)$. Then, given positive $\gamma > 0$, determine whether

$$
\gamma_{\mathrm{opt}} := \inf_{\phi \in \mathcal{H}_\infty} \|\pi^m [W] + m\phi\|_\infty < \gamma. \tag{7.14}
$$

It should be emphasized that this problem is well-posed even for unstable $W$. For $m(s) = e^{-hs}$ and stable $W$, the rank criterion in the Zhou–Khargonekar formula gives a solution for this specific form of one-block problem. Furthermore, this result naturally extends to general inner function $m(s)$ and possibly unstable $W$. The following proposition, a consequence of Lemma 7.2, is used in the solution of the Nehari problem and can be seen as a generalization of Lemma 2.6 in [264].

**Proposition 7.9.** *Let* $W(s) = C(sI - A)^{-1}B$, *and let* $m(s)$ *be an inner function satisfying* $A \in \mathcal{M}_m$. *For two functions* $x \in \mathcal{H}(m)$ *and* $y \in m^\sim \mathcal{H}(m)$, *we have* $y = (\boldsymbol{\Gamma}_{m^\sim \pi^m [W]})x$ *and* $x = (\boldsymbol{\Gamma}_{m^\sim \pi^m [W]})^* y$ *if and only if there exist* $\mathbf{x}, \mathbf{y} \in \mathbb{R}^n$ *satisfying*

$$
y = m^\sim W x - C(sI - A)^{-1}\mathbf{x}, \tag{7.15}
$$
$$
x = m W^\sim y - B^{\mathsf{T}}(sI + A^{\mathsf{T}})^{-1}\mathbf{y}. \tag{7.16}
$$

Then, a solution to the equivalent Nehari problem for $\gamma = 1$ in (7.14) is given as follows.

**Theorem 7.10 (see [124]).** *Let* $m(s)$ *be an inner function, let* $(A, B, C, 0)$ *with* $A \in \mathcal{M}_m$ *be a minimal realization of a rational matrix* $W$, *and let* $F := m^\sim \pi^m [W]$. *Suppose that the essential norm of* $\boldsymbol{\Gamma}_F$ *is less than 1 and that for any* $\rho \geq 1$,

$$
H_\rho := \begin{bmatrix} A & \rho^{-1} B B^{\mathsf{T}} \\ -\rho^{-1} C^{\mathsf{T}} C & -A^{\mathsf{T}} \end{bmatrix} \in \mathcal{M}_m. \tag{7.17}
$$

*Then,* $\gamma_{\mathrm{opt}} < 1$ *if and only if the* $(2,2)$-*block of* $m^\sim(H_\rho)$ *is of full rank for all* $\rho \geq 1$. *In addition, if* $\gamma_{\mathrm{opt}} < 1$, *then* $\Theta$ *in* (7.11) *belongs to* $\mathcal{L}_\infty$ *and is given by*

$$
\Theta = \begin{bmatrix} I & 0 \\ 0 & I \end{bmatrix} + \begin{bmatrix} m^\sim I & 0 \\ 0 & I \end{bmatrix} \pi^m [\Theta_r],
$$

$$
\Theta_r = \left[ \begin{array}{cc|cc} A & BB^{\mathsf{T}} & 0 & B \\ -C^{\mathsf{T}}C & -A^{\mathsf{T}} & \Sigma_{22}^{-1}C^{\mathsf{T}} & -\Sigma_{22}^{-1}\Sigma_{21}B \\ \hline C & 0 & 0 & 0 \\ 0 & B^{\mathsf{T}} & 0 & 0 \end{array} \right],
$$

$$
\Sigma = \begin{bmatrix} \Sigma_{11} & \Sigma_{12} \\ \Sigma_{21} & \Sigma_{22} \end{bmatrix} := m^\sim\left( \begin{bmatrix} A & BB^{\mathsf{T}} \\ -C^{\mathsf{T}}C & -A^{\mathsf{T}} \end{bmatrix} \right).
$$

For arbitrary $\gamma > 0$, checking whether $\gamma_{\text{opt}}$ satisfies (7.14) can be done by making use of the matrix $H_\rho$ defined in Theorem 7.10 and the following equivalences:

$$\gamma_{\text{opt}} < \gamma$$
$$\Leftrightarrow \inf_{\phi \in \mathcal{H}_\infty} \|\pi^m[W'] - m\phi\|_\infty < 1, \quad \text{with } W' := (C/\sqrt{\gamma})(sI - A)^{-1}(B/\sqrt{\gamma})$$
$$\Leftrightarrow \tilde{m}(H_{\rho\gamma})|_{22} \text{ is full rank for all } \rho \geq 1$$
$$\Leftrightarrow \tilde{m}(H_\rho)|_{22} \text{ is full rank for all } \rho \geq \gamma.$$

Therefore,

$$\gamma_{\text{opt}} = \max\{\gamma : \tilde{m}(H_\gamma)|_{22} \text{ is singular}\} \tag{7.18}$$

holds. In other words, as in Chapter 6, we plot the smallest singular value of $\tilde{m}(H_\gamma)|_{22}$ versus $\gamma$ and determine $\gamma_{\text{opt}}$ as the largest $\gamma$ value which makes this matrix singular.

In particular, application of the above result to pseudorational transfer functions, whose factorizations are obtained in Section 4.3, leads to the following.

**Corollary 7.11 (see [121]).** *Consider the stable plant* (4.34) *in the form*

$$P(s) = \frac{N_1(s)N_2(s)}{D(s)},$$

*whose inner-outer factorization $P = P_i P_o$ is given by*

$$P_i(s) = e^{-Ls} \cdot \frac{N_2(s)}{N_2(-s)} \quad and \quad P_o(s) = e^{Ls} \cdot \frac{N_1(s)N_2(-s)}{D(s)},$$

*where $L \geq 0$ and $N_1$, $N_2$, $D$ are as specified in Theorem 4.14. Let $A$ be a Hurwitz matrix and $C(sI - A)^{-1}B$ be a minimal realization of an outer function $W(s)$. Define $H_\gamma$ as in (7.13). Assume that $1/P_i(s)$ is analytic on the set of eigenvalues of $H_\gamma$. Then, we have*

$$\gamma_{\text{opt}} = \inf_{C \in \mathcal{C}(P)} \|W(1 + PC)^{-1}\|_\infty \tag{7.19}$$

$$= \inf_{\phi \in \mathcal{H}_\infty} \|W - P_i \phi\|_\infty \tag{7.20}$$

$$= \max\{\gamma : \tilde{P_i}(H_\gamma)|_{22} \text{ is singular}\}, \tag{7.21}$$

*where*

$$\tilde{P_i}(H_\gamma) = e^{LH_\gamma} N_2(-H_\gamma) N_2(H_\gamma)^{-1}.$$

***Proof.*** The equivalence of (7.20) and (7.21) is deduced from Theorem 7.10 and the ensuing discussion. The equivalence of (7.19) and (7.20) comes from the controller parameterization and outer factor absorption; see [62, 69] for details. □

**Example 7.12.** Consider a pseudorational inner function of the form (4.33), namely

$$P_i(s) = \frac{0.2 - se^{-hs}}{s + 0.2e^{-hs}}, \quad \text{where } h \in [0, \frac{5\pi}{2}).$$

We will find $\gamma_{\text{opt}}$ for the weight $W(s) = \frac{1}{s+\varepsilon}$, where $\varepsilon = 0.01$. By Theorem 7.10, $\gamma_{\text{opt}}$ is the maximal $\gamma$ such that the $(2, 2)$-entry of

$$\tilde{P_i}(H_\rho) = (0.2I + H_\rho e^{hH_\rho})(-H_\rho + 0.2e^{hH_\rho})^{-1},$$

with

$$H_\rho := \begin{bmatrix} -\varepsilon & \rho^{-1} \\ -\rho^{-1} & \varepsilon \end{bmatrix},$$

is nonzero for all $\rho > \gamma$. Note that the eigenvalues of $H_\rho$ are $\pm j\sqrt{\rho^{-2} - \varepsilon^2}$, and they do not intersect with the poles of $1/P_i(s)$ for all $\rho \leq 1/\varepsilon$. By Lemma 7.8, we have that $\gamma_{\mathrm{opt}} \leq \|\pi^m [W]\|_\infty$. Hence, Figure 7.3 tells us that it suffices to check the rank condition for $\rho \leq 10$. In fact, according to Corollary 7.11, $\gamma_{\mathrm{opt}}$ is the largest $\gamma \leq 10$ for which $P_i\tilde{\ }(H_\gamma)|_{2,2}$ is singular; hence, numerically evaluating the singular values of this matrix, we can determine $\gamma_{\mathrm{opt}}$ (see Exercise 7.13 below). The optimal performance level $\gamma_{\mathrm{opt}}$ is computed this way for each value of $h \in [0, \frac{5\pi}{2})$; the results are shown in Figure 7.4.    ■

### Exercise 7.13.

For each fixed $h \in [0, \frac{5\pi}{2})$, verify the numerical value of $\gamma_{\mathrm{opt}}$ displayed in Figure 7.4 by plotting the smallest singular value of $P_i\tilde{\ }(H_\gamma)|_{2,2}$ for $\gamma \in (0, 10)$. Perform an alternative verification of these results by using $M(s) = P_i(s)$ in the approach outlined in part 3 of Exercise 2.16. For this purpose, find the smallest $\omega_\gamma$ satisfying (2.57), call it $\omega_{\mathrm{opt}}$, and then define $\gamma_{\mathrm{opt}} = 1/\sqrt{\omega_{\mathrm{opt}}^2 + \varepsilon^2}$.

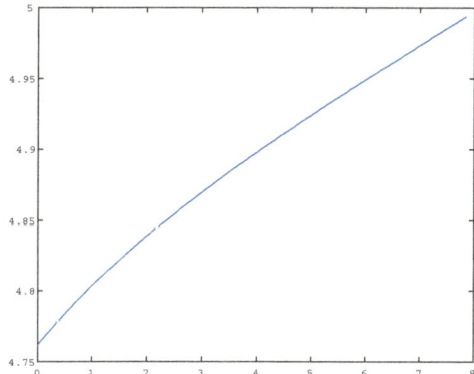

**Figure 7.4.** $\gamma_{\mathrm{opt}}$ *for* $h \in [0, \frac{5\pi}{2})$.

**Example 7.14.** Theorem 7.10 is applicable to MIMO systems. For example, let

$$W(s) = \left(sI - \begin{bmatrix} -1 & 1 \\ 0 & 1 \end{bmatrix}\right)^{-1}$$

and $m(s)$ be as in (7.7) with $h = 1$. In this case, $\|\pi^m [W]\|_\infty \approx 8.5619$. Figure 7.5 shows $\sigma_{\min}(m\tilde{\ }(H_\gamma)|_{22})$ for $\gamma \in (0, 8.6]$, which indicates $\gamma_{\mathrm{opt}} \approx 3.64$ by (7.18).    ■

In the next section, a wide class of $\mathcal{H}_\infty$ control problems is shown to be reducible to the Nehari problem with a generalized $m$-truncation investigated here.

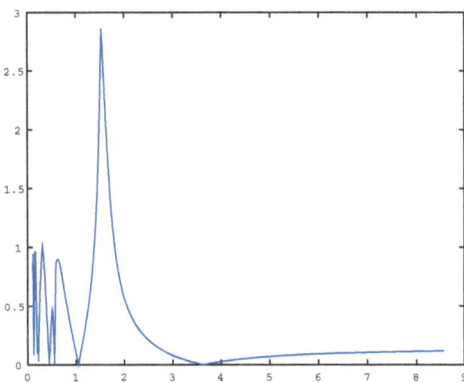

**Figure 7.5.** $\sigma_{\min}(\tilde{m}(H_\gamma)|_{22})$ *for* $\gamma \in (0, 8.6]$.

## 7.4 ▪ Extension to the standard problem

### 7.4.1 ▪ Problem formulation

In Chapters 4–6, the crux of the theory lies in the following assumptions:

- the weighting functions are rational; and

- the plant has a special factorization in the form $P = M_n N_o / M_d$, where $M_n$, $M_d$ are inner, $N_o$ is outer, and either $M_n$ or $M_d$ is rational.

Under these assumptions, a class of $\mathcal{H}_\infty$ control problems was reduced to rank conditions of suitable matrices. To generalize these partial finite dimensionality assumptions, we investigate the following problem (see Figure 7.6): given a rational transfer matrix

$$P(s) = \left[ \begin{array}{c|c} P_{11} & P_{12} \\ \hline P_{21} & P_{22} \end{array} \right] = \left[ \begin{array}{c|cc} A & B_1 & B_2 \\ \hline C_1 & D_{11} & D_{12} \\ C_2 & D_{21} & D_{22} \end{array} \right], \tag{7.22}$$

an inner function $m(s)$, and a prespecified performance level $\gamma > 0$, determine whether there exists a controller $C$ which internally stabilizes the generalized plant $P_{\text{inf}}$ given by

$$P_{\text{inf}}(s) := \left[ \begin{array}{c|c} P_{11} & P_{12} \\ \hline P_{21} & P_{22} \end{array} \right] \left[ \begin{array}{c|c} I & 0 \\ \hline 0 & mI \end{array} \right] \tag{7.23}$$

and guarantees

$$\|\mathcal{F}_l\left(P_{\text{inf}}, C\right)\|_\infty < \gamma. \tag{7.24}$$

If such a controller exists, find all admissible ones.

In this framework, we can formulate various $\mathcal{H}_\infty$ control problems for infinite dimensional systems with a finite number of unstable poles and rational weights (plants with infinitely many unstable poles will be discussed in Section 7.5.2). To see this, let us list a few $\mathcal{H}_\infty$ optimization problems.

**Example 7.15.** Consider a scalar plant which can be factored as

$$G(s) = G_r(s)m(s)G_o(s), \tag{7.25}$$

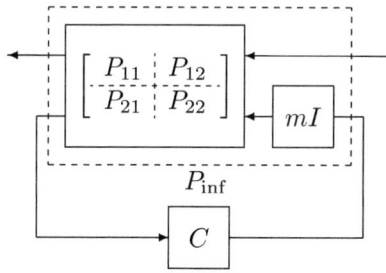

**Figure 7.6.** *Block diagram for problem* (7.24).

where $m$ is inner, $G_o$ is outer, and $G_r$ is rational. This factorization is not necessarily unique, but it does not affect the result. Note that $G_r$ is not necessarily stable but includes all unstable poles of $G$. It is known [62] that the outer function $G_o$ can be absorbed into controllers without degrading the resulting $\mathcal{H}_\infty$ performance. Then, for stable rational weights $W_1$ and $W_2$, the corresponding weighted mixed sensitivity is

$$\begin{bmatrix} W_1(1-GC)^{-1} \\ W_2GC(1-GC)^{-1} \end{bmatrix} = \mathcal{F}_l\left(P_{\mathrm{inf}}, \check{C}\right), \tag{7.26}$$

where $P_{\mathrm{inf}}$ is given by (7.23) with

$$P(s) = \left[ \begin{array}{c|c} W_1 & W_1 G_r \\ 0 & W_2 G_r \\ \hline 1 & G_r \end{array} \right]$$

and $\check{C} = G_o C$.  ■

**Example 7.16.** Consider the control input energy minimization for the same plant as that in the previous example. Given a stable weighting function $W_a$, which characterizes additive plant uncertainty (see Chapter 5) for robust stability, and also controls input energy minimization, we try to minimize

$$\left\| W_a C(1-GC)^{-1} \right\|_\infty = \left\| W_a m C(1-GC)^{-1} \right\|_\infty$$
$$= \mathcal{F}_l\left(P_{\mathrm{inf}}, \check{C}\right),$$

where $P_{\mathrm{inf}}$ is given by (7.23) with

$$P(s) = \left[ \begin{array}{c|c} 0 & W_a G_o^{-1} \\ \hline 1 & G_r \end{array} \right].$$

We here invoked the fact that the multiplication by any inner function preserves the $\mathcal{H}_\infty$ norm. The important point here is that the weighting function $W_a$ must be chosen in such a way that $W_a G_o^{-1}$ is rational but not necessarily so for $W_a$.  ■

**Example 7.17.** In the above examples, typically, the state-space realizations of $P(s)$ do not satisfy orthogonality conditions (A3) appearing in Assumption 7.1. This can be circumvented by augmenting the external input and by redefinition of the controller.

In order to illustrate this point let us consider the generalized feedback system shown in Figure 7.7, part (a). As above, $m$ is a scalar inner function, $G_o$ is outer such that $G_o^{-1} \in \mathcal{H}_\infty$, and $G_r$ is a rational transfer matrix. The weights $W_a$, $W_d$, and $W_n$ are assumed to be rational square outer matrices (in most practical cases they are diagonal matrices and $W_a$, $W_n$ are invertible in $\mathcal{H}_\infty$). By defining $G := G_r m G_o$ and $\check{C} := G_o C$ we have

$$
\mathcal{F}_l\left(P_{\mathrm{inf}}, \check{C}\right) = \left[ \begin{array}{cc} (I - GC)^{-1}W_d & GC(I - GC)^{-1}W_n \\ W_a m G_o C(I - GC)^{-1}W_d & W_a m G_o C(I - GC)^{-1}W_n \end{array} \right] \quad (7.27)
$$
$$
= \left[ \begin{array}{cc} (I - m G_r \check{C})^{-1}W_d & m G_r \check{C}(I - m G_r \check{C})^{-1}W_n \\ W_a \check{C}(I - m G_r \check{C})^{-1}W_d & W_a \check{C}(I - m G_r \check{C})^{-1}W_n \end{array} \right].
$$

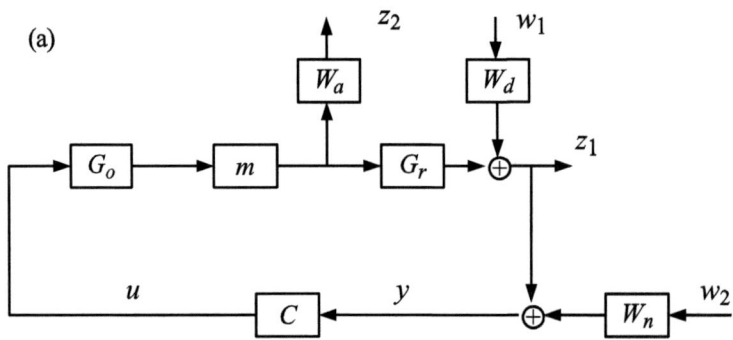

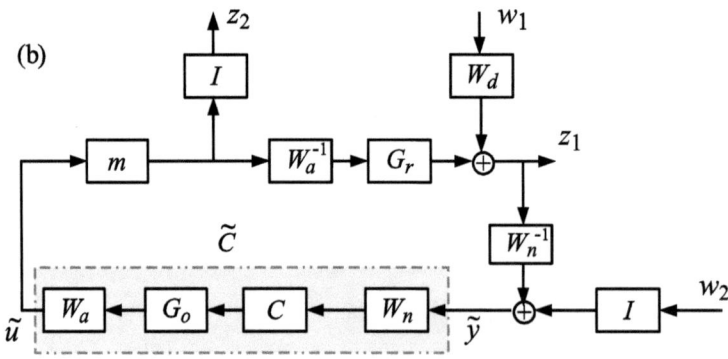

**Figure 7.7.** *Generalized feedback system* (a) *and its equivalent* (b).

Clearly, in the SISO case, defining $W_1 = W_d$ and finding a weight $W_2$ such that $|W_2(j\omega)| = \frac{|W_a(j\omega)W_d(j\omega)|}{|G_r(j\omega)|}$ for all $\omega$, we observe that (7.27) is the same as (7.26) for the special case $W_n = 0$. On the other hand, in order to satisfy (A3) of Assumption 7.1 we need a stable $W_n$ such that $W_n^{-1}$ is stable as well. For simplicity, let us assume that $W_a = \sigma I$ and $W_n = \varepsilon I$ for positive constants $\sigma$ and $\varepsilon$. Consider minimal realizations $W_d(s) = C_d(sI - A_d)^{-1}B_d$ and $G_r(s) = C_r(sI - A_r)^{-1}B_r$. Then, the state-space realization of $P(s)$ for this example is in the form (7.22) with

$$A = \begin{bmatrix} A_r & 0 \\ 0 & A_d \end{bmatrix}, \quad B_1 = \begin{bmatrix} 0 & 0 \\ B_d & 0 \end{bmatrix}, \quad B_2 = \begin{bmatrix} B_r \\ 0 \end{bmatrix}, \tag{7.28}$$

$$C_1 = \begin{bmatrix} C_r & C_d \\ 0 & 0 \end{bmatrix}, \quad D_{11} = \begin{bmatrix} 0 & 0 \\ 0 & 0 \end{bmatrix}, \quad D_{12} = \begin{bmatrix} 0 \\ \sigma I \end{bmatrix}, \tag{7.29}$$

$$C_2 = \begin{bmatrix} C_r & C_d \end{bmatrix}, \quad D_{21} = \begin{bmatrix} 0 & \varepsilon I \end{bmatrix}, \quad D_{22} = 0. \tag{7.30}$$

We assume that all eigenvalues of $A_d$ are in $\mathbb{C}_-$ and $A_r$ does not have any eigenvalues on the imaginary axis. Then, (A1) and (A2) of Assumption 7.1 are satisfied. Also note that in this case, the orthogonality conditions appearing in (A3) of Assumption 7.1 are satisfied, the only missing point is the normalization of the matrices $D_{12}$ and $D_{21}$. That is done by shifting $W_a$ and $W_n$ in the feedback loop as shown in Figure 7.7, part (b).  ∎

**Exercise 7.18.**
1. Assume that $W_a(s) = \sigma I$ and $W_n(s) = \varepsilon I$, with $\sigma > 0$ and $\varepsilon > 0$. Then, let $W_d(s) = C_d(sI - A_d)^{-1}B_d$ and $G_r(s) = C_r(sI - A_r)^{-1}B_r$ be minimal realizations. Obtain a state space realization for the underlying system $P(s)$ corresponding to the feedback system $\mathcal{F}_l(P_{\text{inf}}, \widetilde{C})$ shown in Figure 7.7, part (b). Verify that this system satisfies Assumption 7.1.
2. Now consider $\sigma = 0.4$, $\varepsilon = 0.001$, $W_d(s) = (s + 0.01)^{-1}$,

$$G_o(s) = \begin{bmatrix} \dfrac{4 - e^{-s}}{2 + e^{-s}} & 2 \\ 0 & \dfrac{s + 1}{s + e^{-s}} \end{bmatrix}, \quad G_r(s) = \begin{bmatrix} \dfrac{1}{s^2 - 2s + 2} & \dfrac{1}{s + 1} \\ \dfrac{1}{s^2 - 2s + 2} & 2 \end{bmatrix}.$$

Obtain a state-space realization for $P(s)$ corresponding to Figure 7.7, part (b).

As illustrated by the above examples, the meanings of transfer functions in (7.23) are as follows:
  • $m$ is the plant nonminimum phase factor, possibly infinite dimensional.
  • $P$ contains the weighting functions and a finite dimensional factor of the plant, including all its unstable modes.
Recall that both weighting functions and unstable modes of the plant to be controlled are assumed to be finite-dimensional in the skew Toeplitz approach of [66]; see also Chapter 6 for numerical implementations (infinitely many unstable poles are also allowed, but in that case, $M_n$ must be finite dimensional, and duality is used; this extension is done in Section 7.5.2). In particular, the rationality assumptions of $W_1$, $W_2$, and $M_d$ used in Chapter 6 correspond to the rationality of $P$.

## 7.4.2 ▪ General solution

Before proceeding to the infinite dimensional case, let us review the finite dimensional case, that is, the $\mathcal{H}_\infty$ control problem posed in the previous section with $m(s) = 1$.

**Proposition 7.19 (see [52]).** *Suppose that $P$ in* (7.22) *satisfies Assumption 7.1. Then, a controller $K$ which internally stabilizes $P$ and satisfies $\|\mathcal{F}_l(P, K)\|_\infty < \gamma$ exists if and only if there exists the stabilizing solutions $X \geq 0$ and $Y \geq 0$ to the following two AREs:*

$$XA + A^\mathsf{T}X + C_1^\mathsf{T}C_1 + \gamma^{-2}XB_1B_1^\mathsf{T}X - F^\mathsf{T}F = 0, \tag{7.31}$$

*where $F := -B_2^\mathsf{T}X$, and*

$$AY + YA^\mathsf{T} + B_1B_1^\mathsf{T} + \gamma^{-2}YC_1^\mathsf{T}C_1Y - LL^\mathsf{T} = 0, \tag{7.32}$$

*where $L := -YC_2^\mathsf{T}$, so that $\rho(XY) < \gamma^2$. Moreover, when these conditions hold and $D_{22} = 0$, all such controllers are given by*

$$K = \mathcal{F}_l(M, Q), \tag{7.33}$$

*where $Q \in \mathcal{H}_\infty$ satisfies $\|Q\|_\infty < \gamma$ and*

$$M = \left[\begin{array}{c|cc} \hat{A} & -ZL & ZB_2 \\ \hline F & 0 & I \\ -C_2 & I & 0 \end{array}\right], \tag{7.34}$$

*with*

$$\hat{A} := A + \gamma^{-2}B_1B_1^\mathsf{T}X + B_2F + ZLC_2,$$
$$Z := (I - \gamma^{-2}YX)^{-1}.$$

**Exercise 7.20.**
Consider the system given in Exercise 7.18, and obtain $\gamma_\text{opt}$ for the special case $m(s) = 1$. *Hint*: You may use the MATLAB command `care` and apply a bi-section search; alternatively, you may directly apply the `hinfsyn` command.

**Assumption 7.21.** *The matrix defined by*

$$\grave{A} := \hat{A} - ZLC_2 - ZB_2F \tag{7.35}$$

*belongs to $\mathcal{M}_m$, and the following realization is minimal:*

$$N_{11} := \left[\begin{array}{c|c} \grave{A} & ZL \\ \hline F & 0 \end{array}\right]. \tag{7.36}$$

Even for $\grave{A} \notin \mathcal{M}_m$, the first assumption can be satisfied by slightly changing the inner function $m(s)$. In the case where the second assumption is violated, we can simply replace (7.36) by any minimal realization. In this sense, Assumption 7.21 is not restrictive.

We are now ready to state the main result of this chapter.

**Theorem 7.22 (see [122]).** *Let $\gamma > 0$ be a prespecified performance level. Suppose that $P$ in* (7.22) *and an inner function $m(s)$ satisfy Assumption 7.1 and $A \in \mathcal{M}_m$. The three conditions in Proposition 7.19 are necessary for the existence of a controller $C$ which internally stabilizes $P_\text{inf}$ and satisfies* (7.24). *Suppose that these conditions and*

*Assumption 7.21 are satisfied. Suppose also that for all $\rho \geq \gamma$ we have*

$$H_\rho^{(\gamma)} := \begin{bmatrix} \grave{A} & \rho^{-1}ZLL^\mathsf{T}Z^\mathsf{T} \\ -\rho^{-1}F^\mathsf{T}F & -\grave{A}^\mathsf{T} \end{bmatrix} \in \mathcal{M}_m. \qquad (7.37)$$

*Then, there exists a controller $C$ which internally stabilizes $P_{\mathrm{inf}}$ and satisfies (7.24) if and only if the essential norm of $\Gamma_{m^\sim \pi^m [N_{11}]}$ is less than $\gamma$ and $m^\sim(H_\rho^{(\gamma)})\big|_{22}$ is of full rank for all $\rho \geq \gamma$. Moreover, when these conditions are satisfied, all internally stabilizing controllers $C$ satisfying (7.24) are given by*

$$C = m^\sim \mathcal{F}_l(M, Q) \qquad (7.38)$$

*and*

$$Q = \pi^m[N_{11}] + m\phi \qquad (7.39)$$

*for arbitrary $\phi \in \mathcal{H}_\infty$ satisfying $\|Q\|_\infty < \gamma$.*

Note that the superscript of $H_\rho^{(\gamma)}$ represents the $\gamma$-dependency of $F$, $L$, $Z$, which are associated with the solutions to the AREs. Recall that all $\phi \in \mathcal{H}_\infty$ satisfying $\|Q\|_\infty < \gamma$ are characterized by Theorem 7.10.

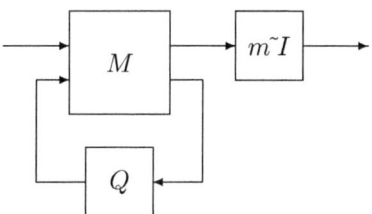

**Figure 7.8.** *Structure of controllers given by* (7.38).

We give some remarks on the underlying idea. Proposition 7.19 motivates us to consider the controllers in Figure 7.8, which automatically satisfies the $\mathcal{H}_\infty$ norm constraint (7.24) for admissible controllers in Figure 7.6. Actually, we can show that all admissible controllers are given in this form [122]. Conversely, controllers in Figure 7.8 internally stabilize the feedback system in Figure 7.6 if and only if (7.39) holds for some $\phi \in \mathcal{H}_\infty$. Finally, the standard $\mathcal{H}_\infty$ control problem reduces to the one-block problem given by (7.39) and $\|Q\|_\infty < \gamma$, which is in the specific form investigated in the previous section.

The additional rank condition on $m^\sim(H_\rho^{(\gamma)})|_{22}$ corresponds to the effect due to the nonminimum phase factor $m(s)$. It is easy to show that Theorem 7.22 is equivalent to Proposition 7.19 when $m(s) = 1$. Recall that $m^\sim(X) = I$ for any square matrix $X$. Thus, it follows that the essential norm of $\Gamma_{m^\sim \pi^m [N_{11}]}$ is zero, because $\pi^m[N_{11}] = 0$ and because $m^\sim(H_\rho^{(\gamma)})|_{22} = I$ is trivially of full rank for any $\rho > 0$.

Note also that in Theorem 7.22, we do not need to verify the rank condition for $\rho > \|\pi^m[N_{11}]\|_\infty$, because $m^\sim(H_\rho^{(\gamma)})|_{22}$ is automatically of full rank for such a $\rho$. Therefore, we can numerically check the additional rank condition by plotting the smallest singular value of $m^\sim(H_\rho^{(\gamma)})|_{22}$, evaluated over a sufficiently fine grid on $\rho$ as in the example given below.

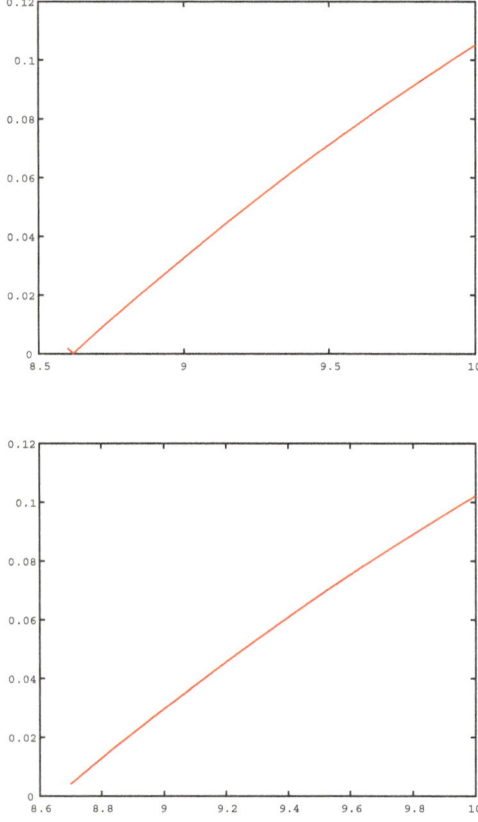

**Figure 7.9.** $\sigma_{\min}(\tilde{m}(H_\rho^{(\gamma)})|_{22})$ *for* $\gamma = 8.6$ *(top) and* $\gamma = 8.7$ *(bottom).*

**Example 7.23.** Let us consider the mixed sensitivity minimization problem in Section 6.2.2. The rational part is given as

$$P_{11} = \begin{bmatrix} W_s \\ 0 \end{bmatrix}, \; P_{12} = \begin{bmatrix} -W_s P_r \\ W_t P_r \end{bmatrix}, \; P_{21} = 1, \; P_{22} = P_r, \; P_r = N_o M_d^{-1},$$

where $W_t P_r$ is biproper. After applying standard transformation such as loop shifting (see, e.g., [83]), we obtain a realization satisfying Assumption 7.21. Figure 7.9 shows the smallest singular value of $\tilde{m}(H_\rho^{(\gamma)})|_{22}$ for $\rho \geq \gamma$. This figure implies that $\gamma = 8.7$ is achievable; however, $\gamma = 8.6$ is not. This is consistent with the fact that the optimal mixed sensitivity is 8.6279, as shown in Section 6.2.2. ∎

In the case of Theorem 7.10, thanks to (7.18), the $\gamma$-bisection search for the computation of $\gamma_{\mathrm{opt}}$ is not needed. Similarly for Theorem 7.22, Figure 7.9 motivates us to evaluate

$$\gamma_{\mathrm{opt}} = \max\{\gamma : \tilde{m}(H_\gamma^{(\gamma)})|_{22} \text{ is singular}\}. \tag{7.40}$$

Actually, this equality provides useful information, although its mathematically rigorous proof requires additional assumptions, e.g., the continuity of $H_\rho^{(\gamma)}$ with respect to

$\gamma$. In particular, in Example 7.29 given in the next section, this numerical equivalence holds.

**Exercise 7.24.**
Consider the system given in Exercise 7.18, and compute $\gamma_{\text{opt}}$ for $m(s) = \frac{0.2 - se^{-s}}{s + 0.2e^{-s}}$. For this purpose, plot the smallest singular value of $m\tilde{}(H_\rho^{(\gamma)})|_{22}$, over a sufficiently fine grid on $\rho$, and determine the smallest $\gamma$ for which $m\tilde{}(H_\rho^{(\gamma)})|_{22}$ is nonsingular for all $\rho \geq \gamma$. Additionally, compute (7.40) to see whether the above discussion holds for this example.

# 7.5 ▪ Duality between unstable zeros and poles

## 7.5.1 ▪ Generalized Smith predictor

A controller of the form in Figure 7.8 cannot be reliably implemented due to unstable pole-zero cancellations and/or noncausality due to $m\tilde{}$. To avoid this issue, an explicit realization of all suboptimal controllers can be obtained by simply replacing $e^{-hs}$ by $m(s)$ in Theorem 1 of [159]: all $\gamma$-suboptimal controllers are parameterized in the form depicted in Figure 7.10 with suitably defined rational transfer matrix $J_r$ and $\Pi \in \mathcal{H}(m)$:

$$J_r := \left[ \begin{array}{c|cc} A_J & -m\tilde{}(A)L & B_2 \\ \hline F & 0 & I \\ \hdashline -C_2 m(A) & I & 0 \end{array} \right],$$

$$A_J := A + B_2 F + m\tilde{}(A) L C_2 m(A),$$

$$\Pi := \pi^m \left[ P_{22}^* \right] \in \mathcal{H}(m), \quad P_{22}^* := \left[ \begin{array}{c|c} A & B_2 \\ \hline -C_2 m(A) & 0 \end{array} \right].$$

Note that both $J_r$ and $\Pi$ are independent of $\gamma$ and the choice of the stable free parameter $\phi$. If we allow $\phi$ to be any function in $\mathcal{H}_\infty$, then Figure 7.10 gives a parameterization of all internally stabilizing controllers.

In Figure 7.10, the only infinite dimensional part is the block represented by $\Pi$ in the feedback path. When $m(s) = e^{-hs}$, this block is a continuous-time FIR system, which can be implemented by finite-time integration. This structure in delay compensating controllers is well known as the modified Smith predictor; see, e.g., [161]. As a natural extension, the effect due to the general nonminimum phase factor, represented by inner function $m$, can be compensated for by the *feedback* component $\Pi \in \mathcal{H}(m)$.

**Example 7.25.** In the Smith predictor–type realization for the controller obtained in Example 7.23, $P_{22}^* = m(1)/(s-1)$. Therefore, the $\gamma$-independent feedback compensator is given by $\Pi = (m(1) - m(s))/(s-1)$, which is stable despite the unstable pole of $P_{22}$. ∎

## 7.5.2 ▪ Systems with infinitely many unstable poles

In this section, we analyze the performance degrading effect caused by unstable modes. The main idea is to formulate a dual problem which can be handled by the methods presented in the previous section. The material presented in this section is taken from

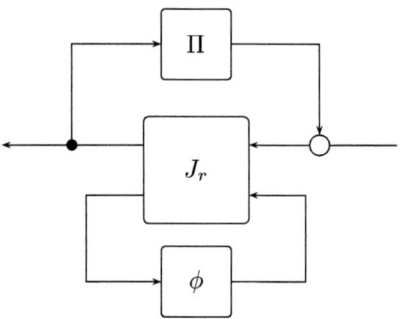

**Figure 7.10.** *Smith predictor–type structure.*

[123], which can be seen as an extension of the SISO case [95], covered in Sections 4.5, 5.1.3, and 6.3.2.

Define the generalized plant $P_{\rm u}$ by

$$\hat{P} := \left[\begin{array}{c:c} \hat{P}_{11} & \hat{P}_{12} \\ \hdashline \hat{P}_{21} & \hat{P}_{22} \end{array}\right] := \left[\begin{array}{c|cc} \hat{A} & \hat{B}_1 & \hat{B}_2 \\ \hline \hat{C}_1 & 0 & \hat{D}_{12} \\ \hat{C}_2 & \hat{D}_{21} & \hat{D}_{22} \end{array}\right], \tag{7.41}$$

$$P_{\rm u} := \left[\begin{array}{c:c} I & 0 \\ \hdashline 0 & \tilde{m}\,I \end{array}\right] \left[\begin{array}{c:c} \hat{P}_{11} & \hat{P}_{12} \\ \hdashline \hat{P}_{21} & \hat{P}_{22} \end{array}\right] \tag{7.42}$$

with an inner function $m$. Denote the set of its internally stabilizing controllers by $\mathscr{C}(P_{\rm u})$. In what follows, for a prespecified performance level $\gamma > 0$, determine whether there exists a controller $C \in \mathscr{C}(P_{\rm u})$ that guarantees

$$\|\mathcal{F}_l(P_{\rm u}, C)\|_\infty < \gamma.$$

If such a controller exists, find all admissible ones.

**Example 7.26.** In comparison to (7.25), we can formulate a control problem for the plant which can be factored as

$$\hat{G} = \hat{G}_r \cdot \hat{G}_o \cdot \tilde{m}, \tag{7.43}$$

where $m$ is inner, $\hat{G}_r$ is rational (possibly with unstable poles and/or unstable zeros), and $\hat{G}_o, \hat{G}_o^{-1} \in \mathcal{H}_\infty$. Since $\tilde{m} = 1/m$, unstable zeros of $m$ represent unstable poles of $G$. Therefore, we can analyze the effect on the performance limit caused by unstable modes by taking $m$ as an appropriate Blaschke product. Again, $m$ is not assumed to be rational. Hence, $m$ can deal with even infinitely many unstable modes; see Section 4.5 for restrictions on the relative degree of the underlying plant and Section 6.3.2 for a SISO design example.  ∎

Fortunately, this problem can be converted to another one for which Theorem 7.22 is applicable.

**Example 7.27.** We again consider the weighted mixed sensitivity minimization problem to explain the basic idea of this problem conversion. For stable rational weights $\hat{W}_s$ and $\hat{W}_t$, let us find an internally stabilizing controller $\hat{C}$ that minimizes the weighted mixed sensitivity

$$\left\| \begin{bmatrix} \hat{W}_s (1 - \hat{G}\hat{C})^{-1} \\ \hat{W}_t (1 - \hat{G}\hat{C})^{-1}\hat{G}\hat{C} \end{bmatrix} \right\|. \tag{7.44}$$

By straightforward computation, it can be verified that $\hat{C}$ internally stabilizes $\hat{G}$ if and only if $C := \hat{C}^{-1}$ internally stabilizes $G := \hat{G}^{-1}$ and

$$\left\| \begin{bmatrix} \hat{W}_s (1 - \hat{G}\hat{C})^{-1} \\ \hat{W}_t (1 - \hat{G}\hat{C})^{-1}\hat{G}\hat{C} \end{bmatrix} \right\| = \left\| \begin{bmatrix} \hat{W}_t (1 - GC)^{-1} \\ \hat{W}_s (1 - GC)^{-1}GC \end{bmatrix} \right\|. \tag{7.45}$$

It readily follows from (7.43) that $G$ has a factorization in the form of (7.25). This problem, therefore, falls into the framework in the previous section. Recall Proposition 4.31 for the SISO version of this "design by duality" approach. See also an example given in Section 6.3.2.  ∎

We generalize this problem conversion technique in Theorem 7.28 below. This is an extension of the results given in Section 4.5. In particular, recall that the plant and the controller must be biproper when dealing with the stabilization of systems with infinitely many poles in $\mathbb{C}_+$; hence, the assumption that $\hat{D}_{22}$ is nonsingular is necessary for this result.

**Theorem 7.28 (see [119]).** *Assume the following:*

1. $(\hat{A}, \hat{B}_2)$ *is stabilizable and* $(\hat{A}, \hat{C}_2)$ *is detectable.*

2. $\hat{D}_{22}$ *is nonsingular.*

3. $A := \hat{A} - \hat{B}_2 \hat{D}_{22}^{-1} \hat{C}_2$ *belongs to* $\mathcal{M}_m$.

4. $D_{12} = 0$.

*Define*

$$P := \left[ \begin{array}{c|cc} A & B_1 & B_2 \\ \hline C_1 & 0 & 0 \\ C_2 & D_{21} & D_{22} \end{array} \right] \tag{7.46}$$

$$:= \left[ \begin{array}{c|cc} \hat{A} - \hat{B}_2 \hat{D}_{22}^{-1} \hat{C}_2 & \hat{B}_1 - \hat{B}_2 \hat{D}_{22}^{-1} \hat{D}_{21} & -\hat{B}_2 \hat{D}_{22}^{-1} \\ \hline \hat{C}_1 & 0 & 0 \\ \hat{D}_{22}^{-1} \hat{C}_2 & \hat{D}_{22}^{-1} \hat{D}_{21} & \hat{D}_{22}^{-1} \end{array} \right].$$

*Then the following statements hold:*

1. $(A, B_2)$ *is stabilizable and* $(A, C_2)$ *is detectable.*

2. $A$ *belongs to* $\mathcal{M}_m$.

3. *When $C\hat{C} = I$, we can show the following two statements:*

- *$C$ internally stabilizes $P$ if and only if $\hat{C}$ internally stabilizes $P_{\mathrm{u}}$.*
- *The two closed-loop systems are the same, i.e.,*

$$\mathcal{F}_l\left(P, C\right) = \mathcal{F}_l\left(P_{\mathrm{u}}, \hat{C}\right).$$

In order to analyze the performance degrading effect caused by unstable poles, we need the solution to the AREs corresponding not to (7.41) but to (7.46). In the dual case investigated in Section 7.4.2, since $C\hat{C} = I$, all suboptimal controllers are given by exchanging the input and output signals in Figure 7.10. As a result, any suboptimal controller is given in the form of a rational system with a common stable and infinite-dimensional *feed-forward* component. This compensates for the effect of unstable modes caused by $\tilde{m}$.

**Example 7.29.** We consider the mixed sensitivity minimization problem (7.44) where $\hat{W}_s := W_s I_2$, $\hat{W}_t := W_m I_2$, and $\hat{G} := P_r(P I_2)$, with $I_2$ being the identity matrix in $\mathbb{R}^{2\times 2}$, the underlying plant $P$ being given by (6.73) and having infinitely many unstable poles, and the weighting functions $W_s$ and $W_m$ being given by (6.78) and (6.77), respectively; in this MIMO example, the plant is augmented and cascaded with

$$P_r(s) := \left(s I_2 - \begin{bmatrix} a_x & 1 \\ 0 & -0.5 \end{bmatrix}\right)^{-1} + I_2, \tag{7.47}$$

where $a_x > 0$ is a parameter whose impact on the optimal performance level $\gamma_{\mathrm{opt}}$ is to be investigated. Theorem 7.22 is applied after converting the problem by Theorem 7.28, where $m(s) := M_d(s)$ is given by (6.74). First, let us take $a_x = 0.5$. Figure 7.11 shows the smallest singular value of $\tilde{m}(H_\gamma^{(\gamma)})|_{22}$ for $\gamma$ larger than 3.747, which is the best achievable performance for $m(s) = 1$. It is expected from (7.40) that the optimal performance is about 4.06. Actually, the smallest singular value of $\tilde{m}(H_\rho^{(\gamma)})|_{22}$ for $\gamma = 4.0, 4.1$ and $\rho \geq \gamma$ justifies this estimation; see Figure 7.12. The optimal mixed sensitivity levels, $\gamma_{\mathrm{opt}}$, for several values of $a_x$ are listed in Table 7.1.  ∎

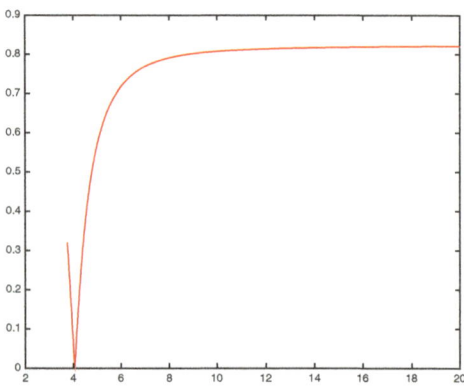

**Figure 7.11.** $\sigma_{\min}(\tilde{m}(H_\gamma^{(\gamma)})|_{22})$ *for* $\gamma > 3.747$.

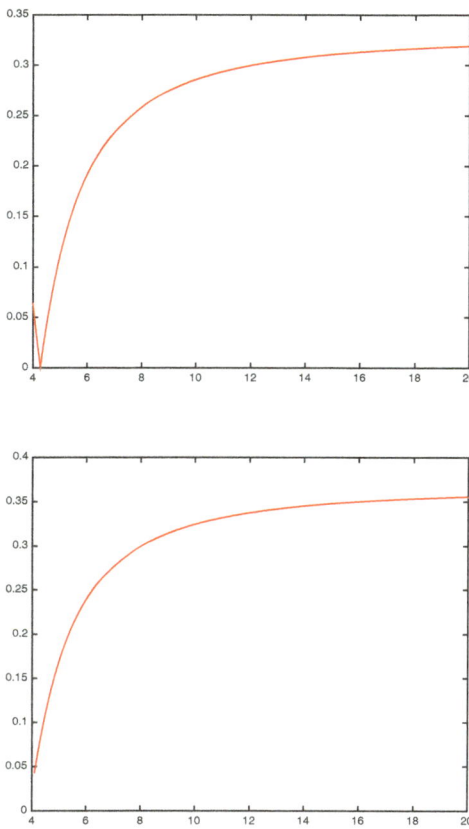

**Figure 7.12.** $\sigma_{\min}(\tilde{m}(H_\rho^{(\gamma)})|_{22})$ *for* $\gamma = 4$ *(top) and* $\gamma = 4.1$ *(bottom).*

**Table 7.1.** *Optimal mixed sensitivity level* $\gamma_{\mathrm{opt}}$ *for several values of* $a_x$.

| $a_x$ | 0.2 | 0.3 | 0.4 | 0.5 | 0.6 | 0.7 | 0.8 | 0.9 |
|---|---|---|---|---|---|---|---|---|
| $\gamma_{\mathrm{opt}}$ | 2.225 | 2.57 | 3.15 | 4.06 | 5.58 | 8.505 | 16.50 | 120.9 |

# Bibliography

[1] M. J. ABLOWITZ AND A. S. FOKAS, *Complex Variables*, Second Edition, Cambridge University Press, New York, 2003. (Cited on p. 73)

[2] V. M. ADAMJAN, D. Z. AROV, AND M. G. KREIN, *Analytic properties of Schmidt pairs for a Hankel operator and the generalized Shur–Takagi problem*, Mathematics of the USSR-Sbornik, vol. 15 (1971), pp. 31–73. (Cited on p. 155)

[3] M. ADIMY, F. CRAUSTE, AND A. EL ABDLLAOUI, *Discrete maturity-structured model of cell differentiation with applications to acute myelogenous leukemia*, Journal of Biological Systems, vol. 16, no. 3 (2008), pp. 395–424. (Cited on pp. 42, 43)

[4] E. J. AKUTOWICZ, *Characterization of Blaschke products in a half-plane*, American Journal of Mathematics, vol. 78, no. 4 (1956), pp. 677–684. (Cited on p. 11)

[5] S. ALCANTARA, W. D. ZHANG, C. PEDRET, R. VILANOVA, AND S. SKOGESTAD, *IMC-like analytical $\mathcal{H}_\infty$ design with S/SP mixed sensitivity consideration: Utility in PID tuning guidance*, Journal of Process Control, vol. 21 (2011), pp. 976–985. (Cited on p. 33)

[6] F. A. ALIEV AND V. B. LARIN, *Optimization of Linear Control Systems: Analytical Methods and Computational Algorithms*, Gordon and Breach, Amsterdam, 1998. (Cited on p. 48)

[7] F. A. ALIEV AND V. B. LARIN, *Errata to: "Comments on optimizing simultaneously over the numerator and denominator polynomials in the Youla-Kučera parameterization,"* IEEE Transactions on Automatic Control, vol. 52 (2007), p. 763. (Cited on p. 48)

[8] Y. ALTINTAS AND M. WECK, *Chatter stability of metal cutting and grinding*, CIRP Annals—Manufacturing Technology, vol. 53 (2004), pp. 619–642. (Cited on p. 31)

[9] R. J. ANDERSON AND M. W. SPONG, *Bilateral control of teleoperators with time delay*, IEEE Transactions on Automatic Control, vol. 34 (1989), pp. 494–501. (Cited on p. 31)

[10] D. AVANESSOFF, C. BONNET, H. CAVALERA, A. R. FIORAVANTI, AND L. H. V. NGUYEN *User document YALTA*, v.1.0.1; https://gforge.inria.fr/projects/yalta-toolbox, September 2015. (Cited on pp. 67, 80, 120)

[11] D. AVANESSOFF, A. R. FIORAVANTI, AND C. BONNET, *YALTA: A MATLAB toolbox for the $H_\infty$-stability analysis of classical and fractional systems with commensurate delays*, Preprints of 5th IFAC Symposium on System, Structure and Control, Grenoble, France, February 2013, pp. 839–844. (Cited on pp. 67, 71, 80, 120)

[12] J. L. AVILA, C. BONNET, J. CLAIRAMBAULT, H. ÖZBAY, S.-I. NICULESCU, F. MERHI, A. BALLESTA, R. TANG, AND J. P. MARIE, *Analysis of a new model of cell population dynamics in Acute Myeloid Leukemia*, in Delay Systems: From Theory to Numerics and Applications, T. Vyhlidal, J.-F. Lafay, and R. Sipahi, eds., Advances in Delays and Dynamics at Springer (ADD@S), vol. 1, pp. 315–328, 2014. (Cited on p. 42)

[13] J. L. AVILA, C. BONNET, H. ÖZBAY, J. CLAIRAMBAULT, S.-I. NICULESCU, P. HIRSCH, AND F. DELHOMMEAU, *A coupled model for healthy and cancerous cells dynamics in Acute Myeloid Leukemia*, Preprints of the IFAC World Congress, Cape Town, South Africa, August 2014, pp. 7529–7534. (Cited on p. 42)

[14] T. AZUMA, S. SAGARA, M. FUJITA, AND K. UCHIDA, *Output feedback control synthesis for linear time-delay systems via infinite-dimensional LMI approach*, Proceedings of the 42nd IEEE Conference on Decision and Control, Maui, Hawaii, December 2003, pp. 4026–4031. (Cited on p. 2)

[15] M. J. BALAS, *Finite-dimensional controllers for linear distributed parameter systems: Exponential stability using residual mode filters*, Journal of Mathematical Analysis and Applications, vol. 133 (1988), pp. 283–296. (Cited on p. 2)

[16] J. A. BALL, I. GOHBERG, AND L. RODMAN, *Interpolation of Rational Matrix Functions*, Birkhäuser, Basel, 1990. (Cited on pp. 15, 17)

[17] B. BAMIEH, F. PAGANINI, AND M. A. DAHLEH, *Distributed control of spatially invariant systems*, IEEE Transactions on Automatic Control, vol. 47 (2002), pp. 1091–1107. (Cited on p. 3)

[18] B. BAMIEH AND J. B. PEARSON, *A general framework for linear periodic systems with applications to $\mathcal{H}_\infty$ sampled-data control*, IEEE Transactions on Automatic Control, vol. 37 (1992), pp. 418–435. (Cited on p. 3)

[19] B. R. BARMISH, *New Tools for Robustness of Linear Systems*, Macmillan, New York, 1994. (Cited on p. 2)

[20] T. BAŞAR AND P. BERNHARD, *$\mathcal{H}_\infty$ Optimal Control and Related Minimax Design Problems: A Dynamic Game Approach*, Second Edition, Birkhäuser, Boston, 1995. (Cited on p. 3)

[21] R. BELLMAN AND K. L. COOKE, *Differential-Difference Equations*, Academic Press, New York, 1963. (Cited on p. 67)

[22] P. BENNER, S. GUGERCIN, AND K. WILLCOX, *A survey of projection-based model reduction methods for parametric dynamical systems*, SIAM Review, vol. 57, no. 4 (2015), pp. 483–531. (Cited on p. 141)

[23] H. BERCOVICI, C. FOIAS, AND A. TANNENBAUM, *On skew Toeplitz operators*, Operator Theory: Advances and Applications, vol. 32 (1988), pp. 21–43. (Cited on p. 2)

[24] S. P. BHATTACHARYYA, H. CHAPELLAT, AND L. H. KEEL, *Robust Control: The Parametric Approach*, Prentice–Hall, Upper Saddle River, NJ, 1995. (Cited on p. 2)

[25] C. BONNET AND J. R. PARTINGTON, *Analysis of fractional delay systems of retarded and neutral type*, Automatica, vol. 38 (2002), pp. 1133–1138. (Cited on p. 3)

[26] J. BONTSEMA AND S. A. DE VRIES, *Robustness of flexible structures against small time delays*, Proceedings of the 27th IEEE Conference on Decision and Control (1988) pp. 1647–1648. (Cited on p. 3)

[27] S. BOYD AND V. BALAKRISHNAN, *A regularity result for the singular values of a transfer matrix and a quadratically convergent algorithm for computing its $\mathcal{L}_\infty$-norm*, Systems & Control Letters, vol. 15 (1990), pp. 1–7. (Cited on p. 27)

[28] S. BOYD, V. BALAKRISHNAN, AND P. KABAMBA, *A bisection method for computing the $\mathcal{H}_\infty$ norm of a transfer matrix and related problems*, Mathematics of Control, Signals, and Systems, vol. 2 (1989), pp. 207–219. (Cited on p. 27)

[29] D. BREDA, S. MASET, AND R. VERMIGLIO, *TRACE-DDE: A tool for robust analysis and characteristic equations for delay differential equations*, in Topics in Time Delay Systems: Analysis, Algorithms, and Control, J.J. Loiseau, W. Michiels, S.-I. Niculescu, and R. Sipahi, eds., Springer, LNCIS, vol. 388 (2009), pp. 183–192. (Cited on p. 71)

[30] N. A. BRUINSMA AND M. STEINBUCH, *A fast algorithm to compute the $\mathcal{H}_\infty$-norm of a transfer function matrix*, Systems & Control Letters, vol. 14 (1990), pp. 287–293. (Cited on p. 27)

[31] R. BYERS, *A bisection method for measuring the distance of a stable matrix to the unstable matrices*, SIAM Journal on Scientific and Statistical Computing, vol. 9 (1988), pp. 875–881. (Cited on p. 27)

[32] C. I. BYRNES, M. W. SPONG, AND T-J. TARN, *A several complex variables approach to feedback stabilization of linear neutral delay-differential systems*, Mathematical Systems Theory, vol 17 (1984), pp. 97–133. (Cited on p. 72)

[33] Y-Y. CAO AND J. LAM, *Robust $\mathcal{H}_\infty$ control of uncertain Markovian jump systems with time-delay*, IEEE Transactions on Automatic Control, vol. 45 (2000), pp. 77–83. (Cited on p. 3)

[34] H. CHE, P. YAN, AND Z. ZHANG, *Robust repetitive control for time delay systems with application to nano manipulations*, Proceedings of the 36th Chinese Control Conference, July 2017, Dalian, China, pp. 3118–3123. (Cited on p. 78)

[35] T. CHEN AND B. A. FRANCIS, *$\mathcal{H}_\infty$-optimal sampled-data control: Computation and designs*, Automatica, vol. 32 (1996), pp. 223–228. (Cited on p. 3)

[36] E. N. CHUKWU, *Stability and Time-Optimal Control of Hereditary Systems*, Academic Press, San Diego, 1992. (Cited on p. 31)

[37] J. A. COOK AND B. K. POWELL, *Modeling of an internal combustion engine for control analysis*, IEEE Control Systems Magazine, vol. 8, no. 4 (1988), pp. 20–26. (Cited on p. 31)

[38] R. CURTAIN AND K. MORRIS, *Transfer functions of distributed parameter systems: A tutorial*, Automatica, vol. 45 (2009), pp. 1101–1116. (Cited on pp. 31, 41)

[39] R. F. CURTAIN, *$H^\infty$ control for distributed parameter systems: A survey*, Proceedings of the 29th CDC, Honolulu, Hawaii, December 1990, pp. 22–26. (Cited on p. 2)

[40] R. F. CURTAIN AND K. GLOVER, *Robust stabilization of infinite dimensional systems by finite dimensional controllers*, Systems & Control Letters, vol. 7 (1986), pp. 41–47. (Cited on p. 2)

[41] R. F. CURTAIN AND A. J. SASANE, *Hankel norm approximation for well-posed linear systems*, Systems & Control Letters, vol. 48 (2003), pp. 407–414. (Cited on p. 2)

[42] R. F. CURTAIN AND Y. ZHOU, *A weighted mixed-sensitivity $\mathcal{H}_\infty$-control design for irrational transfer matrices*, IEEE Transactions on Automatic Control, vol. 41 (1996), pp. 1312–1321. (Cited on p. 2)

[43] R. F. CURTAIN AND H. J. ZWART, *An Introduction to Infinite-Dimensional Linear Systems Theory*, Springer-Verlag, New York, 1995. (Cited on pp. 3, 31, 151, 155)

[44] R. D'ANDREA AND G. E. DULLERUD, *Distributed control design for spatially interconnected Systems*, IEEE Transactions on Automatic Control, vol. 48 (2003), pp. 1478–1495. (Cited on p. 3)

[45] M. A. DAHLEH AND I. J. DIAZ-BOBILLO, *Control of Uncertain Systems: A Linear Programming Approach*, Prentice–Hall, Englewood Cliffs, NJ, 1995. (Cited on p. 2)

[46] C. A. DESOER AND M. VIDYASAGAR, *Feedback Systems: Input-Output Properties*, Academic Press, 1975, reprinted by SIAM, Philadelphia, in 2009. (Cited on pp. 5, 8)

[47] W. DJEMA, C. BONNET, F. MAZENC, J. CLAIRAMBAULT, E. FRIDMAN, P. HIRSCH, AND F. DELHOMMEAU, *Control in dormancy or eradication of cancer stem cells: Mathematical modeling and stability issues*, Journal of Theoretical Biology, vol. 449 (2018), pp. 103–123. (Cited on p. 42)

[48] S. M. DJOUADI AND J. D. BIRDWELL, *On the optimal two-block $\mathcal{H}_\infty$ problem*, Proceedings of the 2005 American Control Conference, Portland, OR, June 2005, pp. 4289–4294. (Cited on p. 2)

[49] W. F. DONOGHUE, *Distributions and Fourier Transforms*, Academic Press, New York, 1969. (Cited on p. 61)

[50] R. C. DORF AND R. H. BISHOP, *Modern Control Systems*, 13th Edition, Pearson, Santa Monica, CA, 2017. (Cited on pp. 3, 46)

[51] J. C. DOYLE, B. A. FRANCIS, AND A. R. TANNENBAUM, *Feedback Control Theory*, Macmillan, New York, 1992. (Cited on pp. 2, 4, 8, 17, 21, 82, 86, 91)

[52] J. C. DOYLE, K. GLOVER, P. P. KHARGONEKAR, AND B. A. FRANCIS, *State-space solutions to standard $\mathcal{H}_2$ and $\mathcal{H}_\infty$ control problems*, IEEE Transactions on Automatic Control, vol. 34 (1989), pp. 831–847. (Cited on pp. 2, 151, 164)

[53] L. DUGARD AND E. I. VERRIEST, EDS., *Stability and Control of Time-Delay Systems*, LNCIS vol. 228, Springer-Verlag, London, 1998. (Cited on p. 3)

[54] G. E. DULLERUD AND F. PAGANINI, *A Course in Robust Control Theory: A Convex Approach*, Springer-Verlag, New York, 2000. (Cited on p. 2)

[55] H. DYM, T. T. GEORGIOU, AND M. C. SMITH, *Explicit formulas for optimally robust controllers for delay systems*, IEEE Transactions on Automatic Control, vol. 40 (1995), pp. 656–669. (Cited on p. 3)

[56] K. ENGELBORGHS, T. LUZYANINA, AND D. ROOSE, *Numerical bifurcation analysis of delay differential equations using DDE-BIFTOOL*, ACM Transactions on Mathematical Software, vol. 28 (2002), pp. 1–21. (Cited on pp. 67, 71)

[57] K. ENGELBORGHS, T. LUZYANINA, AND G. SAMAEY, *DDE-BIFTOOL v.2.00: A MATLAB Package for Bifurcation Analysis of Delay Differential Equations*, Report TW 330, Katholieke Universiteit Leuven, October 2001. (Cited on p. 71)

[58] K. ENGELBORGHS AND D. ROOSE, *On stability of LMS methods and characteristic roots of delay differential equations*, SIAM Journal on Numerical Analysis, vol. 40 (2002), pp. 629–650. (Cited on p. 71)

[59] D. ENNS, H. ÖZBAY, AND A. TANNENBAUM, *Abstract model and controller design for an unstable aircraft*, AIAA Journal of Guidance Control and Dynamics, vol. 15 (1992), pp. 498–508. (Cited on p. 105)

[60] M. FERRARA, L. GUERRINI, AND R. MAVILIA, *Modified neoclassical growth models with delay: A critical survey and perspectives*, Applied Mathematical Sciences, vol. 7 (2013), pp. 4249–4257. (Cited on p. 31)

[61] A. R. FIORAVANTI, C. BONNET, H. ÖZBAY, AND S-I. NICULESCU, *A numerical method for stability windows and unstable root-locus calculation for linear fractional time-delay systems*, Automatica, vol. 48 (2012), pp. 2824–2830. (Cited on pp. 42, 80)

[62] D. S. FLAMM, *Outer factor absorption for $H_\infty$ control problems*, International Journal of Robust and Nonlinear Control, vol 2, no. 1 (1992), pp. 31–48. (Cited on pp. 158, 161)

[63] D. S. FLAMM AND S. K. MITTER, *$\mathcal{H}_\infty$ sensitivity minimization for delay systems*, Systems & Control Letters, vol. 9 (1987), pp. 17–24. (Cited on p. 2)

[64] D. S. FLAMM AND H. YANG, *Optimal mixed sensitivity for SISO distributed plants*, IEEE Transactions on Automatic Control, vol. 39 (1994), pp. 1150–1165. (Cited on p. 2)

[65] C. FOIAS AND A. E. FRAZHO, *The Commutant Lifting Approach to Interpolation Problems*, Birkhäuser, Basel, 1990. (Cited on p. 13)

[66] C. FOIAS, H. ÖZBAY, AND A. TANNENBAUM, *Robust Control of Infinite Dimensional Systems: Frequency Domain Methods*, LNCIS vol. 209, Springer-Verlag, London, 1996. (Cited on pp. xi, 1, 2, 5, 13, 17, 22, 24, 25, 151, 163)

[67] C. FOIAS, A. TANNENBAUM, AND G. ZAMES, *Weighted sensitivity minimization for delay systems*, IEEE Transactions on Automatic Control, vol. 31 (1986), pp. 763–766. (Cited on p. 2)

[68] B. A. FRANCIS, *A Course in $\mathcal{H}_\infty$ Control Theory*, LNCIS, vol. 88, Springer-Verlag, Berlin, 1987. (Cited on pp. 2, 18, 21)

[69] B. A. FRANCIS AND G. ZAMES, *On $\mathcal{H}_\infty$ optimal sensitivity theory for SISO feedback systems*, IEEE Transactions on Automatic Control, vol. 29 (1984), pp. 9–16. (Cited on p. 158)

[70] E. FRIDMAN, *Introduction to Time-Delay Systems: Analysis and Control*, Birkhäuser, Cham, 2014. (Cited on pp. 3, 31, 32)

[71] E. FRIDMAN, A. PILA, AND U. SHAKED, *Regional stabilization and $\mathcal{H}_\infty$ control of time-delay systems with saturating actuators*, International Journal of Robust and Nonlinear Control, vol. 13 (2003), pp. 885–907. (Cited on p. 3)

[72] E. FRIDMAN AND U. SHAKED, *A descriptor system approach to $\mathcal{H}_\infty$ control of linear time-delay systems*, IEEE Transactions on Automatic Control, vol. 47 (2002), pp. 253–270. (Cited on p. 3)

[73] R. FRISCH AND H. HOLME, *The characteristic solutions of a mixed difference and differential equation occurring in economic dynamics*, Econometrica, vol. 3 (1935), pp. 225–239. (Cited on p. 34)

[74] M. FU, A. W. OLBROT, AND M. P. POLIS, *Robust stability for time-delay systems: The edge theorem and graphical tests*, IEEE Transactions on Automatic Control, vol. 34 (1989), pp. 813–820. (Cited on p. 3)

[75] F. R. GANTMACHER, *The Theory of Matrices*, Chelsea, New York, 1960. (Cited on p. 153)

[76] T. T. GEORGIOU AND M. C. SMITH, *Optimal robustness in the gap metric*, IEEE Transactions on Automatic Control, vol. 35 (1990), pp. 673–686. (Cited on pp. 88, 89)

[77] T. T. GEORGIOU AND M. C. SMITH, *Robust stabilization in the gap metric: Controller design for distributed plants*, IEEE Transactions on Automatic Control, vol. 37 (1992), pp. 1133–1143. (Cited on pp. 2, 89)

[78]  T. T. GEORGIOU AND M. C. SMITH, *Robust control of feedback systems with combined plant and controller uncertainty*, Proceedings of the American Control Conference, 1990, pp. 2009–2013. (Cited on p. 89)

[79]  T. GLAD AND L. LJUNG, *Control Theory: Multivariable and Nonlinear Methods*, Taylor & Francis, London, 2000. (Cited on p. 2)

[80]  K. GLOVER, *All optimal Hankel-norm approximations of linear multivariable systems and their $L_\infty$-error bounds*, International Journal of Control, vol. 39 (1984), pp. 1115–1193. (Cited on p. 18)

[81]  K. GLOVER AND D. MCFARLANE, *Robust stabilization of normalized coprime factor plant descriptions with $\mathcal{H}_\infty$ bounded uncertainty*, IEEE Transactions on Automatic Control, vol. 34 (1989), pp. 821–830. (Cited on p. 88)

[82]  G. H. GOLUB AND C. F. VAN LOAN, *Matrix Computations*, The Johns Hopkins University Press, Baltimore, MD, 1989. (Cited on p. 153)

[83]  M. GREEN AND D. J. N. LIMEBEER, *Linear Robust Control*, Prentice–Hall, Englewood Cliffs, NJ, 1995. (Cited on pp. 2, 166)

[84]  C. GU, *Eliminating the genericity conditions in the skew Toeplitz operator algorithm for $H_\infty$ optimization*, SIAM Journal on Mathematical Analysis, vol. 23 (1992), pp. 1623–1636. (Cited on p. 2)

[85]  C. GU, O. TOKER, AND H. ÖZBAY, *On the two block $\mathcal{H}_\infty$ problem for a class of unstable distributed systems*, Linear Algebra and Its Applications, vol. 234 (1996), pp. 227–244. (Cited on p. 2)

[86]  D-W. GU, P. H. PETKOV, AND M. M. KONSTANTINOV, *Robust Control Design with MATLAB®*, Second Edition, Springer-Verlag, London, 2013. (Cited on p. 82)

[87]  G. GU AND P. P. KHARGONEKAR, *Linear and nonlinear algorithms for identification in $\mathcal{H}_\infty$ with error bounds*, IEEE Transactions on Automatic Control, vol. 37 (1992), pp. 953–963. (Cited on p. 82)

[88]  G. GU, P. P. KHARGONEKAR, AND E. B. LEE, *Approximation of infinite-dimensional systems*, IEEE Transactions on Automatic Control, vol. 34 (1989), pp. 610–618. (Cited on pp. 112, 141)

[89]  K. GU, V. L. KHARITONOV, AND J. CHEN, *Stability of Time-Delay Systems*, Birkhäuser, Boston, 2003. (Cited on pp. 3, 31)

[90]  C. GUIVER AND M. R. OPMEER, *Model reduction by balanced truncation for systems with nuclear Hankel operators*, SIAM Journal on Control and Optimization, vol. 52, no. 2 (2014), pp. 1366–1401. (Cited on p. 141)

[91]  S. GUMUSSOY, *Coprime-inner/outer factorization of SISO time-delay systems and FIR structure of their optimal $\mathcal{H}_\infty$ controllers*, International Journal of Robust and Nonlinear Control, vol. 22 (2012), pp. 981–998. (Cited on p. 110)

[92]  S. GUMUSSOY, D. HENRION, M. MILLSTONE, AND M. L. OVERTON, *Multiobjective robust control with HIFOO 2.0*, IFAC Proceedings Volumes (Proceedings of the 6th IFAC Symposium on Robust Control Design), vol. 42, no. 6 (2009), pp. 144–149. (Cited on p. 135)

[93]  S. GUMUSSOY AND W. MICHIELS, *Fixed-order H-infinity control for interconnected systems using delay differential algebraic equations*, SIAM Journal on Control and Optimization, vol. 49 (2011), pp. 2212–2238. (Cited on pp. 29, 135)

[94] S. GUMUSSOY AND W. MICHIELS, *Computation of extremum singular values and the strong H-infinity norm of SISO time-delay systems*, Automatica, vol. 54 (2015), pp. 266–271. (Cited on p. 28)

[95] S. GUMUSSOY AND H. ÖZBAY, *On the mixed sensitivity minimization for systems with infinitely many unstable modes*, Systems & Control Letters, vol. 53 (2004), pp. 211–216. (Cited on p. 168)

[96] S. GUMUSSOY AND H. ÖZBAY, *Stable $\mathcal{H}_\infty$ controller design for time delay systems*, International Journal of Control, vol. 81 (2008), pp. 546–556. (Cited on p. 94)

[97] S. GUMUSSOY AND H. ÖZBAY, *Sensitivity minimization by strongly stabilizing controllers for a class of unstable time-delay systems*, IEEE Transactions on Automatic Control, vol. 54 (2009), pp. 590–595. (Cited on p. 94)

[98] S. GUMUSSOY AND H. ÖZBAY, *On feedback stabilization of neutral time delay systems with infinitely many unstable poles*, Proceedings of the 14th IFAC Workshop on Time Delay Systems, Budapest, Hungary, June 2018, to appear. (Cited on pp. 72, 75)

[99] L. GÜVENÇ, *Repetitive controller design in parameter space*, Proceedings of the 2001 American Control Conference, Arlington, VA, June 2001, pp. 2749–2754. (Cited on p. 3)

[100] I. GYÖRI, F. HARTUNG, AND J. TURI, *Preservation of stability in delay equations under delay perturbations*, Journal of Mathematical Analysis and Applications, vol. 220 (1998), pp. 290–312. (Cited on p. 3)

[101] J. K. HALE AND S. M. VERDUYN LUNEL, *Introduction to Functional Differential Equations*, Series in Applied Mathematical Sciences, vol. 99, Springer-Verlag, 1993. (Cited on pp. 67, 68)

[102] Y. HALEVI, *On control of flexible structures*, Proceedings of the 41st IEEE Conference on Decision and Control, Las Vegas, NV, December 2002, pp. 232–237. (Cited on p. 3)

[103] Y. HALEVI, *Control of flexible structures governed by the wave equation using infinite dimensional transfer functions*, ASME Journal of Dynamic Systems, Measurement, and Control, vol. 127 (2005), pp. 579–588. (Cited on p. 3)

[104] S. HARA, Y. YAMAMOTO, AND H. FUJIOKA, *Modern and classical analysis/synthesis methods in sampled-data control—A brief overview with numerical examples*, Proceedings of the 35th IEEE Conference on Decision and Control, Kobe, Japan, December 1996, pp. 1251–1256. (Cited on p. 3)

[105] S. HARA, Y. YAMAMOTO, T. OMATA, AND M. NAKANO, *Repetitive control system—A new-type servo system*, IEEE Transactions on Automatic Control, vol. 33 (1988), pp. 659–668. (Cited on pp. 3, 76, 78)

[106] A. HAURANIA, H. H. MICHALSKA, AND B. BOULET, *Delay-dependent robust stabilization of uncertain neutral systems with saturating actuators*, Proceedings of the 2003 American Control Conference, Denver, CO, June 2003, pp. 509–514. (Cited on p. 3)

[107] H. HEMAMI AND H. ÖZBAY, *Modeling and control of biological systems with multiple afferent and efferent transmission delays*, Journal of Robotic Systems, vol. 17 (2000), pp. 609–622. (Cited on p. 31)

[108] K. HIRATA, Y. YAMAMOTO, AND A. TANNENBAUM, *A Hamiltonian-based solution to the two block $\mathcal{H}_\infty$ problem for general plants in $\mathcal{H}_\infty$ and rational weights*, Systems & Control Letters, vol. 40 (2000), pp. 83–95. (Cited on p. 2)

[109] K. HOFFMAN, *Banach Spaces of Analytic Functions*, Prentice–Hall, Englewood Cliffs, NJ, 1962. (Cited on pp. 58, 63)

[110] C. V. HOLLOT, V. MISRA, D. TOWSLEY, AND W. B. GONG, *Analysis and design of controllers for AQM routers supporting TCP flows*, IEEE Transactions on Automatic Control, vol. 47 (2002), pp. 945–959. (Cited on pp. 31, 34, 35, 36)

[111] Y-P. HUANG AND K. ZHOU, *Robust stability of uncertain time-delay systems*, IEEE Transactions on Automatic Control, vol. 45 (2000), pp. 2169–2173. (Cited on p. 3)

[112] M. HUZMEZAN, W. A. GOUGH, G. A. DUMONT, AND S. KOVAC, *Time delay integrating systems: A challenge for process control industries. A practical solution*, Control Engineering Practice, vol. 10 (2002), pp. 1153–1161. (Cited on p. 31)

[113] O. IFTIME, M. KAASHOEK, AND A. SASANE, *A Grassmannian band method approach to the Nehari–Takagi problem*, Journal of Mathematical Analysis and Applications, vol. 310 (2005), pp. 97–115. (Cited on p. 2)

[114] O. IFTIME AND H. ZWART, *Nehari problems and equalizing vectors for infinite-dimensional systems*, Systems & Control Letters, vol. 45 (2002), pp. 217–225. (Cited on p. 22)

[115] A. ISIDORI, *Lectures in Feedback Design for Multivariable Systems*, Advanced Textbooks in Control and Signal Processing. Springer, Cham, 2017. (Cited on p. 46)

[116] K. ITO AND K. A. MORRIS, *An approximation theory of solutions to operator Riccati equations for $H_\infty$ control*, SIAM Journal on Control and Optimization, vol. 36 (1998), pp. 82–99. (Cited on p. 2)

[117] M. KALECKI, *A macrodynamic theory of business cycles*, Econometrica, vol. 3 (1935), pp. 327–344. (Cited on p. 34)

[118] A. E. KARAGÜL, O. DEMIR, AND H. ÖZBAY, *Computation of optimal $\mathcal{H}_\infty$ controllers and approximations of fractional order systems: A tutorial review*, Applied and Computational Mathematics, vol. 12 (2013), pp. 261–288. (Cited on p. 80)

[119] K. KASHIMA, *A new expression for the $H^2$ performance limit based on state-space representation*, Automatica, vol. 45 (2009), pp. 283–390. (Cited on pp. 151, 169)

[120] K. KASHIMA, H. ÖZBAY, AND Y. YAMAMOTO, *A Hamiltonian-based solution to the mixed sensitivity optimization problem for stable pseudorational plants*, Systems & Control Letters, vol. 54 (2005), pp. 1063–1068. (Cited on pp. 2, 108)

[121] K. KASHIMA AND Y. YAMAMOTO, *A new characterization of invariant subspaces of $\mathcal{H}_2$ and applications to the optimal sensitivity problem*, Systems & Control Letters, vol. 54 (2005), pp. 539–545. (Cited on pp. 62, 158)

[122] K. KASHIMA AND Y. YAMAMOTO, *Finite rank criteria for $\mathcal{H}_\infty$ control of infinite-dimensional systems*, IEEE Transactions on Automatic Control, vol. 53 (2008), pp. 881–893. (Cited on pp. 151, 164, 165)

[123] K. KASHIMA AND Y. YAMAMOTO, *On standard $\mathcal{H}_\infty$ control problems for systems with infinitely many unstable poles*, Systems & Control Letters, vol. 57 (2008), pp. 309–314. (Cited on p. 168)

[124] K. KASHIMA, Y. YAMAMOTO, AND H. ÖZBAY, *Parameterization of suboptimal solutions of the Nehari problem for infinite-dimensional systems*, IEEE Transactions on Automatic Control, vol. 52 (2007), pp. 2369–2374. (Cited on pp. 151, 156, 157)

[125] A. KATARIA, H. ÖZBAY, AND H. HEMAMI, *Controller design for natural and robotic systems with transmission delays*, Journal of Robotic Systems, vol. 19 (2002), pp. 231–244. (Cited on p. 31)

[126] H. K. KHALIL, *Nonlinear Systems*, Third Edition, Prentice–Hall, 2002. (Cited on p. 46)

[127] P. P. KHARGONEKAR AND K. POOLLA, *Robust stabilization of distributed systems*, Automatica, vol. 22 (1986), pp. 77–84. (Cited on p. 3)

[128] V. L. KHARITONOV, *Robust stability analysis of time delay systems: A survey*, Annual Reviews in Control, vol. 23 (1999), pp. 185–196. (Cited on p. 3)

[129] V. L. KHARITONOV, *Time-Delay Systems: Lyapunov Functionals and Matrices*, Birkhäuser, New York, 2013. (Cited on pp. 3, 31)

[130] H. KIMURA, *Chain-Scattering Approach to $\mathcal{H}_\infty$ Control*, Birkhäuser, Boston, 1997. (Cited on p. 2)

[131] C. KNOSPE AND L. ZHU, *Performance limitations of non-laminated magnetic suspension systems*, IEEE Transactions on Control Systems Technology, vol. 19 (2011), pp. 327–336. (Cited on p. 42)

[132] A. KOJIMA AND S. ISHIJIMA, *Robust controller design for delay systems in the gap-metric*, IEEE Transactions on Automatic Control, vol. 40 (1995), pp. 370–374. (Cited on p. 3)

[133] V. B. KOLMANOVSKII AND V. R. NOSOV, *Stability of Functional Differential Equations*, Academic Press, London, 1986. (Cited on p. 3)

[134] M. KRSTIC, *Delay Compensation for Nonlinear, Adaptive and PDE Systems*, Birkhäuser, Boston, 2009. (Cited on p. 3)

[135] V. KUČERA, *Author's reply to comments on "Optimizing simultaneously over the numerator and denominator polynomials in the Youla-Kucera parameterization,"* IEEE Transactions on Automatic Control, vol. 52 (2007), p. 763. (Cited on p. 48)

[136] D. K. LE AND A. E. FRAZHO, *A numerical procedure for a non-rational $\mathcal{H}_\infty$ optimization problem in control design*, Systems & Control Letters, vol. 16 (1991), pp. 9–15. (Cited on pp. 2, 22)

[137] K. LENZ AND H. ÖZBAY, *Analysis and robust control techniques for an ideal flexible beam*, in Multidisciplinary Engineering Systems: Design and Optimization Techniques and their Applications, C. T. Leondes, ed., Academic Press, 1993, pp. 369–421. (Cited on pp. 2, 40, 41, 79)

[138] K. LENZ, H. ÖZBAY, A. TANNENBAUM, J. TURI, AND B. MORTON, *Frequency domain analysis and robust control design for an ideal flexible beam*, Automatica, vol. 27 (1991), pp. 947–961. (Cited on pp. 2, 41, 79)

[139] X. LI AND C. E. DE SOUZA, *Delay-dependent robust stability and stabilization of uncertain linear delay systems: A linear matrix inequality approach*, IEEE Transactions on Automatic Control, vol. 42 (1997), pp. 1144–1148. (Cited on p. 3)

[140] H. LOGEMANN, R. REBARBER, AND G. WEISS, *Conditions for robustness and nonrobustness of the stability of feedback systems with respect to small delays in the feedback loop*, SIAM Journal on Control and Optimization, vol. 34 (1996), pp. 572–600. (Cited on p. 3)

[141] J. J. LOISEAU, M. CARDELLI, AND X. DUSSER, *Neutral time-delay systems that are not formally stable are not BIBO stabilizable*, IMA Journal of Mathematical Control and Information, vol. 19 (2002), pp. 217–227. (Cited on p. 72)

[142] J. LOUISELL, *Stability exponent and eigenvalue abscissas by way of the imaginary axis eigenvalues*, in Advances in Time-Delay Systems, S. Niculescu and K. Gu, eds., Springer-Verlag, LNCSE, vol. 38 (2004), pp. 193–206. (Cited on p. 71)

[143] J. J. LOISEAU AND H. MOUNIER, *Stabilisation de l'équation de la chaleur commande en flux*, ESAIM: Proceedings-Fractional Differential Systems: Models, Methods and Applications, vol. 5 (1998), pp. 131–144. (Cited on p. 41)

[144] T. A. LYPCHUK, M. C. SMITH, AND A. TANNENBAUM, *Weighted sensitivity minimization: General plants in $\mathcal{H}_\infty$ and rational weights*, Linear Algebra and Its Applications, vol. 109 (1988), pp. 71–90. (Cited on pp. 2, 22)

[145] J. M. MACIEJOWSKI, *Multivariable Feedback Design*, Addison–Wesley, Workingham, England, 1989. (Cited on p. 2)

[146] M. C. MACKEY, *Unified hypothesis for the origin of aplastic anaemia and periodic hematopoiesis*, Blood, vol. 51 (1978), pp. 941–956. (Cited on p. 42)

[147] P.M. MÄKILÄ, *Comments on "Robust, fragile, or optimal?"* IEEE Transactions on Automatic Control, vol. 43 (1998), pp. 1265–1267. (Cited on p. 89)

[148] S. MASCOLO, *Congestion control in high-speed communication networks using the Smith principle*, Automatica, vol. 35 (1999), pp. 1921–1935. (Cited on p. 31)

[149] D. MATIGNON, *Stability properties for generalized fractional differential systems*, ESAIM: Proceedings, vol. 5 (1998), pp. 145–158. (Cited on pp. 42, 129)

[150] A. MEGRETSKI, *H-Infinity model reduction with guaranteed suboptimality bound*, Proceedings of the 2006 American Control Conference, Minneapolis, MN, June 2006, pp. 448–453. (Cited on p. 141)

[151] G. MEINSMA, M. FU, AND T. IWASAKI, *Robustness of the stability of feedback systems with respect to small time delays*, Systems & Control Letters, vol. 36 (1999), pp. 131–134. (Cited on p. 3)

[152] G. MEINSMA AND L. MIRKIN, *$\mathcal{H}_\infty$ control of systems with multiple I/O delays via decomposition to adobe problems*, IEEE Transactions on Automatic Control, vol. 50 (2005), pp. 199–211. (Cited on pp. 3, 107, 151)

[153] G. MEINSMA, L. MIRKIN, AND Q-C. ZHONG, *Control of systems with I/O delay via reduction to a one-block problem*, IEEE Transactions on Automatic Control, vol. 47 (2002), pp. 1890–1895. (Cited on pp. 151, 154, 155)

[154] G. MEINSMA AND H. ZWART, *On $H^\infty$ control for dead-time systems*, IEEE Transactions on Automatic Control, vol. 45 (2000), pp. 272–285. (Cited on pp. 111, 151)

[155] W. MICHIELS AND S. GUMUSSOY, *Characterization and computation of $\mathcal{H}_\infty$ norms for time-delay systems*, SIAM Journal on Matrix Analysis and Applications, vol. 31 (2010), pp. 2093–2115. (Cited on pp. 27, 28)

[156] W. MICHIELS AND S.-I. NICULESCU, *Stability and Stabilization of Time-Delay Systems: An Eigenvalue Based Approach*, Advances in Design and Control, vol. 12, SIAM, Philadelphia, 2007. (Cited on pp. 31, 66, 68)

[157] W. MICHIELS AND H. U. ÜNAL, *Evaluating and approximating FIR filters: An approach based on functions of matrices*, IEEE Transactions on Automatic Control, vol. 60, no. 2 (2015), pp. 463–468. (Cited on p. 141)

[158] R. H. MIDDLETON AND D. E. MILLER, *On the achievable delay margin using LTI control for unstable plants*, IEEE Transactions on Automatic Control, vol. 52 (2007), pp. 1194–1207. (Cited on pp. 83, 86)

[159] L. MIRKIN, *On the extraction of dead-time controllers and estimators from delay-free parameterizations*, IEEE Transactions on Automatic Control, vol. 48 (2003), pp. 543–553. (Cited on pp. 107, 154, 167)

[160] L. MIRKIN, *On the approximation of distributed-delay control laws*, Systems & Control Letters, vol. 51, no. 5 (2004), pp. 331–342. (Cited on p. 111)

[161] L. MIRKIN AND N. RASKIN, *Every stabilizing dead-time controller has an observer-predictor-based structure*, Automatica, vol. 39 (2003), pp. 1747–1754. (Cited on pp. 53, 107, 167)

[162] L. MIRKIN AND G. TADMOR, $\mathcal{H}_\infty$ *control of system with I/O delay: A review of some problem-oriented methods*, IMA Journal of Mathematical Control and Information, vol. 19 (2002), pp. 185–199. (Cited on pp. 3, 151)

[163] V. MISRA, W. B. GONG, AND D. TOWSLEY, *Fluid-based analysis of a network of AQM routers supporting TCP flows with an application to RED*, Proceedings of ACM/SIGCOMM, (2000), pp. 151–160. (Cited on p. 34)

[164] C. A. MONJE, Y. Q. CHEN, B. M. VINAGRE, D. XUE, AND V. FELIU, *Fractional-Order Systems and Controls: Fundamentals and Applications*, Springer, London, 2010. (Cited on pp. 41, 80)

[165] M. MORARI AND E. ZAFIRIOU, *Robust Process Control*, Prentice–Hall, Englewood Cliffs, NJ, 1989. (Cited on p. 3)

[166] Ö. MORGÜL, *On the stabilization and stability robustness against small delays of some damped wave equations*, IEEE Transactions on Automatic Control, vol. 40 (1995), pp. 1626–1630. (Cited on p. 3)

[167] K. MORRIS, *Introduction to Feedback Control*, Harcourt/Academic Press, San Diego, 2001. (Cited on pp. 3, 17)

[168] K. A. MORRIS, *Convergence of controllers designed using state-space techniques*, IEEE Transactions on Automatic Control, vol. 39 (1994), pp. 2100–2104. (Cited on p. 2)

[169] K. A. MORRIS, $\mathcal{H}_\infty$*-output feedback of infinite-dimensional systems via approximation*, Systems & Control Letters, vol. 44 (2001), pp. 211–217. (Cited on p. 2)

[170] M. NAGAHARA AND Y. YAMAMOTO, $\mathcal{H}_\infty$*-optimal fractional delay filters*, IEEE Transactions on Signal Processing, vol. 61, no. 18 (2013), pp. 4473–4480. (Cited on p. 141)

[171] M. NAGAHARA AND Y. YAMAMOTO, *FIR Digital Filter Design by Sampled-Data* $\mathcal{H}_\infty$ *Discretization*, Preprints of IFAC World Congress, Cape Town, South Africa, August 2014, pp. 3110–3115. (Cited on pp. 111, 136, 141)

[172] M. NAGAHARA AND Y. YAMAMOTO, *Digital repetitive controller design via sampled-data delayed signal reconstruction*, Automatica, vol. 65 (2016), pp. 203–209. (Cited on p. 78)

[173] K. M. NAGPAL AND R. RAVI, $\mathcal{H}_\infty$ control and estimation problems with delayed measurements: State-space solutions, SIAM Journal on Control and Optimization, vol. 35 (1997), pp. 1217–1243. (Cited on p. 2)

[174] Z. NEHARI, On bounded bilinear forms, Annals of Mathematics, vol. 65 (1957), pp. 153–162. (Cited on p. 18)

[175] L. H. V. NGUYEN AND C. BONNET, Stabilization of some fractional neutral delay systems which possibly possess an infinite number of unstable poles, In: A. Seuret et al. (eds.), Low-Complexity Controllers for Time-Delay Systems. Advances in Delays and Dynamics, vol. 2, pp. 47–60, Springer, Cham, 2014. (Cited on p. 75)

[176] L. H. V. NGUYEN AND C. BONNET $\mathcal{H}_\infty$-stability analysis of various classes of neutral systems with commensurate delays and with chains of poles approaching the imaginary axis, Proceedings of the 54th IEEE Conference on Decision and Control (CDC), Osaka, Japan, December 2015, pp. 6416–6421 (Cited on pp. 9, 67)

[177] L. H. V. NGUYEN, C. BONNET, AND A. R. FIORAVANTI $H_\infty$-stability analysis of fractional delay systems of neutral type, SIAM Journal on Control and Optimization, vol. 54, no. 2 (2016), pp. 740–759. (Cited on pp. 9, 67)

[178] S.-I. NICULESCU, Delay Effects on Stability: A Robust Control Approach, LNCIS, vol. 269, Springer-Verlag, London, 2001. (Cited on pp. 3, 31)

[179] S.-I. NICULESCU, J.-M. DION, AND L. DUGARD, Robust stabilization for uncertain time-delay systems containing saturating actuators, IEEE Transactions on Automatic Control, vol. 41 (1996), pp. 742–747. (Cited on p. 3)

[180] J. E. NORMEY-RICO AND E. F. CAMACHO, Dead-time compensators: A survey, Control Engineering Practice, vol. 16 (2008), pp. 407–428. (Cited on p. 33)

[181] J. E. NORMEY-RICO AND E. F. CAMACHO, Unified approach for robust dead-time compensator design, Journal of Process Control, vol. 19 (2009), pp. 38–47. (Cited on p. 31)

[182] Y. OHTA, Hankel singular values and vectors of a class of infinite-dimensional systems: Exact Hamiltonian formulas for control and approximation problems, Mathematics of Control, Signals, and Systems, vol. 12 (1999), pp. 361–375. (Cited on pp. 2, 22)

[183] Y. OHTA, A study on the norm of mixed Hankel-Toeplitz operator, Proceedings of the 2000 American Control Conference, Chicago, IL, June 2000, pp. 2765–2769. (Cited on p. 2)

[184] N. OLGAÇ AND R. SIPAHI, A practical method for analyzing the stability of neutral type LTI-time delayed systems, Automatica, vol. 40 (2004), pp. 847–853. (Cited on p. 71)

[185] M. OLIVI, F. SEYFERT, AND J.-P. MARMORAT, Identification of microwave filters by analytic and rational $H_2$ approximation, Automatica, vol. 49, no. 2 (2013), pp. 317–325. (Cited on p. 141)

[186] M. R. OPMEER, Model order reduction by balanced proper orthogonal decomposition and by rational interpolation, IEEE Transactions on Automatic Control, vol. 57, no. 2 (2012), pp. 472–477. (Cited on p. 141)

[187] G. OROSZ AND G. STÉPÁN, Subcritical Hopf bifurcations in a car-following model with reaction-time delay, Proceedings of the Royal Society A, vol. 462, no. 2073 (2006), pp. 2643–2670. (Cited on p. 31)

[188] A. OUSTALOUP, F. LEVRON, B. MATHIEU, AND F. M. NANOT, Frequency-band complex noninteger differentiator: Characterization and synthesis, IEEE Transactions on Circuit and Systems—I: Fundamental Theory and Applications, vol. 47, no. 1, (2000), pp. 25–39. (Cited on p. 80)

[189] H. ÖZBAY, *A simpler formula for the singular values of a certain Hankel operator*, Systems & Control Letters, vol. 15 (1990), pp. 381–390. (Cited on p. 22)

[190] H. ÖZBAY, *Controller reduction in the 2-block $\mathcal{H}_\infty$ optimal design for distributed plants*, International Journal of Control, vol. 54 (1991), pp. 1291–1308. (Cited on p. 111)

[191] H. ÖZBAY, *Introduction to Feedback Control Theory*, CRC Press, Boca Raton, FL, 1999. (Cited on pp. 3, 4, 27, 49, 82)

[192] H. ÖZBAY, *Coping with time delays in networked control systems*, Proceedings of the 19th International Symposium on Mathematical Theory of Networks and Systems (MTNS 2010), July 2010, Budapest, Hungary, pp. 1333–1338. (Cited on pp. vii, 32)

[193] H. ÖZBAY, *Stable $\mathcal{H}_\infty$ controller design for systems with time delays*, in Perspectives in Mathematical System Theory, Control, and Signal Processing, J. C. Willems et al. (eds.), LNCIS, vol. 398, Springer-Verlag, Berlin, Heidelberg, 2010, pp. 105–113. (Cited on p. 94)

[194] H. ÖZBAY, *Computation of $\mathcal{H}_\infty$ controllers for infinite dimensional plants using numerical linear algebra*, Numerical Linear Algebra with Applications, vol. 20 (2013), pp. 327–335. (Cited on pp. 97, 101)

[195] H. ÖZBAY, *Robust control of infinite dimensional systems*, in Encyclopedia of Systems and Control, J. Baillieul and T. Samad (eds.), Springer-Verlag, London, 2014. (Cited on p. 135)

[196] H. ÖZBAY, C. BONNET, H. BENJELLOUN, AND J. CLAIRAMBAULT, *Stability analysis of cell dynamics in leukemia*, Mathematical Modelling of Natural Phenomena, vol. 7 (2012), pp. 203–234. (Cited on pp. 31, 43)

[197] H. ÖZBAY, C. BONNET, AND A. R. FIORAVANTI, *PID controller design for fractional-order systems with time delays*, Systems & Control Letters, vol. 61 (2012), pp. 18–23. (Cited on p. 80)

[198] H. ÖZBAY, H. Ç. SAĞLAM, AND M. K. YÜKSEL, *Hopf cycles in one sector optimal growth models with time delay*, Macroeconomic Dynamics, vol. 21 (2017), pp. 1887–1901. (Cited on p. 31)

[199] H. ÖZBAY, M. C. SMITH, AND A. TANNENBAUM, *Mixed sensitivity optimization for a class of unstable infinite dimensional systems*, Linear Algebra and Its Applications, vol. 178 (1993), pp. 43–83. (Cited on p. 2)

[200] H. ÖZBAY AND A. TANNENBAUM, *A skew Toeplitz approach to the $\mathcal{H}_\infty$ control of multivariable distributed systems*, SIAM Journal on Control and Optimization, vol. 28 (1990), pp. 653–670. (Cited on p. 2)

[201] H. ÖZBAY AND A. TANNENBAUM, *On the structure of suboptimal $H_\infty$ controllers in the sensitivity minimization problem for distributed stable plants*, Automatica, vol. 27 (1991), pp. 293–305. (Cited on p. 134)

[202] H. ÖZBAY AND J. TURI, *On input/output stabilization of singular integro-differential systems*, Applied Mathematics and Optimization, vol. 30 (1994), pp. 21–49. (Cited on p. 2)

[203] A. A. OZDEMIR AND S. GUMUSSOY, *Transfer function estimation in system identification toolbox via vector fitting*, IFAC-PapersOnLine, vol. 50, no. 1 (2017), pp. 6232–6237. (Cited on p. 141)

[204] F. PADULA, S. ALCÁNTARA, R. VILANOVA, AND A. VISIOLI, *$H_\infty$ control of fractional linear systems*, Automatica, vol. 49 (2013), pp. 2276–2280. (Cited on p. 42)

[205] J. R. PARTINGTON, *Linear Operators and Linear Systems: An Analytical Approach to Control Theory*, Cambridge University Press, Cambridge, UK, 2004. (Cited on p. 34)

[206] J. R. PARTINGTON, *Some frequency-domain approaches to the model reduction of delay systems*, Annual Reviews in Control, vol. 28 (2004), pp. 65–73. (Cited on pp. 111, 141)

[207] J. R. PARTINGTON AND C. BONNET, $\mathcal{H}_\infty$ *and BIBO stabilization of delay systems of neutral type*, Systems & Control Letters, vol. 52 (2004), pp. 283–288. (Cited on pp. 3, 75)

[208] J. R. PARTINGTON AND K. GLOVER, *Robust stabilization of delay systems by approximation of coprime factors*, Systems & Control Letters, vol. 14 (1990), pp. 325–331. (Cited on p. 3)

[209] T. E. PEERY AND H. ÖZBAY, $\mathcal{H}_\infty$ *optimal repetitive controller design for stable plants*, Transactions of the ASME Journal of Dynamic Systems, Measurement, and Control, vol. 119 (1997), pp. 541–547. (Cited on p. 3)

[210] I. R. PETERSEN, V. A. UGRINOVSKII, AND A. V. SAVKIN, *Robust Control Design Using* $\mathcal{H}_\infty$ *Methods*, Springer-Verlag, London, 2000. (Cited on p. 2)

[211] T. QI, J. ZHU, AND J. CHEN, *Fundamental limits on uncertain delays: When is a delay system stabilizable by LTI controllers?*, IEEE Transactions on Automatic Control, vol. 62, no. 3 (2017), pp. 1314–1328. (Cited on pp. 84, 86)

[212] P.-F. QUET, B. ATASLAR, A. IFTAR, H. ÖZBAY, S. KALYANARAMAN, AND T. KANG, *Rate-based flow controllers for communication networks in the presence of uncertain time-varying multiple time-delays*, Automatica, vol. 38 (2002), pp. 917–928. (Cited on pp. 31, 32)

[213] P.-F. QUET AND H. ÖZBAY, *On the design of AQM supporting TCP flows using robust control theory*, IEEE Transactions on Automatic Control, vol. 49 (2004), pp. 1031–1036. (Cited on pp. 35, 36, 117, 120)

[214] R. RABAH, G. M. SKLYAR, AND A. V. REZOUNENKO, *On strong regular stabilizability for linear neutral type systems*, Journal of Differential Equations, vol. 245 (2008), pp. 569–593. (Cited on p. 72)

[215] T. REYA, S. J. MORRISON, M. F. CLARKE, AND I. L. WEISSMAN, *Stem cells, cancer, and cancer stem cells*, Nature, vol. 414, no. 6859 (2001), pp. 105–111. (Cited on p. 42)

[216] A. A. RODRIGUEZ AND M. A. DAHLEH, *Weighted* $\mathcal{H}_\infty$ *optimization for stable infinite dimensional systems using finite dimensional techniques*, Proceedings of the 29th IEEE Conference on Decision and Control, Honolulu, HI, December 1990, pp. 1814–1819. (Cited on p. 2)

[217] A. A. RODRIGUEZ AND M. A. DAHLEH, *On the computation of induced norms for noncompact Hankel operators arising from distributed control problems*, Systems & Control Letters, vol. 19 (1992), pp. 429–438. (Cited on p. 2)

[218] D. ROOSE, T. LUZYANINA, K. ENGELBORGHS, AND W. MICHIELS, *Software for stability and bifurcation analysis of delay differential equations and applications to stabilization*, in Advances in Time-Delay Systems, S. Niculescu and K. Gu, eds., Springer-Verlag, LNCSE, vol. 38 (2004), pp. 167–181. (Cited on p. 67)

[219] W. RUDIN, *Real and Complex Analysis*, Third Edition, McGraw–Hill, New York, 1987. (Cited on p. 115)

[220] H. SANO, *Finite-dimensional* $\mathcal{H}_\infty$ *control of flexible beam equation systems*, IMA Journal of Mathematical Control and Information, vol. 19 (2002), pp. 477–491. (Cited on p. 2)

[221] D. SARASON, *Generalized interpolation in* $\mathcal{H}_\infty$, Transactions of the American Mathematical Society, vol. 127 (1967), pp. 179–203. (Cited on p. 13)

[222] A. SASANE, *Hankel Norm Approximation for Infinite-Dimensional Systems*, Springer-Verlag, Berlin, 2002. (Cited on p. 2)

[223] L. SCHWARTZ, *Méthodes Mathématiques pour les Sciences Physiques*, Hermann, Paris, 1961. (Cited on pp. 36, 37)

[224] L. SCHWARTZ, *Théorie des Distributions*, Hermann, Paris, 1966. (Cited on pp. 36, 37, 39)

[225] R. SIPAHI, S-I. NICULESCU, C. T. ABDALLAH, W. MICHIELS, AND K. GU, *Stability and stabilization of systems with time delay*, IEEE Control Systems Magazine, vol. 31, no. 1 (2011), pp. 38–65. (Cited on p. 117)

[226] S. SKOGESTAD AND I. POSTLETHWAITE, *Multivariable Feedback Control: Analysis and Design*, John Wiley & Sons, Chichester, England, 1996. (Cited on p. 2)

[227] M. C. SMITH, *On stabilization and existence of coprime factorizations*, IEEE Transactions on Automatic Control, vol. 34 (1989), pp. 1005–1007. (Cited on p. 48)

[228] O. J. STAFFANS, *On the distributed stable full information* $\mathcal{H}_\infty$ *minimax problem*, International Journal of Robust and Nonlinear Control, vol. 8 (1998), pp. 1255–1305. (Cited on p. 3)

[229] O. J. STAFFANS, *Well-Posed Linear Systems*, Cambridge University Press, Cambridge, UK, 2005. (Cited on p. 2)

[230] G. STEIN, *Respect the unstable*, IEEE Control Systems Magazine, vol. 23, no. 4 (2003), pp. 12–25. (Cited on p. 105)

[231] M. STEINBUCH, *Repetitive control for systems with uncertain period-time*, Automatica, vol. 38 (2002), pp. 2103–2109. (Cited on p. 3)

[232] B. SZ.-NAGY AND C. FOIAS, *Harmonic Analysis of Operators on Hilbert Space*, North–Holland, Amsterdam, 1970. (Cited on p. 13)

[233] G. TADMOR, *The Nehari problem in systems with distributed input delays is inherently finite dimensional*, Systems & Control Letters, vol. 26 (1995), pp. 11–16. (Cited on p. 22)

[234] G. TADMOR, *Weighted sensitivity minimization in systems with a single input delay: A state space solution*, SIAM Journal of Control and Optimization, vol. 35 (1997), pp. 1445–1469. (Cited on pp. 2, 151)

[235] S. TARBOURIECH AND J. M. G. DA SILVA, JR., *Synthesis of controllers for continuous-time delay systems with saturating controls via LMIs*, IEEE Transactions on Automatic Control, vol. 45 (2000), pp. 105–111. (Cited on p. 3)

[236] S. TARBOURIECH, P . L. D. PERES, G. GARCIA, AND I. QUEINNEC, *Delay-dependent stabilization of time-delay systems with saturating actuators*, Proceedings of the 39th IEEE Conference on Decision and Control, Sydney, Australia, December 2000, pp. 3248–3253. (Cited on p. 3)

[237] R. TEMPO, G. CALAFIORE, AND F. DABBENE, *Randomized Algorithms for Analysis and Control of Uncertain Systems*, Springer-Verlag, London, 2005. (Cited on p. 2)

[238] O. TOKER AND H. ÖZBAY, $\mathcal{H}_\infty$ *Optimal and suboptimal controllers for infinite dimensional SISO plants*, IEEE Transactions on Automatic Control, vol. 40 (1995), pp. 751–755. (Cited on pp. 97, 98, 99, 101)

[239] O. TOKER AND H. ÖZBAY, *Gap metric problem for MIMO delay systems: Parameterization of all sub-optimal controllers*, Automatica, vol. 31 (1995), pp. 931–940. (Cited on p. 89)

[240] O. TOKER AND H. ÖZBAY, *On the rational $\mathcal{H}_\infty$ controller design for infinite dimensional plants*, International Journal of Robust and Nonlinear Control, vol. 6 (1996), pp. 383–397. (Cited on p. 111)

[241] O. TOKER AND H. ÖZBAY, *HINFCON: A MATLAB based program for $\mathcal{H}_\infty$ optimal/suboptimal controller design*, 1996, http://www.ee.bilkent.edu.tr/~ozbay/HINFCON.zip (Cited on p. 99)

[242] K. UCHIDA, K. IKEDA, T. AZUMA, AND A. KOJIMA, *Finite-dimensional characterizations of $\mathcal{H}_\infty$ control for linear systems with delays in input and output*, International Journal of Robust and Nonlinear Control, vol. 13 (2003), pp. 833–843. (Cited on p. 2)

[243] V. UTKIN, J. GULDNER, AND J. SHI, *Sliding Mode Control in Electromechanical Systems*, Taylor & Francis, London, 1999. (Cited on p. 3)

[244] B. VAN KEULEN, *$\mathcal{H}_\infty$ Control for Distributed Parameter Systems: A State Space Approach*, Birkhäuser, Boston, 1993. (Cited on pp. 2, 151)

[245] M. VIDYASAGAR, *Control System Synthesis: A Factorization Approach*, MIT Press, Cambridge, MA, 1985 (reprinted by Morgan & Claypool in 2011). (Cited on pp. 2, 4)

[246] M. VIDYASAGAR, *Nonlinear Systems Analysis*, Second Edition, Classics in Applied Mathematics 42, SIAM, Philadelphia, 2002. (Cited on p. 46)

[247] A. VISIOLI AND Q-C. ZHONG, *Smith predictor based control*, in Control of Integral Processes with Dead Time, Advances in Industrial Control, pp. 141–185, Springer, London, 2011. (Cited on pp. 33, 49)

[248] T. VYHLÍDAL AND P. ZÍTEK, *Mapping based algorithm for large-scale computation of quasi-polynomial zeros*, IEEE Transactions on Automatic Control, vol. 54, no. 1 (2009), pp. 171–177. (Cited on pp. 67, 71, 120)

[249] T. VYHLÍDAL AND P. ZÍTEK, *QPmR-quasi-polynomial root-finder: Algorithm update and examples*, in Delay Systems: Advances in Delays and Dynamics, vol. 1, T. Vyhlidal, J. F. Lafay, R. Sipahi, eds., Springer, Cham, 2014, pp. 299–312. (Cited on pp. 9, 67, 120)

[250] M. WAKAIKI, Y. YAMAMOTO, AND H. ÖZBAY, *Sensitivity reduction by strongly stabilizing controllers for MIMO distributed parameter systems*, IEEE Transactions on Automatic Control, vol. 54 (2012), pp. 2089–2094. (Cited on p. 94)

[251] K. WATANABE, E. NOBUYAMA, AND A. KOJIMA, *Recent advances in control of time delay systems: A tutorial review*, Proceedings of the 35th IEEE Conference on Decision and Control, Kobe, Japan, 1996, pp. 2083–2089. (Cited on p. 3)

[252] G. WEISS AND M. HÄFELE, *Repetitive control of MIMO systems using $\mathcal{H}_\infty$ design*, Automatica, vol. 35 (1999), pp. 1185–1199. (Cited on p. 3)

[253] G. WEISS, Q-C. ZHONG, T. C. GREEN, AND J. LIANG, *$\mathcal{H}_\infty$ repetitive control of DC-AC converters in microgrids*, IEEE Transactions on Power Electronics, vol. 19, no. 1 (2004), pp. 219–230. (Cited on p. 78)

[254] K. WILLCOX AND A. MEGRETSKI, *Fourier series for accurate, stable, reduced-order models in large-scale applications*, SIAM Journal on Scientific Computing, vol. 26, no. 3 (2005), pp. 944–962. (Cited on p. 141)

[255] J. J. WINKIN, F. M. CALLIER, B. JACOB, AND J. R. PARTINGTON, *Spectral factorization by symmetric extraction for distributed parameter systems*, SIAM Journal on Control and Optimization, vol. 43, no. 4 (2005), pp. 1435–1466. (Cited on p. 59)

[256] Z. WU AND W. MICHIELS, *Reliably computing all characteristic roots of delay differential equations in a given right half plane using a spectral method*, Journal of Computational and Applied Mathematics, vol. 236 (2012), pp. 2499–2514. (Cited on p. 71)

[257] S. XU, J. LAM, S. HUANG, AND C. YANG, $\mathcal{H}_\infty$ *model reduction for linear time-delay systems: Continuous-time case*, International Journal of Control, vol. 74, no. 11 (2001), pp. 1062–1074. (Cited on p. 141)

[258] Y. YAMAMOTO, *Pseudo-rational input/output maps and their realizations: A fractional representation approach to infinite-dimensional systems*, SIAM Journal on Control and Optimization, vol. 26 (1988), pp. 1415–1430. (Cited on pp. 31, 37, 38, 39)

[259] Y. YAMAMOTO, *Reachability of a class of infinite-dimensional linear systems: An external approach with applications to general neutral systems*, SIAM Journal on Control and Optimization, vol. 27 (1989), pp. 217–234. (Cited on pp. 39, 61)

[260] Y. YAMAMOTO, *On the state space and frequency domain characterization of $\mathcal{H}_\infty$ norm of sampled-data systems*, Systems & Control Letters, vol. 21 (1993), pp. 163–172. (Cited on p. 3)

[261] Y. YAMAMOTO, *From Vector Spaces to Function Spaces: Introduction to Functional Analysis with Applications*, SIAM, Philadelphia, 2012. (Cited on pp. 3, 4, 5, 7, 9, 13, 17, 36, 37, 58)

[262] Y. YAMAMOTO AND S. HARA, *Relationships between internal and external stability for infinite-dimensional systems with applications to a servo problem*, IEEE Transactions on Automatic Control, vol. 33 (1988), pp. 1044–1052. (Cited on pp. 75, 78)

[263] Y. YAMAMOTO AND S. HARA, *Internal and external stability and robust stability condition for a class of infinite-dimensional systems*, Automatica, vol. 28 (1992), pp. 81–93. (Cited on p. 78)

[264] Y. YAMAMOTO, K. HIRATA, AND A. TANNENBAUM, *Some remarks on Hamiltonians and the infinite dimensional one block $H^\infty$ problem*, Systems & Control Letters, vol. 29 (1996), pp. 111–117. (Cited on p. 157)

[265] M. O. YEĞIN, *Numerical Computation and Implementation of $\mathcal{H}_\infty$ Controllers for Retarded and Neutral Time Delay Systems*, MSc Thesis, Bilkent University, Ankara, Turkey, November 2017. (Cited on pp. xii, 99, 135)

[266] M. O. YEĞIN AND H. ÖZBAY, *Numerical computation of $\mathcal{H}_\infty$ optimal controllers for time delay systems using YALTA*, IFAC-PapersOnLine, vol. 49, no. 10 (2016), pp. 182–187. (Cited on pp. 99, 135)

[267] S. YI, P. W. NELSON, AND A. G. ULSOY, *Time-Delay Systems: Analysis and Control Using the Lambert W Function*, World Scientific, Singapore, 2010. (Cited on p. 31)

[268] D. C. YOULA, H. A. JABR, AND J. J. BONGIORNO, JR., *Modern Wiener Hopf design of optimal controllers: Part II*, IEEE Transactions on Automatic Control, vol. 21 (1976), pp. 319–338. (Cited on p. 48)

[269] V. YÜCESOY AND H. ÖZBAY, *On the optimal Nevanlinna-Pick interpolant and its applications to robust control of systems with time delays*, Proceedings of the 22nd International Symposium on Mathematical Theory of Networks and Systems (MTNS2016), July 2016, Minneapolis, MN, pp. 515–516. (Cited on p. 17)

[270]  V. YÜCESOY AND H. ÖZBAY, *Optimal Nevanlinna Pick interpolant and its application to robust repetitive control of time delay systems*, Applied and Computational Mathematics, vol. 17, no. 1, (2018), pp. 96–108. (Cited on p. 17)

[271]  M. K. YÜKSEL, *Capital dependent population growth induces cycles*, Chaos, Solitons & Fractals, vol. 44 (2011), pp. 759–763. (Cited on p. 31)

[272]  P. ZAK, *Kaleckian lags in general equilibrium*, Review of Political Economy, vol. 11 (1999), pp. 321–330. (Cited on pp. 31, 34)

[273]  G. ZAMES, *Feedback and optimal sensitivity: Model reference transformations, multiplicative seminorms and approximate inverses*, IEEE Transactions on Automatic Control, vol. 26 (1981), pp. 301–320. (Cited on p. 2)

[274]  G. ZAMES AND S. K. MITTER, *A note on essential spectrum and norms of mixed Hankel–Toeplitz operators*, Systems & Control Letters, vol. 10 (1988), pp. 159–165. (Cited on p. 24)

[275]  M. ZEREN AND H. ÖZBAY, *On the strong stabilization and stable $\mathcal{H}_\infty$ controller design problems for MIMO systems*, Automatica, vol. 36 (2000), pp. 1675–1684. (Cited on p. 94)

[276]  M. ZEREN AND H. ÖZBAY, *Comments on: "Solutions to the combined sensitivity and complementary sensitivity problem in control systems,"* IEEE Transactions on Automatic Control, vol. 43 (1998), p. 724. (Cited on p. 17)

[277]  Q. C. ZHONG, *On distributed delay in linear control Laws—Part I: Discrete-delay implementations*, IEEE Transactions on Automatic Control, vol. 49, no. 11 (2004), pp. 2074–2080. (Cited on p. 111)

[278]  Q. C. ZHONG, *On distributed delay in linear control laws—Part II: Rational implementations inspired from the $\delta$-operator*, IEEE Transactions on Automatic Control, vol. 50, no. 5 (2005), p. 729–734. (Cited on p. 111)

[279]  Q.-C. ZHONG, *Robust Control of Time-Delay Systems*, Springer-Verlag, London, 2006. (Cited on pp. 3, 33, 107, 141, 151)

[280]  K. ZHOU, WITH J. C. DOYLE, *Essentials of Robust Control*, Prentice–Hall, Upper Saddle River, NJ, 1998. (Cited on p. 2)

[281]  K. ZHOU, J. C. DOYLE, AND K. GLOVER, *Robust and Optimal Control*, Prentice–Hall, Upper Saddle River, NJ, 1996. (Cited on pp. 2, 25, 58, 153)

[282]  K. ZHOU AND P. P. KHARGONEKAR, *On the weighted sensitivity minimization problem for delay systems*, Systems & Control Letters, vol. 8 (1987), pp. 307–312. (Cited on pp. 2, 108, 151)

[283]  J. ZHU, T. QI, D. MA, AND J. CHEN, *Limits of Stability and Stabilization of Time-Delay Systems*, Springer, Cham, 2018. (Cited on pp. 3, 86)

[284]  H. ZWART, *Transfer functions for infinite-dimensional systems*, Systems & Control Letters, vol. 52 (2004), pp. 247–255. (Cited on p. 31)

# Index